Methods in Cell Biology

VOLUME 37
Antibodies in Cell Biology

Series Editors

Leslie Wilson
Department of Biological Sciences
University of California
Santa Barbara, California

Paul Matsudaira
Whitehead Institute for Biomedical Research and
Department of Biology
Massachusetts Institute of Technology
Cambridge, Massachusetts

Methods in Cell Biology

Prepared under the Auspices of the American Society for Cell Biology

VOLUME 37
Antibodies in Cell Biology

Edited by

David J. Asai
Department of Biological Sciences
Purdue University
West Lafayette, Indiana

ACADEMIC PRESS, INC.
A Division of Harcourt Brace & Company

San Diego   New York   Boston   London   Sydney   Tokyo   Toronto

Cover photograph: Second meiotic spindle in a *Xenopus* egg stained with an antibody to tubulin and viewed by confocal immunofluorescence microscopy (in false color). Photograph by David L. Gard, University of Utah.

Academic Press, Inc.
1250 Sixth Avenue, San Diego, California 92101-4311

United Kingdom Edition published by
Academic Press Limited
24–28 Oval Road, London NW1 7DX

International Standard Book Number: 0-12-564137-0 (Hardcover)

International Standard Book Number: 0-12-064520-3 (Paperback)

PRINTED IN THE UNITED STATES OF AMERICA
93 94 95 96 97 98   EB   9 8 7 6 5 4 3 2 1

I dedicate this volume to Ray D. Owen and June W. Owen of Caltech.
Ray conveyed to me the wonder of the immune response.
June provided valuable perspectives on life and science.
And together they convinced me that work such as what went into the
editing of this volume is only worthwhile if it benefits young people.

CONTENTS

CONTRIBUTORS

Numbers in parentheses indicate the pages on which the authors' contributions begin.

Caroline E. Alfa (201), Department of Biology, University College London, London WC1E 6BT, England

David J. Asai (57, 441), Department of Biological Sciences, Purdue University, West Lafayette, Indiana 47907

Detlev Drenckhahn (7), Department of Anatomy, Julius-Maximilians University, D-97070 Würzburg, Germany

Page A. Erickson (283), Neuroscience Research Institute, and Department of Biological Sciences, University of California at Santa Barbara, Santa Barbara, California 93106

Steven K. Fisher (283), Neuroscience Research Institute, and Department of Biological Sciences, University of California at Santa Barbara, Santa Barbara, California 93106

Imelda M. Gallagher (201), Department of Biology, University College London, London WC1E 6BT, England

David L. Gard (147), Department of Biology, University of Utah, Salt Lake City, Utah 84112

Spyros D. Georgatos (407), Programme of Cell Biology, European Molecular Biology Laboratory, D-6900 Heidelberg, Germany

Edward S. Golub (1), The Pacific Center for Ethics and Applied Biology, San Diego, California 92121

Erika L. F. Holzbaur (361), University of Pennsylvania, School of Veterinary Medicine, Philadelphia, Pennsylvania 19104

Jeremy S. Hyams (201), Department of Biology, University College London, London WC1E 6BT, England

Brigitte M. Jockusch (343), Cell Biology Group, University of Bielefeld, DW-4800 Bielefeld, Germany

Thomas Jöns (7), Department of Anatomy, Julius-Maximilians University, D-97070 Würzburg, Germany

Mary Ann Jordan (129), Department of Biological Sciences, University of California, Santa Barbara, Santa Barbara, California 93106

Harish C. Joshi (259), Department of Anatomy and Cell Biology, Emory University School of Medicine, Atlanta, Georgia 30322

John Z. Kiss (311), Department of Botany, Miami University, Oxford, Ohio 45056

Darryl L. Kropf (147), Department of Biology, University of Utah, Salt Lake City, Utah 84112

Seung-won Lee (119), Department of Biological Sciences, Purdue University, West Lafayette, Indiana 47907

Jan L. M. Leunissen (241), AURION, Immuno Gold Reagents and Accessories, 6702 AA Wageningen, The Netherlands

Geoffrey P. Lewis (283), Neuroscience Research Institute, and Department of Biological Sciences, University of California at Santa Barbara, Santa Barbara, California 93106

Kent McDonald (311), Laboratory for Three-Dimensional Fine Structure, Molecular, Cellular, and Developmental Biology, University of Colorado, Boulder, Colorado 80309

Robert A. Obar (361), Alkermes, Inc., Cambridge, Massachusetts 02139

Joann J. Otto (105, 119), Department of Biological Sciences, Purdue University, West Lafayette, Indiana 47907

Frank Schmitz (7), Department of Anatomy, Julius-Maximilians University, D-97070 Würzburg, Germany

Jonathan M. Scholey (223), Section of Molecular and Cellular Biology, Division of Biological Sciences, University of California at Davis, Davis, California 95616, and Department of Cellular and Structural Biology, University of Colorado Health Science Center, Denver, Colorado 80262

Thomas J. Smith (75), Department of Biological Sciences, Purdue University, West Lafayette, Indiana 47907

Wen-Jing Y. Tang (95), Kewalo Marine Laboratory, Pacific Biomedical Research Center, University of Hawaii, Honolulu, Hawaii 96813

Constance J. Temm-Grove (343), Cell Biology Group, University of Bielefeld, DW-4800 Bielefeld, Germany

Douglas Thrower (129), Department of Biological Sciences, University of California, Santa Barbara, Santa Barbara, California 93106

Peter van de Plas (241), AURION, Immuno Gold Reagents and Accessories, 6702 AA Wageningen, The Netherlands

Susan M. Wick (171), Department of Plant Biology, Biological Sciences Center, University of Minnesota, St. Paul, Minnesota 55108

John K. Wilder (57), Department of Biological Sciences, Purdue University, West Lafayette, Indiana 47907

Leslie Wilson (129), Department of Biological Sciences, University of California, Santa Barbara, Santa Barbara, California 93106

Brent D. Wright (223), Section of Molecular and Cellular Biology, Division of Biological Sciences, University of Calfornia at Davis, Davis, California 95616, and Department of Cellular and Structural Biology, University of Colorado Health Science Center, Denver, Colorado 80262

PREFACE

The most striking feature of the vertebrate immune system may be its ability to produce an antibody whose binding sites are complementary to virtually any biological molecule. Cell and developmental biologists have long appreciated this and have pioneered ways to exploit the antibody response in order to generate customized reagents. The purpose of this volume is to build upon this rich tradition by focusing on practical applications of antibodies to address current problems in cell, developmental, and molecular biology.

The volume is loosely organized into three parts. After an immunologist's introductory perspective, the first few chapters describe standard methods used to produce, purify, and fractionate antibodies from conventional antisera and hybridomas. The second part of the volume summarizes important aspects of immunochemistry including the quantification of the immune response: the affinity purification of antibodies, immunoblotting, immunoprecipitations, and solid-phase binding assays. The last part is the largest. Here, several authors have shared their protocols and strategies in the application of antibodies to specific biological problems.

At present, one of the most powerful applications of antibodies is the subcellular localization of antigens. The immunofluorescence localization of antigens in flat, cultured cells, a method developed nearly two decades ago (summarized in the chapter by Mary Osborn and Klaus Weber in Volume 24 of *Methods in Cell Biology*), is a standard technique in many laboratories and does not require a reintroduction here. Instead, the focus of this volume is the next generation of immunolocalization protocols. In these applications, the critical feature is how the cell or tissue is fixed and permeabilized, and the current challenge is to apply these techniques to more intractable samples. Chapters describe the light microscopic immunolocalization of antigens in very large cells, pigmented cells, plant cells, and fungi. Other chapters describe the utilization of gold particles in immunoelectron microscopy and special methods of fixation and permeabilization of tissues for immunolocalization with the electron microscope.

Antibodies can be applied in a dynamic way to problems needing far more than a description of where an antigen is located in a cell. The final three chapters in this volume represent a departure from immunolocalization studies and summarize other important applications of antibodies in biology: the microinjection of antibodies into living cells to perturb cellular function; the use of antibodies to identify cDNA clones expressing antigenic determinants; and the exploitation of antisense antibody strategies to identify complementary structures in the cell. These topics foretell future emphases in the utilization of antibodies.

A theme that recurs in each of the chapters requires an explicit comment: *An antibody-based protocol is only as reliable as the antibody is specific.* The immune system that produces the antibodies is incapable of distinguishing a "good" antibody from one that cross-reacts with high affinity to the "wrong" molecule. The investigator must convince herself that the appropriate controls for specificity (including but not limited to use of nonimmune and preimmune antibodies, absorption of antibodies with a defined determinant to eliminate reactivity, and affinity-purification of antibodies) are applied objectively and that their interpretation is not circularly argued.

I thank the contributors. Their writing detailed protocols in a way that is understandable to a wide readership and their coming through with the chapters indicate a commitment to sharing and teaching that goes far beyond payment or glory. For their patient guidance, I thank the series editors, Les Wilson and Paul Matsudaira, and the acquisitions editor at Academic Press, Phyllis Moses. Most of the initial editing of this volume occurred while I was a guest in the laboratory of Ian Gibbons, whom I warmly thank. I am indebted to Dan Campbell, Justine Garvey, Natalie Cremer, and Dieter Sussdorf; their editions of *Methods in Immunology* (W. A. Benjamin, Inc.) served as the model for this volume and remain key references in my laboratory.

David J. Asai

CHAPTER 1

The Immune Response: An Overview

Edward S. Golub

The Pacific Center for Ethics and Applied Biology
San Diego, California 92121

It will probably not have escaped the notice of the readers of this volume that on those occasions that they go to their colleagues who are card-carrying immunologists for advice about some technique that requires antibodies, the immunologist may not know as much as the cell biologist who is asking the question. This is because immunologists study the mechanism of the immune response to find how the immune system works and are often indifferent to the uses to which the immune system can be applied. Ironically, the immunologist is probably using the tools and methods of cell and molecular biology more than those of applied immunology. It is worth remembering that in carrying out the experiments to determine how the immune system works, immunologists first used antibodies as reagents to identify and eliminate cells by virtue of unique antigens expressed on the cell surface.

The two all-encompassing questions that immunologists have studied for decades, that is, how immunological diversity is generated and what are the cellular events in an immune response, are probably of little interest to the readers of this volume. It is, of course, not essential that someone who wants to use antibodies to identify the location or temporal appearance of a given antigen in a tissue be fully informed about the scientific basis for the development of the reagent. But the understanding of some basic immunological principles and facts can lead to the more intelligent use of immunological reagents and make cell biology as rigorous and almost as much fun as immunology.

In this brief overview I will give an immunologist's view of some of the interesting questions, problems, facts, and theories that the users of antibodies may want to keep in the back of their minds when they use antibodies as reagents.

I. Clonal Selection

The idea of clonal selection, which is the underlying paradigm of modern immunology, is that the immune repertoire of an animal is formed by random rearrangements of roughly 10^4 gene segments. Each of the genes that results from this rearrangement encodes an antibody of a given specificity. This antibody is expressed on the surface of the B lymphocyte that produces it as a surface receptor. When the antigen with which the antibody reacts enters the system, it reacts with the receptor, activating the B cell to proliferate and actively secrete antibody molecules. The facts that there is a clonal expansion of B lymphocytes that have been selected by the specific antigen through its surface immunoglobulin receptor is the reason the phenomenon is called clonal selection, and all of the antibodies that the cell and its progeny produce are the same.

Inherent in the clonal selection paradigm is the counterintuitive notion of the preexistence of a cell containing the rearranged gene encoding the specific antibody. When injecting the mouse or rabbit with antigen, the investigator is not instructing the animal about what antibody specificity to produce; rather, he or she is causing the selection of preexisting cells that contain genes encoding the antibody of the desired specificity. The collection of B lymphocytes containing the rearranged antibody genes constitutes the immune repertoire of the animal.

II. Size of the Immune Repertoire

The fact that the preexisting immune repertoire determines whether an animal will make an antibody of a given specificity raises the question of the size of the immune repertoire. The 10^4 gene segments can rearrange to a theoretical 10^{10} or even 10^{11} different antibody specificities. This is a comforting notion because it means that there is a high chance that the animal will produce antibody of the specificity that the immunizer wants. However, a human has about 10^{12} lymphocytes and a mouse has only about 10^{10}. Given the fact that much of the repertoire is already taken up with cells that have come in contact with the appropriate antigen and have already been engaged in clonal expansion, a mouse may have fewer than 10^8 available B lymphocytes to be selected. This poses a theoretical conundrum and raises questions about the extent of immunological specificity, which will be discussed below. Because the cell biologists reading this volume intend to use antibodies as specific reagents, the question of just what is meant by immunological specificity is not a bit of arcana best left to immunologists. If there are not enough cells for all of the specificities, then perhaps there is not as much specificity as was thought to be present.

III. Specificity and Affinity Maturation

Specificity is defined arbitrarily. We usually think of it as the property of reacting with the antigen against which the antibody was raised, and with no other antigen, but ultimately that distinction is one of technological limitations of the sensitivity of the assay. For example, when the Ly1 antigen was first discovered in the mouse using fluorescence microscopy, it was thought to exist on a distinct and unique subset of T cells. With the advent of flow cytometric methods of analysis it was found that there are low levels of the antigen on other T cells and that there are even some B cells that express it. The antibody used was still reacting with the same antigens that it always did, but we were able to detect more of the low-level reactivity and so our idea of specificity, in this context, changed.

The affinity of the interaction between antigen and antibody is also an integral part of the question of specificity. *Affinity* is the strength of the interaction between the antigen and the antibody and can be measured as an affinity constant in those cases in which the antigen is in solution. The first antibodies that are made after antigen is injected are of the IgM class, and generally have a low affinity for antigen. As the immune response continues (or matures), and especially after a second injection of antigen, the class of antibodies switches from IgM to IgG. During this time the affinity of the antibodies increases. It was originally thought that the increase was due to the selection of those B cells expressing the highest-affinity receptors for antigen, but this has given way to the idea that the change in affinity is due to mutations that result in better binding of antigen to antibody. The general rule, then, is that IgG antibodies of higher affinity appear as the immune response progresses.

But as the affinity increases, structures closely related to the antigen but that may have bound with too low an affinity to be detected with an IgM antibody may now be detected. The use of IgG may therefore broaden the range of closely related antigenic structures that can now be detected, giving a different picture of the presence or quantity of the antigenic determinant being studied.

IV. Idea of Cross-Reactivity

By definition, antibodies have the ability to bind specifically to antigens; but as we have just seen, this can be a somewhat arbitrary definition. Historically it was noted that recovery from a disease protected one from further infection with the same disease but not others. Later it was shown that antisera that neutralized diphtheria toxin did not neutralize tetanus toxin. But Landsteiner showed, contrary to what is generally believed, that antisera can show a great deal of cross-reactivity, that is, react with more antigens than the one that was used to immunize the animal. The strongest reactions are with the homologous antigen, but in the synthetic antigen systems that Landsteiner studied, reactions with chemically related molecules was the rule rather than the exception. That being the case, one must use caution in interpreting data in which an antibody is used to identify the presence of a given molecule, because there is the possibility that the antibody is reacting with a chemically related molecule. With some good luck, this kind of cross-reactivity may not be important in the particular experiment in which it is being used, but the investigator needs to be aware of it at all times.

V. Monoclonality and Multispecificity

Given the above, it is no surprise that one of the most common errors made by those who use antibodies as reagents is to equate monoclonality and monospecificity. Because the immune response is clonal, an antiserum contains the product of all of the B cells that were activated by all of the antigenic determinants (epitopes) on the antigen. With the advent of monoclonal antibodies, investigators have available to them reagents that react with only one of the epitopes on a large antigen. However, that small chemical configuration with which the monoclonal antibody reacts may appear on unrelated antigens in other structures. If the determinant is in low concentration on other structures in a cell it might not have been detected with a conventional antiserum because the concentration of that particular specificity of antibody may have been too low in the serum. But it might now be seen with a monoclonal antibody. Therefore, paradoxically, using a monoclonal antibody may introduce more cross-reactivity.

Another problem with the monoclonal antibody as a reagent is that, because it is directed at only one epitope on an antigen, an alteration in that epitope that reduces the binding to antigen can be read as the absence of the entire antigenic structure. Glycosylation patterns, molecular association, and a host of other factors may alter the availability of the epitope, leading to erroneous information about the presence of the antigen. Unfortunately, there are no rules to choose a particular monoclonal antibody over another.

VI. A Cautionary Note

Immunologists and cell biologists can be seduced with equal ease into using monoclonal antibodies, often because of the common insecurity that drives people to use the most recent technological advance. There are times, of course, when it is imperative to use monoclonal antibodies; but there are times when the low-tech conventional antiserum may be preferable. The crucial fact to remember when using antibodies as reagents is that they are not very smart molecules. They will react with a range of molecular structures, so the cell biologist who is using them must be the clever one and know the limitations as well as the virtues of the reagent.

Suggested Readings

Cohn, M., and Langman, R. E. (1990). The protecton: The unit of humoral immunity selected by evolution. *Immunol. Rev.* **115,** 7.

Freitas, A. A., Rocha, B., and Coutinho, A. A. (1986). Lymphocyte population kinetics in the mouse. *Immunol. Rev.* **91,** 5.

Golub, E. S. (1987). Somatic mutation: Diversity and regulation of the immune repertoire. *Cell* **48,** 732.

Golub, E. S. (1991). The selfish id: An alternative view of the idiotypic network. *Scand. J. Immunol.* **33,** 489.

Golub, E. S., and Green, D. R. (1991). "Immunology: A Synthesis," 2nd Ed. Sinauer Associates, Sunderland, Massachusetts.

Kindt, T. J., and Capra, J. D. (1984). "The Antibody Enigma." Plenum, New York.

Landsteiner, K. "The Specificity of Serological Reactions," Dover, New York, 1936.

Tonegawa, S. (1983). Somatic generation of antibody diversity. *Nature (London)* **302,** 575.

CHAPTER 2

Production of Polyclonal Antibodies against Proteins and Peptides

Detlev Drenckhahn, Thomas Jöns, and Frank Schmitz

Department of Anatomy
Julius-Maximilians University
D-97070 Würzburg, Germany

I. Introduction

During the past two decades immunological laboratory methods have become powerful and indispensable research tools in cell biology, biochemistry, and molecular biology. The still increasing importance of immunoglobulins in cell research can be concluded from the percentage of articles published in the *Journal of Cell Biology* in which antibodies were used to address problems that would have been difficult or impossible to approach by other methods. In 1970 only about 1% of the studies relied on immunological methods, those using antibodies reached about 30% in 1980 and 70% in 1990. Immunological techniques enjoy broad application in cell research mainly because of their high specificity and sensitivity, which allow selective detection, isolation, and quantification of molecules down to the picogram level. The main advantage of antibodies is that they can be induced (raised) against virtually any desired macromolecule and even to substances below M_r 1000 if these are presented to the immune system in an appropriate way.

The most frequently used antibody-dependent techniques in modern cell biology are (1) visualization of macromolecules in cells and tissues by immunocytochemical techniques (73%), (2) identification of proteins in electropherograms by immunoblotting techniques (46%), immunoisolation of proteins by immunoprecipitation (22%), and quantitative determination of particular proteins by immunoassays (3%). (Numbers in parentheses refer to articles published in the *Journal of Cell Biology* in 1991). Other important immunological techniques in cell research are microinjection of antibodies into cells to disrupt or block the activities of individual proteins and their supramolecular assemblies, and the use of antibodies in binding studies and to map and define functional sites on proteins. Antibodies have also become important tools in molecular biology. Not only are antibodies used to screen expression libraries, but they also provide the only link between cDNA sequences and the nature and subcellular location of the encoded proteins. The underlying rationale is to produce fusion proteins in bacteria or to synthesize peptide sequences deduced from the known cDNA sequences and then to use these protein fragments and peptides to elicit antibodies for studying the many cell biological and biochemical aspects of the corresponding protein.

The success of immunological approaches in cell research is based largely on the specificity, cross-reactivity, and affinity of the antibodies available. Depending on the cell biological question to be solved with antibodies one must make certain decisions with respect to the antigen to be used for immunization and the procedures for testing titer and specificity of the antibodies elicited; furthermore, one must decide whether monoclonal rather than polyclonal antibodies may be the more promising tools for addressing a given problem. This article is restricted to the most commonly used procedures for production, purification, and characterization of polyclonal antibodies against proteins and peptides raised in rabbits and mice. The main emphasis will be on step-by-step protocols, most of which are being routinely used in our laboratory.

In addition, comments will be given concerning the purpose of the techniques described and the information provided by them. Moreover, we will briefly discuss when one should consider applying a given technique, and alternative methodological approaches. Some of these questions will be addressed in an introductory survey of immunglobulins. This article provides essential background information for those who are not familiar with the principles of practical immunology. For more detailed information the reader is referred to excellent reviews and monographs (Weir, 1978; Mishell and Shiigi, 1980; Harlow and Lane, 1988; Peters and Baumgarten, 1992).

II. Properties of Immunoglobulins

A. General

1. Structure of Immunoglobulins

The main purpose of the immune system is to bind and eliminate foreign molecules (antigens), such as bacterial toxins, that have invaded an individual. This task is mainly accomplished by immunoglobulins, which are water-soluble glycoproteins that carry the specific antibody activity (for reviews, see Davies and Metzger, 1983; Goodman, 1987; Darnell *et al.*, 1990). Immunoglobulins comprise about 20% of total protein in blood serum, with a concentration of about 15–20 mg/ml.

Figure 1 depicts the general structure of IgG, the most abundant immunoglobulin class in serum. All immunoglobulins are composed of two heavy chains and two light chains linked together by interchain disulfide bonds. Intrachain disulfide bonds give rise to loop domains. Under reducing conditions (presence of 2-mercaptoethanol or dithioerythritol) immunoglobulins are inactivated and fall apart into two heavy chains (50 kDa in IgG) and two light chains (23 kDa). Under nonreducing conditions the immunoglobulin tetramer migrates in sodium dodecyl sulfate-polyacrylamide gel electrophoresis (SDS-PAGE) at about 160 kDa. The two major immunoglobulin classes found in serum are IgG (constituting about 75% of serum immunoglobulins) and IgM (10%). IgM differs from IgG with respect to the heavy chain, which contains an additional constant loop domain (C_H4), is bigger (70 kDa), and bears more carbohydrate residues (18% in IgM, 4% in IgG). Moreover, IgM tends to form pentameric oligomers with a total mass of about 900,000. IgGs occur as four different variants (subclasses) of the heavy chains (IgG 1–4). IgG_1 is the predominant subclass (60–70% of total IgG) in serum.

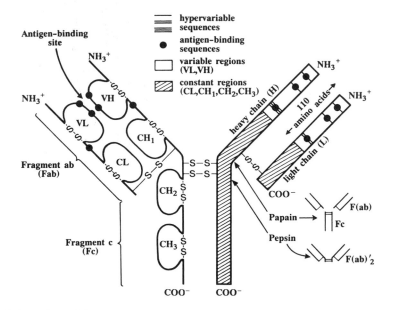

Fig. 1 Schematic model of an IgG molecule. Fragments generated by pepsin and papain treatment are also shown.

The variable regions contain the antigen-binding sites and consist of approximately 120 amino acids at the N terminus of both the heavy and light chains. During a process called *isotype switching* the gene sequences encoding the variable region of the heavy chain can combine with those of the constant regions of any of the heavy chains, thus giving rise to the different classes and subclasses of immunoglobulins. Because in polyclonal sera IgG (IgG$_1$) represents the predominant immunoglobulin class the term *antibody* or *immunoglobulin* as used in this chapter refers to IgG unless otherwise stated.

2. Survey of Cellular Mechanisms Involved in Immune Response

The process by which an individual can generate as many as 10^6–10^8 different variants of immunoglobulins (with respect to the sequence of the variable region) has been clarified to a large extent: antibodies are secreted by plasma cells that arise from precursor cells, the B lymphocytes. Each individual contains as many different subtypes of B cells (10^6–10^8), as immunoglobulins with different antigen specificities can be generated. Each B cell can synthesize and express only a single type of antibody, the antigen specificity of which is unique to this particular B cell. This striking heterogeneity of B cells is generated during fetal and perinatal development by random rearrangement and subsequent trimming (splicing, deletion) of the genes encoding the variable regions of the heavy and light chains. When a foreign molecule (antigen) is introduced into an individual the antigen will be bound by immunoglobulins (mainly of the monomeric IgM type) exposed on the surface of B cells. Because the specificity of immunoglobulins exposed on a given B cell differs from that of other B cells with respect to the variable region, the antigen will be bound by only those B cells that bear immunoglobulins on their surface with sufficient high affinity for the injected antigen (K_D 10^{-4}–10^{-12} M). The locus on an antigen that is bound by the antibody is defined as the *epitope*, whereas the complementary binding site on the immunoglobulin molecule is named the *paratope*. The paratope is formed by six of the seven different hypervariable stretches (complementarity determining regions, CDRs) present in the Fab portion of the antibody molecule (Fig. 1). Binding of an antigen to surface immunoglobulins of a B cell is the initial event that will eventually lead to proliferation of this particular B cell and subsequent differentiation to plasma cells, which will secrete the same type of antibody (with respect to its paratope) that had originally been expressed on the surface of the antigen-binding B cell.

After binding the antigen, B cells internalize (endocytose) the antigen into their lysosomal system, where the antigen is degraded (processed) into fragments. Some of these fragments will be bound by the major histocompatibility complex II (MHC II) (repertoire of 20–40 different variants). Binding to MHC II molecules occurs within the lysosomal system. Subsequently the MHC II–antigen complex will be exposed at the cell surface by exocytotic externalization. This process is defined as antigen presentation. If the presented antigen contains epitopes that differ from fragments of the animal's own proteins (that are also permanently exposed by MHC II molecules) these epitopes will be recognized by thymus-derived lymphocytes, the helper T cells. Helper T cells that bind to the antigen–MHC II complex will secrete specific growth hormones (interleukins) that stimulate the presenting B cells to replicate (clonal proliferation) and to develop into antibody-secreting plasma cells. Because only those helper T cells that do not bind to the body's own proteins have been selected during embryonic development, helper T cells are important for the control of immunological self-tolerance.

The extent of an immune response depends largely on the number and activity of committed helper T cells ready for stimulation of antigen-presenting B cells. This task is done primarily by macrophages, a population of leukocytes specialized for ingestion (phagocytosis) of particulate matter such as dead cells, cell fragments, or macromolecular complexes. Macrophages will nonspecifically remove such materials, including foreign antigens, and will present fragments of the ingested materials bound on their surface by MHC II molecules. Binding of helper T cells to these MHC II–bound antigen fragments is a strong stimulant for proliferation of the helper T cells. Thus antigen-presenting macrophages will effectively support the immune response to a certain antigen by causing proliferation and activation of a fraction of helper T cells specific for a particular antigenic site.

Clonal proliferation and activation of B lymphocytes takes place primarily in lymphatic tissues, such as lymph nodes or the lymph follicles present in the spleen or in the mucosa of the alimentary tract. Injection of antigens into lymph nodes or the spleen takes advantage of this situation. The clonally proliferated B lymphocytes will eventually differentiate into plasma cells, which begin to secrete antibodies within a period of usually 5 to 20 days after application of the immunogen. This primary immune response is characterized by secretion of IgM. Activated B lymphocytes gradually begin switching over to synthesize IgG-specific heavy chains (γ chains) instead of IgM heavy chains (μ chains) and differentiate to plasma cells secreting IgG instead of IgM. This phenomenon is called isotype switching. The variable region of the secreted immunoglobulins remains unchanged by this switch. A fraction of the B lymphocyte clones (as well as clones of helper T cells) will differentiate into memory cells that spread over the lymphatic tissues throughout the body. When the same antigen is reintroduced weeks, months, or even years later memory cells will be able to proliferate immediately and develop into antibody-secreting plasma cells, thus providing the cellular basis for the rapid and heightened secondary (anamnestic) antibody response following a secondary exposure to the antigen (booster immunization). This secondary response leads to antibody levels in serum (titers) that are higher than the titers of the primary response. The predominant immunoglobulin class secreted during the second response is IgG.

B. Polyepitopic, Monoepitopic, and Monoclonal Antibodies

Most proteins are complex antigens containing two or more epitopes (antigenic determinants) that will cause stimulation of different B lymphocyte clones. The resulting antibody response is polyclonal and, in addition, polyepitopic (Benjamin, 1984; Goodman, 1987). The fraction of immunoglobulins that binds to the same stretch of sequence (or conformational site) is defined as the monoepitopic antibody fraction. This fraction is a polyclonal mixture of immunoglobulins with different affinities for this site (Berzowsky and Berkower, 1984). The lymphocyte clone with the best complementary fit for a given epitope will evoke immunoglobulins that bind with highest affinity to this epitope (K_D of down to $10^{-12} M$). In addition, other clones will also be stimulated. These may bind to the epitope less tightly and, accordingly, will lead to immunoglobulins with lower affinities (K_D up to $10^{-4} M$).

Monoclonal antibodies are defined by their origin from a single B lymphocyte or plasma cell (see Article 3). Such antibodies are obtained by *in vitro* fusion of activated B lymphocytes (or plasma cells) with a plasmacytoma tumor cell line. The resulting hybrid cells (hybridomas) behave like tumor cells and undergo continuous proliferation. The immunoglobulins secreted by a clone derived from a single hybridoma cell are monoclonal by definition.

C. Specificity, Cross-Reactivity

1. Species Specificity, Isoform Specificity

A poly- or monoclonal antibody is defined as specific when it reacts selectively with the antigen used for immunization and not with any other antigen. Often antibodies show selective specificity only for the protein of the animal species from which it was isolated (species specificity). But in many cases a specific antibody will be species nonspecific and react with a given protein in many species, sometimes throughout the entire eukaryotic kingdom. Species specificity cannot be predicted. It depends largely on the degree of evolutionary conservation of a given epitope or protein. A further type of specificity is known as tissue specificity or isoform specificity. Isoform-specific antibodies react more or less selectively with the protein isoform that is present in the tissue from which the protein was purified. At the same time such isoform-specific antibodies may be species nonspecific and react in a tissue-specific manner in a variety of animal species.

Isoform specificity is more frequently observed in monoepitopic or monoclonal antibodies but may also occur in polyepitopic antibodies. For instance, polyclonal antibodies raised against myosin of cross-striated muscle will normally not react with myosins of smooth muscle or nonmuscle cells in the same animal, but will react with cross-striated types of myosin in many other animal species (Gröschel-Stewart and Drenckhahn, 1982). This kind of isoform specificity may even occur at the cellular level. In the chicken intestinal epithelium the brush border (terminal web, TW) isoform of β-spectrin (TW 260) differs immunologically and by sequence from the β-spectrin isoform restricted to the basolateral cell surface of the same cell type (235-kDa spectrin) (Glenney and Glenney, 1983). A single neuron may contain at least three immunologically different isoforms of ankyrin, one present in nodes of Ranvier (erythrocyte-related ankyrin), one in the cell body (erythroid ankyrin), and a third isoform, brain ankyrin, along the remaining plasma membrane of an individual neuron (Kordeli and Bennett, 1991).

Such restricted specificities of antibodies are often observed only with the native proteins (e.g., are seen in immunoprecipitation, immunoassays, and immunostaining of whole cells and tissue sections) but may not be detected in immunoblots of electrophoretically separated proteins. Denaturation of proteins by SDS-PAGE may expose conserved epitopes shared by different isoforms. Such hidden (cryptic) epitopes may not be accessible to antibodies on the native protein. For instance, polyclonal antibodies to platelet myosin do not react with native smooth muscle myosin, but display binding to SDS-denatured smooth muscle myosin in immunoblots (Larson *et al.*, 1984). The reverse may also be the case, namely, no detectable signal in immunoblots but binding of the antibody to the native protein.

These examples have been mentioned to stress that it is often important to know the kind of specificity of the antibodies applied before drawing conclusions as to the occurrence of a certain protein in a certain cell type or subcellular structure.

2. Cross–Reactivity

Antibodies may cross-react with protein bands and cellular structures that do not contain the protein against which the antibody has been raised. One possibility for cross-reactivity is that the antiserum is contaminated with antibody fractions not related to the immunogen. But even affinity-purified and monoclonal antibodies may show strong cross-reactivity. In these cases the cross-reacting protein most likely contains a shared epitope related or identical to that present on the immunogen. It is often difficult to eliminate the cross-reacting fraction of immunoglobulins from polyclonal antisera (impossible for monoclonal antibodies).

3. Contaminating Immunoglobulins

The most frequent reason for reaction of polyclonal antisera with more than one protein is contamination of the immunogen with other proteins. Because the strength of the immune response to a given antigen does not necessarily depend on the dose of antigen administered it is possible that a minor contaminant may be a strong immunogen causing high-affinity antibodies that dominate the immune response.

4. Autoantibodies, Spontaneous Antibodies

The other class of contaminating immunoglobulins are autoantibodies, that is, immunoglobulins directed against proteins (molecules) of the animal itself (anti-self antibodies). Such autoantibodies are normally already present in the serum sample taken from the animal prior to immunization (preimmune serum). Frequently observed autoantibodies in rabbits are directed against nuclear proteins (Nigg, 1988). But spontaneous antibodies against cytoskeletal proteins, particularly intermediate filament proteins, are also rather frequent (Osborn *et al.*, 1977; Osborn and Weber, 1982). Intermediate filament antibodies rarely occur in guinea pigs. Rabbits often contain antibodies reacting with the Golgi apparatus and centrosome (Conolly *et al.*, 1979). The apical membrane of collecting duct epithelia of the kidney is another structure to which autoantibodies occur in rabbits (unpublished observations).

Further types of spontaneous antibodies frequently observed in rabbits are directed against arthropods (probably against proteins of the host's parasites). The possibility of pre-existing antibodies should be excluded before starting to immunize an animal by screening the preimmune serum by immunoblotting and immunostaining. In any case samples of preimmune serum should be saved (at least 10 ml per animal).

5. Antiidiotypic Antibodies

An antibody developed against an epitope of an immunogen is termed idiotypic antibody or antibody 1. The antigen-binding site of the idiotypic antibody (its paratope) is by definition complementary to the sequence of the immunogenic epitope. Overshooting immune response against the immunogen may be modulated (reduced) by spontaneous generation of antibodies directed against the variable region of the idiotype antibody. Such antiidiotypic antibodies (antibody 2) (Jerne, 1974; Köhler *et al.*, 1989) will form complexes with the idiotypic antibodies and may cause a fall in the titer of an idiotypic antibody. A fraction of the antiidiotypic antibodies (antibody 2b) may be perfectly complementary to the paratope of the idioptypic antibodies. Thus, antibodies 2b will mimic the physicochemical properties of the epitope of the antigen to which the paratope of the idiotypic antibody is complementary.

Antiidiotypic antibodies may be the reason for the occurrence of antibodies in serum that cross-react with proteins known or suspected to serve as physiological binding sites for the protein used as immunogen. For instance, antibodies raised against insulin may induce antiidiotypic antibodies directed against the insulin-binding site (paratope) on the first, idiotypic antibody. These antiidiotypic antibodies may not only recognize the insulin-binding site on the idiotypic antibody but may also recognize the insulin-binding site on the insulin receptor. Thus the antiidiotypic antibody can behave like insulin (the primary immunogen) in that it binds to the insulin receptor (Sege and Petersen, 1978; Gaulton and Greene, 1986).

Affinity purification of the antiserum to the immunogen will eliminate the antiidiotypic immunoglobulin fraction. On the other hand, the antiidiotypic antibodies can be purified by affinity purification to the cross-reacting protein (or protein band in immunoblots). To prove that these antibodies are anti-idioptypic, that is, directed against the antigen-binding sequence of the idiotypic antibody, addition of the idiotypic antibody [F(ab) fragments] to the purified antiidiotypic antibody should gradually cause neutralization (inhibition) of binding of the antiidiotypic antibody to the cross-reacting protein (protein band). Antiidiotypic antibodies may be valuable for the identification of binding sites between protein 1 (the primary immunogen) and protein 2 (the protein to which the antiidiotypic antibody binds) (Djabali *et al.*, 1991; Jöns and Drenckhahn, 1992). Article 19 describes the characterization and use of anti-idiotypic antibodies.

III. Polyclonal versus Monoclonal Antibodies

Although the production of monoclonal antibodies has become a routine procedure during the past 10 years the use of polyclonal primary antibodies in cell research has not changed much. About two-thirds of the studies using antibodies published in the *Journal of Cell Biology* in 1990 applied polyclonal primary antibodies, the remaining third used monoclonals (25%) or a combination of monoclonal and polyclonal antibodies (10%). The main reason for the preferential use of polyclonal antibodies appears to be trivial: their generation is rather simple, quick, and inexpensive whereas the production of monoclonal antibodies requires equipment and skills in cell culture techniques and is relatively time consuming and expensive. In addition, a suitable test system must be established to screen hundreds of cell culture supernatants in order to identify and select the clones producing the monoclonal antibodies of interest. But even if facilities, time, and money are available for the production of monoclonal antibodies it depends on the particular research problem pursued whether a monoclonal approach is worthwhile. Because conventional production of monoclonal antibodies requires immunization of mice (rats), small samples of polyclonal antibodies will be obtained during test bleeds and exsanguination as the by-products of the approach.

A. Properties and Applications of Polyclonal Antibodies

As outlined above, polyclonal antibodies raised against a protein are usually directed against more than one epitope. This is the reason why polyclonal antibodies can often be used for direct immunoprecipitation without the need for addition of immobilized protein A (see Article 7). A further advantage of polyepitopic specificity is that denaturation of proteins by SDS or by fixation with aldehydes usually will not destroy all epitopes so that polyclonal antibodies can be used in most instances for both immunoblotting and for immunostaining of chemically fixed cells and tissues. Fixation is particularly important for immunocytochemical localization of soluble proteins, which may otherwise be extracted during permeabilization or undergo artifactual redistribution.

Polyclonal antibodies against proteins are often not isoform specific and not sensitive to posttranslational modifications of the molecule. This broad reactivity of polyclonal antibodies may be of advantage in many cases. Isoform specificity of polyclonal antisera may be obtained later by affinity purification to peptide sequences or fusion proteins that contain isoform-specific sequences. This kind of antibody selection is possible only if the isoform-specific sequence is contained in the original antigen (see Section XII).

Affinity purification of polyclonal antibodies to SDS-denatured protein bands in immunoblots or to aldehyde-fixed antigens (dotted on nitrocellulose or coupled to column material) may allow the selection and concentration of IgG clones that can be used for detection of the antigen even under harsh conditions as, for example, in aldehyde-fixed cells and tissue embedded in plastic or paraffin (see Section XIII).

B. Properties and Applications of Monoclonal Antibodies

Monoclonal antibodies enjoy a constant specificity and affinity and are theoretically available in unlimited amounts.

By definition monoclonal antibodies are monoepitopic. This property allows mapping of functional sites on proteins. However, this monoepitopic specificity often makes monoclonal antibodies sensitive to any kind of modification of proteins (epitopes) such as denaturation by SDS, fixation with aldehydes, or posttranslational modifications. Therefore, it should be borne in mind that the absence of reactivity in a certain assay (e.g., in immunoblots) may be a result of inaccessibility of the epitope under the given conditions. A polyclonal mixture of different monoclonal antibodies specific for a given protein may help to overcome this problem.

Monoclonal antibodies are often specific for isoforms and spliced variants. This may be of advantage in certain cases but may be a disadvantage in cases in which more general aspects of the protein in question are to be addressed.

Cross-reactivity is rare but if it occurs it is due primarily to shared epitopes of otherwise unrelated proteins. Cross-reactivity of monoclonal antibodies cannot be eliminated by absorption or affinity purification.

Nonspecific cross-reactivity is often observed with IgM-type monoclonal antibodies. Even weak affinities of a single paratope for an unrelated protein may result in significant nonspecific binding caused by the several binding sites present on a typical pentameric IgM molecule. This is the reason for the nonspecific coimmunoprecipitation of unrelated proteins often observed with IgMs.

Production of monoclonal antibodies is the way of choice to obtain immunoglobulins against proteins that cannot be purified or sequenced or that are available in low amounts (microgram level).

Monoclonal antibodies are powerful tools when used to identify proteins by blocking their specific activities. In a kind of shotgun approach cellular fragments or even whole cells can be used for immunization and the resulting monoclonal antibodies are then assayed in an appropriate test system for blocking activities. In a similar approach new protein components of whole organelles, membrane domains, or supramolecular assemblies can be identified. In these cases immunostaining and immunoblotting may be the assay system of choice.

Cell culture systems have been developed that allow *in vitro* immunization and production of monoclonal antibodies. Low amounts of antigen (below 1 μg) are sufficient for immunization (for reviews see Reading, 1982; Peters and Baumgarten, 1992).

C. General Recommendation for Production of Polyclonal versus Monoclonal Antibodies

1. If the purpose of the antibody to be raised is more general, that is, if the antibody is to be used for immunolocalization as well as for immunoblotting, immunoprecipitation, and immunosorbent assays, one should try to elicit polyclonal antibodies in rabbits. The amount of serum available is generally high enough (100 ml or more) for all needs, including affinity purification. A specific high-affinity antiserum can be extremely valuable.

If the protein cannot be obtained with sufficient high purity (>90%) further purification of the antigen by SDS-PAGE and excision of the corresponding protein band is the way of choice. A minimum of 100–200 μg of protein is needed as immunogen per rabbit. If the protein (protein band) is adsorbed on nitrocellulose filters even a few micrograms may be sufficient to elicit antibodies in rabbits, provided the excised nitrocellulose material is injected into the spleen.

Two or more rabbits should be immunized at the same time. If further progress in research depends urgently on a good antibody a double strategy is recommended, namely to immunize rabbits and mice (rats). The genetic background of one of the immunized species may be more susceptible to the immunogen applied and give a better immune response to it than that of another species. If the polyclonal sera obtained do not suffice one can still decide, even several months later, to boost the mice or rats and to use them for production of monoclonal antibodies. In any case it is good to possess both poly- and monoclonal antibodies.

2. If the protein is difficult to purify and obtain in sufficient high quantities there are various possibilities to obtain antibodies, depending on whether the sequence is known or not.

a. If the sequence is known, fusion proteins made by molecular cloning in a procaryotic expression system, or synthetic peptides, can be used for immunization. It is best to try both approaches and to immunize rabbits and mice (rats) at the same time. Generally, the peptide approach is less expensive and quicker but its success is uncertain. In our hands only 15% of the peptides used for immunization will evoke antibodies recognizing the full protein (experience with 50 different peptide sequences). Depending on the affinity and specificity of the polyclonal antibody raised one still has the option to take the mice (rats) immunized with fusion proteins and try to produce monoclonal antibodies.

b. If the sequence is not known one should try to obtain sequence data by microsequencing of the protein band blotted on appropriate materials [e.g., polyvinylidene difluoride (PVDF) membranes]. If stretches of the sequence are obtained by this approach peptides can be synthesized for immunization. The corresponding nucleotide sequence (if not too degenerate) may allow a molecular biology approach to determine the full- or partial-length sequence and to produce fusion proteins or further peptides for immunization.

c. In any case, the inhomogeneous/rare protein preparation can be used for immunization of mice (rats) or, if experience is available, for "*in vitro*" immunization, hoping to elicit monoclonal antibodies reacting with the protein in question.

d. As outlined in Section VI, a few micrograms of a homogeneous protein band blotted on nitrocellulose and implanted into the spleen may evoke a significant antibody response.

IV. Processing of Antigens for Immunization

A. Antigenicity of Proteins and Peptides

1. Foreignness

Proteins purified from a given animal species are often nonimmunogenic or are poor antigens when used as immunogen in the same animal species. This is particularly true of ectodomains of integral membrane proteins or serum proteins. Because these types of proteins are permanently exposed to the immune system of the animal they are potent candidates for induction of self-tolerance. Yet when introduced into other vertebrate species these proteins are more likely to evoke production of specific antibodies. In general terms, immunogenicity of a protein tends to correlate positively with increased evolutionary distance between the species chosen for immunization and the animal source used for immunogen preparation. However, there may be exceptions to this rule: in our hands, certain cytoplasmic proteins of rabbits turned out to be good antigens when injected in rabbits (e.g., myosin and α-actinin from skeletal muscle). Proteins with a high degree of evolutionary conservation, such as calmodulin, actin, tubulin, or peptide hormones, are generally bad antigens in any mammalian species. Therefore, injection of higher doses of such proteins may be required (see below). However, denaturation of such proteins by boiling, treatment with SDS, derivatization with dinitrophenol, cross-linking with glutaraldehyde, or modification with other chemicals may improve immune response. Most likely these denaturing conditions cause exposure of hidden epitopes and thus circumvent self-tolerance.

2. Genetic Background

The genetic constitution of an individual has substantial influence on the immune response. Some individuals of an animal species may produce specific antibodies against an immunogen, whereas other individuals do not produce antibodies, or they produce cross-reacting antibodies. For example, strain 2 guinea pigs produces antibodies against poly-L-lysine but strain 13 guinea pigs do not (Goodman, 1987). It is therefore largely a question of the genetic background and luck whether the individual animal immunized will produce a good antibody. If possible, one should immunize several individuals at the same time to be sure to obtain, at least in one animal, a good antibody. If rabbits and mice do not develop antibodies against a given protein other species may be tried for immunization, such as guinea pigs or chickens (for details of antigen application and bleeding in these species see Herbert, 1978).

3. Molecular Structure and Size of Epitopes

Immunogens are normally large molecules and immunogenicity is mainly a function of molecular size and complexity. In general, peptides below a length of 10 amino acids ($M_r \sim 1000$) are nonimmunogenic when injected in soluble form. Even larger peptides (M_r 1000–10,000) are normally less immunogenic than large proteins above M_r 40,000. However, even single amino acids may act as strong immunogens when bound to carrier proteins (see, e.g., Meyer *et al.*, 1991). Such immunogens that require coupling to carrier proteins are termed haptens.

Several studies have shown that the size of an antigenic determinant (epitope) of a protein is on the order of four to eight amino acids (Getzoff *et al.*, 1987; Geysen *et al.*, 1987). Three features are important for immunogenicity: chemical complexity, hydrophilicity, and accessibility. Homopolymers (e.g., polylysine, polyalanine) are generally nonimmunogenic regardless of size. Copolymers of two or more different amino acids are much better immunogens. Antigenicity correlates positively with hydrophilicity of the epitope (Kyte and Doolittle,

1982). Hydrophobic peptides or protein stretches (e.g., transmembrane domains) are normally poor immunogens.

Antigenic epitopes are mostly located on the free surface of a protein, where it can be readily recognized by surface-bound IgM molecules of B lymphocytes. Two types of antigenic determinants can be distinguished: (1) discontinuous (conformational) epitopes and (2) continuous (sequential) epitopes. Discontinuous epitopes consist of residues that are separated from each other in the amino acid sequence but are brought into proximity by the tertiary structure (folding) of the protein. Continuous epitopes correspond to a stretch of a sequence in the primary structure of a protein. Conformation-dependent (discontinuous) epitopes may be unfolded and destroyed during the denaturing conditions of SDS-PAGE and will no longer be detectable when the protein is electroblotted on nitrocellulose paper. This is probably the main reason why monoclonal antibodies often do not bind to protein bands in immunoblots. Antibodies directed against sequential epitopes will normally bind to both native and denatured proteins. If the main reason for raising the antibody is to use the antibody during immunoblotting experiments, it is recommended that the protein be denatured with SDS prior to immunization. Under these conditions epitopes may be exposed that remain preserved in immunoblots.

4. Number of Epitopes

The number of epitopes recognized by the immune system of an individual depends mainly on the molecular size of the protein. Large proteins such as thyroglobulin (700 kDa) may evoke antibodies against as many as 40 different epitopes, whereas the number of epitopes on hen egg albumin (45 kDa) is about 5. The location and chemical nature of epitopes may partially differ between individuals, but the number of epitopes (on a given protein) to which antibodies are developed remains rather constant (Benjamin, 1984; Goodman, 1987). Certain epitopes may be strongly immunogenic and evoke high-affinity antibodies that dominate the immune response (immunodominance). Often the terminal sequences (C or N terminus) of proteins contain dominant epitopes. To a certain degree an immunodominant epitope may interfere with formation of immunoglobulins against other epitopes of the same molecule. If the immunodominant epitope is located on portions of a protein that belong to conserved domains within a given protein family these dominant epitopes may prevent the generation of isoform-specific antibodies. If the aim of immunization is to obtain isoform-specific antibodies, methods must be applied to avoid or suppress the formation of immunodominant immunoglobulins (see Section XII).

B. Native Proteins as Antigens

As outlined above, proteins used for immunization must be of high purity to avoid generation of antibodies against contaminating antigens. To control purity, 10 μg of the protein antigen should be loaded on a single lane of a 10% SDS slab gel. If no contaminating bands are seen by Coomassie blue staining purity of the protein is better than 90%. Contaminating glycoproteins may not be seen by Coomassie blue staining. In all cases in which glycoproteins can be expected as contaminants, for instance, when working with noncytosolic proteins, silver staining is recommended. Irrespective of the presence of minor contaminants one should try the native antigen. The immune response to native proteins is often better than to denatured proteins. In most cases (but not always) the amounts of minor contaminants are too low to evoke significant immune responses. Depending on the specificity of the antiserum one still has the option to affinity purify the specific immunoglobulins at the respective protein bands blotted on nitrocellulose.

To be safe it is recommended that two rabbits be immunized with the native protein fraction and one or two additional animals with the protein band excised from SDS-PAGE. Alternatively it is possible to immunize with a mixture of native and SDS-denatured proteins.

C. Fusion Proteins

Proteins expressed partially or full length as fusion proteins in *Escherichia coli* frequently form cytoplasmic aggregates termed inclusion bodies. Inclusion bodies are relatively insoluble compared to the majority of the endogeneous bacterial proteins, so the strategy for purification of fusion proteins is to solubilize the bulk of bacterial proteins by detergents or urea and to sediment and further purify the inclusion bodies (for review see Marston, 1986). If the fusion protein is not incorporated into inclusion bodies, partial purification of the fusion protein by conventional procedure (gel filtration, ammonium sulfate precipitation, affinity purification) is necessary. Immunization may also be performed with protein bands excised from SDS slab gels or from nitrocellulose strips.

1. Purification of Inclusion Bodies

a. General Procedure

1. Induce expression of fusion proteins in the log phase of bacterial growth. If *lac* promoter-based expression vectors are used expression of fusion proteins is optimal 2–3 hr after induction of the *lac* promotor with 2 mM (isopropyl-β-D-thiogalactoside) (IPTG). Test expression at various times after induction by SDS-PAGE of bacterial pellets. The pellet (1.5 ml of *E. coli* culture) is dissolved in 100 μl of SDS sample buffer; 2–5 μl of sample buffer-lysed *E. coli* proteins is loaded per lane of a minigel.

2. Induce 100 ml of the *E. coli* culture and sediment bacteria at 3000 g (20 min, 4°C).

3. Wash the pellet once for 20 min in 50 ml of ice-cold PBS, containing 0.3 mM phenylmethylsulfonyl fluoride (PMSF) and 10 U Trasylol (aprotinin)/ml.

Several ways to purify fusion proteins and/or inclusion bodies are possible; as described in the following sections.

b. Separation of Fusion Proteins by SDS-PAGE.

Run the bacterial pellet on SDS-PAGE as described above. Excise the fusion protein band from the stained gels or nitrocellulose blots and use them for immunization as described in Section IV,D.

c. Purification of Inclusion Bodies (First Option)

1. Resuspend the pellet in 10 ml of ice-cold 1 M urea (in H$_2$O) and disrupt bacteria by sonification for 1–2 min.

2. Centrifuge inclusion bodies and cellular fragments and resuspend the pellet in 10 ml of ice-cold 2 M urea, sonify for 1–2 min, and repeat the procedure by increasing the urea concentration stepwise to 3, 4, 5, 6, 7, and 8 M, respectively.

3. Inclusion bodies (pellet) usually become dissolved in 6–7 M urea.

4. Check all steps by SDS-PAGE.

5. Dialyze the fraction(s) containing the dissolved inclusion proteins against phosphate-buffered saline (PBS). Occasionally fusion proteins will precipitate during dialysis.

d. Purification of Fusion Proteins (Second Option).

If the first option (see above) does not work efficiently, try the following procedure.

1. Resuspend the bacterial pellet in 10 ml of a solution containing to 10 mM ethylenediaminetetraacetic acid (EDTA) in 100 mM NaCl, 50 mM Tris (pH 8.0, 4°C) to destabilize the outer bacterial membrane.

2. Centrifuge at 300 g (20 min, 4°C) and dissolve the pellet in 4 ml of 15% sucrose [in 10 mM EDTA, 50 mM Tris (pH 8.0)] to which 1 ml of freshly prepared lysozyme (10 mg/ml in H$_2$O) is added. Incubate for 15 min on ice, prior to addition of 1 mg DNase I (in 100 μl H$_2$O) and 1 ml of MgCl$_2$ (1 M in H$_2$O). Then let stand on ice for another 5 min. Afterward add 3 ml of a solution containing 1% (w/v) Triton X-100, 0.5% (w/v) deoxycholate, 0.1 M NaCl, and 10 mM Tris-HCl (pH 7.4). Mix and incubate on ice for another 15 min.

3. Centrifuge at 3000 g (20 min, 4°C) and wash the pellet three times with 5 ml of 1.75 M guanidinium-HCl (in 1 M NaCl, 1% Triton X-100, pH 8.0) to remove contaminating proteins and membrane fragments. Inclusion bodies will normally remain intact under these conditions.

4. Wash the pellet once with 10 ml of mM Tris-HCl (pH 8.0) and dissolve inclusion bodies in 7–8 M urea, followed by dialysis against PBS as described before.

Procaryotic expression vectors (e.g., pRSET; Invitrogen, San Diego, CA) have been described that contain several histidine residues in the amino-terminal portion of the fusion protein. This histidine-rich stretch can be used to purify the fusion protein by immobilized metal ion affinity chromatography.

For immunization either the soluble fusion proteins or the pellet of inclusion bodies can be used. Further purification by SDS-PAGE and immunization with excised, electroeluted or electroblotted gel bands (nitrocellulose blots) is recommended.

D. SDS-Denatured Proteins Separated by Polyacrylamide Gel Electrophoresis

Proteins that are difficult to purify in native form may be separated by SDS-PAGE. The protein band of interest can be excised and homogenized for immunization. In some instances SDS-denatured proteins may be better antigens than the same protein in its native state. But the reverse is also often the case (see Section IV,B).

Excised Bands of Stained and Fixed Acrylamide Gels

1. Run SDS-PAGE as normal.

2. Stain and fix the gel in 0.25% Coomassie blue G250, 45% isopropanol (propane-2-o1), 9.25% glacial acetic acid in H_2O for 15–30 min.

3. Destain in 5% glacial acetic acid, 7.5% isopropanol in H_2O for 30 min.

4. Carefully cut out the stained protein bands with a sharp scalpel or razor blade. Excised bands can be stored in destaining solution at 4°C or frozen.

5. Equilibrate the gel slices with PBS (pH 7.5) (three times, 5 min each) to neutralize and partially remove acetic acid.

6. Mince the gel bands with a razor blade and with a minimum volume of PBS and homogenize with a Dounce (Wheaton, Millville, NJ) homogenizer or with a loose-fitting motor-driven Teflon pestle. This may take 20 or more passes.

7. Emulsify the homogenate with an equal volume of Freund's complete adjuvant by sonication (three times, 10 sec each, highest setting). A water (gel)-in-oil emulsion will form (see Section V,A).

8. Inject subscapularly in rabbits or intraperitonally in mice. For booster injections use Freund's incomplete adjuvant.

Fixation of the gel by isopropanol (propane-2-ol) and acetic acid probably helps to retard the release of the proteins from the gel (depot effect). The disadvantage of this method is that polyacrylamide may cause granulomas and sterile abscesses at the sites of injection. To avoid this possibility, the proteins must be electroeluted from the gel slices. However, the loss of proteins during electroelution is normally rather high (>70%), and in our hands the eluted proteins do not give better immune responses.

Fixation and Coomassie blue staining of proteins can be avoided when using $CuCl_2$ as negative stain (Lee *et al.*, 1987). This procedure is recommended when trying to elute the protein but may be also used for immunization with excised gel bands.

1. Run SDS-PAGE as normal.

2. Place the gel in prestaining solution [0.1% SDS, 190 mM Tris (pH 8.8)] for 10 min.

3. Rinse briefly (few seconds) with H_2O.

4. Place in an excess volume of 0.3 M $CuCl_2$ in H_2O. $CuCl_2$ will form whitish-blue ternary complexes with Tris and SDS (if Tris and SDS have been washed out during step 3 reequilibrate the gel with prestaining solution and repeat steps 3 and 4). Protein bands remain unstained against a whitish background (takes 10–30 min).

5. Cut out the protein bands (using a dark background to visualize the bands) and remove Cu^{2+} by three incubation steps (10 min each) with 0.25 M EDTA in 0.25 M Tris (pH 9). Bands can be processed either for electroelution of proteins or for homogenization and immunization as described above for Coomassie blue-stained bands.

E. Proteins Transferred to Nitrocellulose

Instead of using excised polyacrylamide bands for immunization proteins can be spotted or electroblotted on nitrocellulose filters containing the adsorbed protein and used for immunization (Knudsen, 1985). When injected into animals nitrocellulose acts as a depot adjuvant. The adsorbed proteins will be slowly released from the nitrocellulose material (about 50% release within 2 weeks). The nitrocellulose strips (fragments) may be either administered (implanted) subcutaneously or injected into the spleen. When injected into the spleen only a single protein band containing 0.1–1 μg (mice) or 1–10 μg (rabbits) of protein adsorbed to nitrocellulose may evoke significant antibody titers (Nilsson *et al.*, 1987). Alternatively, nitrocellulose material may be dissolved in dimethyl-sulfoxide (DMSO) and the protein–nitrocellulose solution emulsified with Freund's adjuvant and used for immunization. But in this case the total amount of protein used for a single immunization step should be much higher (at least 10–20 μg for mice, 50–100 μg for rabbits).

1. Transfer the proteins to nitrocellulose sheets either by direct spotting (a vacuum dot-blot apparatus is recommended) or by electroblotting of SDS-slab gels as normal.

2. In the case of electroblotting visualize the blotted protein bands with 0.5% Ponceau S in 3% (w/v) trichloroacetic acid (TCA) in H_2O.

3. Destain the background with several washes of H_2O.

4. Localize the desired protein band with the help of molecular weight markers or by immunostaining of a side strip cut from the margins of the sheet. Excise the protein bands with a sharp scalpel.

5. Remove Ponceau S stain by several washes with PBS.

6. Cut the nitrocellulose into fine pieces, add 1 ml of PBS, and sonicate several times (e.g., six times, 10 sec each). The material (particles of 0.5 mm^2 or smaller) is now ready for subcutaneous administration or injection into the spleen. Suspend in 200 μl (mice) or 1 ml (rabbits) of PBS. For injection into the spleen much lower amounts of absorbed protein are needed.

7. If they are to be dissolved in DMSO, nitrocellulose strips must be completely dry (dry at 70°C for several hours).

8. Dissolve the excised band in a minimal volume of DMSO (0.5–1 ml). Sonification may be necessary.

9. Add an equal volume of PBS and emulsify with Freund's adjuvant as described in Section V,A.

F. Peptides as Antigens

The main reasons for using synthetic peptides as immunogens are the following:

1. The peptide itself may be a physiologically occurring molecule (e.g., a transmitter, growth factor, or hormone).

2. The peptide sequence may be part of an isoform-specific portion of a protein. Antibodies against this sequence would allow discrimination among different isoforms of that protein.

3. The peptide may be part of the sequence of a protein that is difficult to purify. The sequence to be used for immunization may be deduced from either the known nucleotide sequence or obtained by microsequencing of blotted protein bands. Antibodies against the synthetic peptide may be used for screening expression libraries or for other kinds of immunodetection and purification of the protein in question.

1. Choice of Sequence

As outlined above, antigenicity of a polypeptide correlates positively with hydrophilicity and complexity of the sequence. Hydrophobic sequences are poor antigens. Therefore, if one has the choice in selecting a sequence the following points should be considered.

1. The sequence of the peptide should be as long as possible. We usually take a stretch of 10–20 amino acid residues for immunization. The chance of obtaining an antibody reacting with the native protein increases with the length of the peptide.

2. A terminally located single cysteine residue allows coupling of one end of the peptide sequence to a carrier protein by *m*-maleimidobenzoyl-*N*-hydroxysuccinimide ester (MBS) (see Section IV,F,2,b below). If cysteine is absent from the sequence a cysteine residue may be added to one end of the peptide. If lysine is absent from the sequence the peptide can be coupled via its free N-terminal amino group to carrier proteins using glutaraldehyde as cross-linker (see also point 4 below). But even if lysine is present in the sequence glutaraldehyde coupling to a carrier protein will work in most cases. However, the way of coupling and the effects of formation of intermolecular cross-links on antibody specificity are difficult to predict.

3. Be sure that the sequence lacks potential glycosylation sites [Asn-X-Ser(Thr)]. The antibody generated may not bind to the native protein if it is glycosylated at this site. The same may be true for other posttranslational modifications such as fatty acylation (see below).

4. If the peptide sequence forms a continuous epitope on the protein surface (see above) the corresponding antibody will most likely recognize the native protein. Several criteria have been published that may allow prediction of continuity of an epitope (Hopp, 1986).

5. If possible, use the free N-terminal or C-terminal end of a protein sequence; the termini are often particularly immunogenic and evoke antibodies recognizing the native protein. Be sure that the N terminus of an N-terminal peptide remains free or blocked with acetic acid but is not coupled to the carrier. In this case addition of cysteine to the C-terminal end of the peptide and coupling via MBS is recommended. If SH groups are present in the peptide sequence taken from the cytoplasmic domains of membrane proteins consider the possibility of fatty acid acylation or prenylation at cystein residues. In this case the antibody may not bind to the native protein (as in our own experience with antibodies against the cytoplasmic domain of influenzavirus hemagglutinin).

2. Coupling of Synthetic Peptides to Carrier Proteins

a. Coupling with Glutaraldehyde

1. Dissolve (suspend) carrier proteins, such as keyhole limpet hemocyanin (KLH) or hen egg albumin (chicken ovalbumin), in PBS at a concentration of 1 mg/ml. We do not use bovine serum albumin as carrier for immunization to avoid the possibility of generation of antibodies reacting with serum albumin of other species and with albumin present in tissue culture media (e.g., fetal calf serum).

2. Add high-performance liquid chromatography (HPLC)-purified peptide (10–30 amino acids) at the ratio of 1 mg of peptide per mg protein (KLH or ovalbumin). Cool to 4°C. The number of lysines is high enough in these proteins to couple 1 mg of peptide to 1 mg of protein.

3. Add an equal volume of freshly prepared 2% glutaraldehyde in H_2O dropwise with constant stirring.

4. Stop the reaction after 1 hr by addition of solid sodium borohydride (NaBH$_4$) to a final concentration of 10 mg/ml (250 mM).

5. Incubate for 1 hr at 4°C. Afterward dialyze against PBS overnight.

b. Coupling with m-Maleimidobenzoyl-N-hydroxysuccinimide Ester

In the first step MBS is given to the carrier protein, to the free amino groups (e.g., lysines) of which MBS will bind via its active N-hydroxysuccinimide ester. After addition of peptides containing SH groups a thioether bond is formed between the free thiol group and the double bond of the maleimide moiety of MBS.

1. Dissolve 1 mg carrier protein/ml 50 mM sodium phosphate buffer, pH 8.

2. Add 100 ml of MBS solution containing 10 mg MBS/ml dimethylformamide.

3. Stir the solution for 30 min at room temperature.

4. Remove the excess MBS by gel filtration through Sephadex G25 equilibrated with 50 mM phosphate buffer, pH 6.0 (typical size of the column to be used for filtration of 1–2 ml is 1 $\times$ 10 cm).

5. Pool the fractions containing 0.5–1 mg carrier protein (now activated with MBS).

6. Adjust the pH to 7–7.5 and add 1 mg peptide (10–30 amino acids) per milligram carrier protein.

7. Stir the solution for 3 hr at room temperature and dialyze afterward against PBS (4°C).

c. Multiple Antigen Peptide Method

A new approach to generate antibodies against peptides has been described (Tam and Lu, 1989). Branching heptameric lysine cores were used onto which the peptide of interest was synthesized. The resulting macromolecule consists of multiple copies of a single peptide on all arms of the small branched oligolysine core: these structures were termed multiple antigenic peptides (MAPs) and were shown to generate strong immune responses in both rabbits and mice. It appears that the antibody titers elicited may not only be higher but may also occur faster than those obtained with protein-conjugated peptides (McLean *et al.*, 1991; Forssmann, W.-G., personal communication). The potential advantages of this method are the following (McLean *et al.*, 1991).

1. Most of the immunogen consists of the amino acid sequence against which the antibody is to be raised.

2. Quantitation of the immunizing dose is easier.

3. Preparation of the immunogen is quicker and more convenient in that the conjugation of the peptide to a carrier protein is avoided, because the molecular weight of the branched peptides is large enough (M_r 13,000–15,000 from a 15-amino acid sequence) to elicit an immune response.

The heptameric lysine core can be synthesized onto an alanine residue coupled via an acid-labile bond to the resin by using Fluorenylmethoxycarbonyl (Fmoc)-Lys pentafluorophenyl ester. Peptides can be directly synthesized onto the branched lysine core.

Although the first results obtained with the MAP method are promising, it is too early to recommend it as a standard procedure. If, for instance, the peptide used for immunization does not contain a T-helper cell epitope it will be nonimmunogenic when coupled to oligolysine. But the same peptide may be immunogenic when coupled to a carrier protein that contains numerous T-helper cell epitopes. Diepitopic MAP constructs that contain T-helper cell epitopes (sequences) at some of the NH_2 groups may circumvent the need for a carrier protein (Tam and Lu, 1989).

V. Adjuvants

Several compounds have been described that are capable of potentiating an immune response. The working principles of most of the commonly used adjuvants are as follows.

1. Depot action: Liberation of the injected immunogen is delayed by entrapping it into water-in-oil emulsions (e.g., Freund's adjuvant) or by adsorption of the immunogen to a variety of different materials, such as bentonite, polyacrylamide, nitrocellulose, $Al(OH)_3$, or $Zn(OH)_2$.

2. Presentation of the immunogen by macrophages: Phagocytosis of soluble antigens is made possible by their adsorption to the aforementioned particulate matter and macromolecules.

3. Attraction and activation of macrophages and secretion of interleukins by addition of inflammatory agents: Inflammatory agents include heat-killed bacteria or components of their membrane (*Mycobacteria tuberculosis* and *smegmatis*, *Bordetella pertussis*, muramyl-dipeptide, lipopolysaccharide, lipid A).

A. Freund's Adjuvant

The water-in-oil adjuvant most commonly used for generation of antibodies in animals is Freund's adjuvant. Freund's incomplete adjuvant consists of light mineral oil to which mannitane mono-oleate is added as emulsifier (0.85:1.5, v/v) (Freund and McDermott, 1942). The addition of inactivated and dried mycobacteria (*M. tuberculosis*) at 1 mg/ml is used as inflammatory agent in Freund's complete adjuvant. The working principle of Freund's adjuvant and other oil-based adjuvants is to produce a water-in-oil emulsion in which hydrophilic antigens (proteins) are entrapped in small water droplets (discontinuous phase) suspended in an oil environment (continuous phase). Emulsions are mixtures of immiscible fluids, one of which is suspended as small droplets inside the other. The success of adjuvant application depends largely on the generation of a stable water-in-oil emulsion. If instead an oil-in-water emulsion is produced the adjuvant may be much less effective, because the antigen will rapidly diffuse away from the site of immunization.

Principles of Emulsification

The aqueous antigen solution must be mixed in a way that induces the oil to surround water droplets and inhibits production of oil droplets surrounded by water. Several methods have been described, but we prefer the following procedure.

1. Measure the volume of the aqueous antigen solution. Phosphate-buffered saline or any other nontoxic aqueous solution may be used. A typical volume used for rabbits is 0.5–1.5 ml and for mice is 0.1–0.2 ml, containing 100–200 μg (rabbits) or 10–20 μg (mice) of proteinaceous antigen.

2. Use the same volume of Freund's adjuvant for emulsification. Freund's complete adjuvant is used for the first antigen injection and Freund's incomplete adjuvant is used for all further (booster) injections.

3. Give the adjuvant in a conical plastic tube or any other tube with hydrophobic surface (i.e., siliconized glass tubes, plastic syringes).

4. Add antigen solution in small fractions (50–100 μl). Inject the water phase forcefully into the oil phase with a single stroke of the syringe. Use a 19- to 21-gauge needle. Bending or breaking off the needle may improve breaking up the water phase into small droplets (less than 1 μm in diameter). Vortex vigorously in between.

5. If the volume is small (i.e., less than 0.5 ml) avoid vortexing but instead draw the antigen–adjuvant mixture into a syringe (1-ml all-plastic tuberculin syringe) and express it back into the tube or into another syringe several times until a thick white emulsion has formed. A double-headed syringe needle can also be used to pass the emulsion back and forth between two syringes. Add an additional 50 μl of antigen and repeat the process until all antigen has been emulsified.

6. Sonicate the emulsion (e.g., in a closed plastic tube or closed plastic syringe) six times (10 sec each) at maximal setting.

7. The emulsion is stable if a drop placed on water surface does not disperse.

8. If not stable add more adjuvant and sonicate again.

B. Other Hydrophobic Adjuvants

Freund's adjuvant is still the golden standard of adjuvants. However, it may be harmful if injected accidently into the experimenter's tissue. Long-lasting granulomas and autoimmune responses may occur. Therefore the use of Freund's adjuvant is prohibited in a few countries. A possible alternative is the synthetic polyoxyethylene (hydrophilic)–polyoxypropylene (hydrophobic) copolymer, TiterMax (CytRx Corporation, Atlanta, GA), which appears to be much less harmful but as effective as Freund's adjuvant (Hunter *et al.*, 1989). RAS adjuvant (Fitzgerald, 1991) (RIBI Immunochem. Res. Inc., Hamilton, ME) and ABM adjuvant (Linaris, Bettingen, Germany) are also based on oily substances (but at lower concentrations) with the addition of the endotoxin compound lipid A as an immunostimulant.

Another adjuvant is the highly branched isoparaffinic polyalphaolefin, PAO (USDA clearance for use in foods) (e.g., Conoco and Jet gas stations). Mix 4 ml PAO with 0.9 ml Tween-81 and 0.3 ml Tween-80. This mixture forms stable water-in-oil emulsions even at 5% (v/v) PAO. For immunization mix 8 vol of the antigen solution with 2 vol of the PAO mixture. The antigen partitions into the aqueous phase and an emulsion with low viscosity will form (Enders *et al.*, 1990).

C. Bentonite

Bentonite is a clay originally found near Fort Benton, Wyoming. It consists of layered $Al \cdot Mg \cdot Si(OH)$ lattices with an absorptive surface area of $100-300 \ m^2/g$. The size of particles used for immunization is below 1 μm and thus they are the right size for phagocytosis by macrophages (Potter and Stollerman, 1961; Carpenter and Barsales, 1967; Gallily and Garvey, 1968). Bentonite may be used for intravenous injection, mixed with immunostimulants, or emulsified with Freund's adjuvant for subcutaneous administration.

1. Preparation of Bentonite

1. Suspend and blend 1 g bentonite (Cat. No. B3378; Sigma, St. Louis, MO) in 200 ml of H_2O.
2. Dilute to 1000 ml with H_2O.
3. Remove large particles by centrifuging at 1000 g (10 min).
4. Recentrifuge the supernatant at 2000 g (10 min).
5. Suspend the pellet in an equal volume of PBS. The pellet consists of small particles (≤ 1 μm), visible by phase-contrast microscopy.
6. Freeze into 100-μl aliquots and store frozen.

2. Adsorption of Proteins to Bentonite

1. Thaw 100 μl of the bentonite stock solution, resuspend vigorously in 10 ml PBS, and sediment at 2000 g for 10 min. Use 100 μl of the stock solution (50-μl pellet) to bind 200 μg of protein.
2. Add antigen (dissolved in PBS or saline) to the bentonite pellet while vortexing or drawing the suspension into a syringe and injecting it back into the tube (several cycles). The final volume of the mixture should be 1–2 ml (rabbit) or 0.1–0.5 ml (mice).

3. For subcutaneous or intraperitoneal immunization emulsify with an equal volume of Freund's adjuvant (Freund's complete adjuvant for the first injection and Freund's incomplete adjuvant for the following injections).

4. The antigen–bentonite mixture can also be used for intravenous injections. But in this case emulsification with Freund's adjuvant should be omitted.

D. Aluminium Hydroxide

$Al(OH)_3$ is often used as adsorbent (adjuvant) for intravenous booster injections of protein antigens. $Al(OH)_3$ is nontoxic (although it may cause sterile abscesses and granulomas when injected subcutaneously, rather than intramuscularly) and is the most widely used adjuvant licenced for human use. $Al(OH)_3$ can be purchased as a suspension (Alhydrogel; Superfos, Vedbaek, Denmark) but it is also quickly prepared from soluble aluminium salts, as described in the next section.

Preparation of Aluminium Hydroxide

1. Prepare 10 ml of a 0.2 M potassium alum sulfate [0.94 g $KAl(SO_4)_2 \cdot 12H_2O$] and add 30 ml of 0.2 M NaOH. Vortex. A cloudy, whitish $Al(OH)_3$ precipitate will form.

2. Let stand at room temperature for 10 min and then centrifuge at 1000 g for 10 min. The volume of the resulting pellet (stiff, gellike) is 6–7 ml. Discard the supernatant and resuspend the pellet thoroughly in 40 ml H_2O.

3. Centrifuge at 1000 g for 10 min and discard the supernatant. Repeat two times. The pellet (6–7 ml) contains about 150 mg $Al(OH)_3$ (calculated dry weight).

4. Add H_2O to the pellet to give a final volume of 10 ml. Stir and suspend the pellet thoroughly using a glass pipette until a homogeneous suspension has formed [1.5% $Al(OH)_3$ suspension].

5. Mix equal volumes of the 1.5% $Al(OH)_3$ suspension with the antigen solution. A volume of 100 μl of 1.5% $Al(OH)_3$ will bind 100–200 μg protein. If free of toxic components (e.g., detergents) the suspension is ready for intravenous injection.

6. Otherwise the $Al(OH)_3$–protein mixture can be pelleted at $\geq$1000 g for 10 min. Discard the supernatant, which should be free of proteins (screen by Bradford staining). Resuspend the pellet in an appropriate volume of physiological saline (0.9% NaCl) for intravenous injection (rabbits, 1 ml; mice, 0.2 ml).

E. Liposomes

Adjuvant properties of liposomes (consisting of phosphatidylcholine with or without addition of cholesterol) are probably based primarily on their preferential phagocytosis by macrophages (for reviews see van Rooijen and van Nieuw-megen, 1983; Davis *et al.*, 1987; Alving, 1991). Liposomes can be used for administration of hydrophobic antigens (e.g., membrane proteins) that will partition into the hydrophobic lipid phase and expose their hydrophilic domains on the surface. Adsorption of hydrophilic antigens on the surface of liposomes and even their inclusion into the interior of liposomes may result in strong immune responses (Allison and Byars, 1986). The latter possibility has been used to prepare antibodies against toxic molecules, which are thought to become slowly released when entrapped into liposomes or that may become biologically inactivated when phagocytosed by macrophages. Protein–liposome mixtures suspended in aqueous solutions may be injected without adjuvant (e.g., intravenously) but adsorption to $0.1\ M\ Zn(OH)_2$ or emulsification with oily adjuvants (e.g., polyalphaolefin or Freund's adjuvant) has been reported to potentiate the immune response significantly (Enders *et al.*, 1990).

To 200 mg lecithin add PBS or water to 1 ml. Sonicate vigorously to obtain a 20% (w/v) suspension of lecithin. Add 5% (v/v) of this liposome suspension to the antigen–adjuvant solution (emulsion) and mix thoroughly. Liposomes will further stabilize water-in-oil emulsions and improve immunogenicity.

Liposomes can also be added to other kinds of antigen preparations, for example, proteins adsorbed to bentonite, $Al(OH)_3$, or $Zn(OH)_2$. In the latter cases intraveneous application is possible.

Liposomes have also been successfully used as carriers (adjuvants) for immunization with hydrophobic proteins such as integral membrane proteins (Eidelman *et al.*, 1984).

F. Bacterial and Synthetic Immunostimulants

The potency of the aforementioned adjuvants can be considerably increased by addition of heat-killed bacteria or components of their cell wall that cause a local inflammatory response, including activation of macrophages and secretion of interleukins. These immunostimulants are normally used only for the primary immunization step and not for booster injections (to avoid overshooting inflammatory reactions and dominance of immune responses against bacterial proteins). In all procedures that are based on emulsification with Freund's complete adjuvant (contains 1 mg heat-inactivated mycobacteria per milliliter) addition of bacterial immunostimulants is not necessary. Bacterial stimulants should not be injected intravenously because of the possibility of severe systemic side effects (but see below for derivatives of lipid A). The following bacterial stimulants are frequently used.

1. Heat-killed *M. tuberculosis*, *M. smegmatis*, or *B. pertussis* (Johnson *et al.*, 1956) (10^9–10^{10} organisms for one immunization step, approx. 1 mg dry weight).

2. Muramyl dipeptide (and its derivatives) is the main immunostimulatory component of *M. tuberculosis* (50–100 μg for one immunization step) (for a review see Arnon *et al.*, 1983). This component can also be coupled covalently

to the antigen (Bessler and Hauschildt, 1987). Muramyl dipeptide may bypass the need for some of the factors produced by T-helper cells in the generation of immunity.

3. Lipid A, the major inflammatory component of muramyl dipeptide, is a strong lipophilic immunostimulant (Qureshi *et al.,* 1982). Derivatives can be injected intravenously.

4. Branched and linear polysaccharides such as dextran sulfate and a variety of different unmodified polysaccharides are often strong immunostimulatory agents and can be used as adjuvants (Waksman, 1979; Gurvich and Korukova, 1986; Dalsgaard, 1987; Nilsson *et al.,* 1987; Vanselow, 1987). Therefore, it sometimes may be of advantage to use proteins bound to column materials for immunization, either as suspension or emulsified with Freund's adjuvant. The amount of protein bound should be in the range indicated in Section VII,B.

The list of additional, mostly synthetic, immunostimulants is rapidly growing. However, the use of most of these stimulants is still uncommon in cell biology. Therefore a description is omitted here.

VI. Routes of Immunization

The immune response depends to a large extent on the way the antigen is administered. Before considering this aspect the major routes for immunization will be briefly described. Only adult animals should be taken for immunization.

A. Subcutaneous Injection

Subcutaneous injection is the most commonly used route for immunization in rabbits. The antigen–adjuvant mixture is injected into the loose connective tissue located between the skin and the underlying musculature. Typically, four or more sites along both sides of the back are used. The needle (20 gauge) is inserted into a fold of the skin produced by lifting the skin pinched between the thumb and the forefinger. The volume normally injected per site is 50 μl for mice and 0.5–1 ml for rabbits. The antigen is released relatively slowly and will be drained to the axillary and inguinal lymph nodes. In addition, presentation of the antigen to skin-associated lymphoid cells is possible. Formation of large granulomas and even ulcers is occasionally seen when Freund's adjuvant is used for subcutaneous injection [occasionally $Al(OH)_3$ may also cause ulcers].

B. Subscapular Injection

Subscapular immunization is a special case of subcutaneous administration of antigens. In our hands this is the most effective and convenient way of local administration of an antigen. The site of injection is close to the axillary lymph nodes. Ulcers do not occur. Up to 5 ml can be injected at one site. A safe and easy way for subscapular injection is to kneel down on the floor and fix the rabbit gently between the experimenter's thighs so that the rabbit cannot retreat or escape. The scapula is a triangular bone sliding on the lateral wall of the thorax. The upper edge of the scapula can be easily palpated. Injection is made into the space between the inner (medial) surface of the scapula and the wall of the thorax (consisting of ribs plus covering layer of muscle tissue) (see Fig. 2). Quickly pierce the skin close to the inner surface of the scapula (18- to 20-gauge needle). After having overcome the resistance of the skin the needle should proceed without any significant resistance. Tilt the tip of the needle slightly outward to avoid hurting the wall of the thorax. Proceed as far as possible (3– 5 cm) and then express the syringe. In most cases rabbits are hardly irritated by this procedure and may even continue eating.

C. Intravenous Injection

Intravenous injection of an antigen will carry it to macrophages of the spleen and capillaries in the lung and liver. The spleen is the specific lymphatic tissue largely responsible for the immune response to substances entering the blood circulation. If the antigen is adsorbed to small particulate matter (e.g., to bentonite) the spleen is possibly the major site of antigen trapping. Intravenous administration of antigen is used primarily for booster injections. Generally, intravenous injection will cause a more rapid but less sustained immune response, with peak titers at days 4 to 7 after booster injection, followed by a rapid fall of the titer. Antibodies that are capable of precipitating soluble antigens are

Fig. 2 Subscapular immunization of a rabbit.

obtained with highest probability of intravenous application of the antigen. In rabbits intravenous injection is made into the dorsal marginal ear vein (see Section VIII). In mice intravenous injection (into the tail vein) is not routinely done for production of polyclonal antibodies.

D. Intraperitoneal Injection

Injection of antigens into the peritoneal cavity will cause spreading and dilution of the antigen over the large peritoneal surface. This form of application is less effective in rabbits but the most commonly used site of immunization in mice (Fig. 3). The antigen will be presented to peritoneal macrophages and lymphocytes and will be drained via mesenteric lymph nodes and intestinal lymphatics to the thoracic duct and then will be delivered to the blood circulation. Thus, intraperitoneal application is thought to have both local and systemic effects. Anesthesize mice with methoxyfluorane, or, if not available, with ether (if experienced, intraperitoneal injection can be carried out without anesthesia). Hold the skin of the neck between thumb and forefinger (Fig. 3). Turn the animal with the belly up and fix the base of the tail and the left hind limb between the fourth and fifth finger. Injection should be made in the left midportion of the abdomen (left of the umbilicus) to avoid puncture of the liver and caecum, which are located in the right half of the abdomen. The needle (25 gauge) should be inserted to a depth of 0.5 cm. Up to 2 ml (mice) can be injected at once. Use of Freund's complete adjuvant is recommended for both primary and booster injections because it may cause production of ascites fluid (10 ml or more) with immunoglobulin titers similar to serum levels.

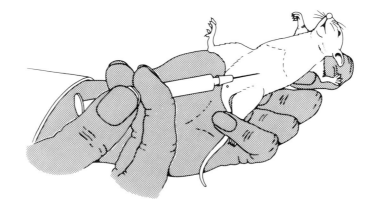

Fig. 3 Intraperitoneal injection of a mouse.

E. Intrasplenic Immunization

Application of antigens directly into the spleen is probably the most effective way to immunize rabbits and mice, and may require only minute amounts of antigens (around 1 μg). The antigen should be immobilized on nitrocellulose for immunization (Spitz, 1986; Nilsson *et al.*, 1987; Grohmann *et al.*, 1991).

Anesthesize animals with an appropriate anaesthetic [e.g., pentobarbitone sodium (Nembutal), 50–100 mg/kg body wt] which should be given slowly and intravenously in rabbits (marginal ear vein) and injected intraperitoneally in mice. Place animals on their right side. Shave a 2-cm strip (mice) or a 5- to 10-cm strip (rabbits) of the abdominal skin parallel to the inferior half of the rib arch (endings of abdominal ribs). Sterilize the area with 70% ethanol. A longitudinal incision (1–2 cm in mice, 5 cm in rabbits) is made through the abdominal skin parallel to the left caudal third (half) of the rib arch (rib endings). The distance between rib arch and incisure should be about 0.5–1 cm in mice and 1–2 cm in rabbits. The edges of the incision are pulled aside to expose the abdominal muscle. These are then incised to open the peritoneal cavity. The spleen is a dark red ribbon-shaped organ that should lie immediately under the incision on the left side of the stomach (i.e., its lower third). In rabbits the spleen is 6–7 cm long, 1 cm wide, and 4–5 mm thick. In mice the spleen is much smaller (1–2 cm long, 3 mm broad, and 2–3 mm thick). Take the free caudal end of the spleen with a pair of blunt forceps and inject the nitrocellulose fragments suspended in 0.5–1 ml of PBS (rabbits) or 0.1 ml (mice). Injection should be made close to the free end of the spleen and the needle (20 gauge) should proceed a few millimeters (mice) or 0.5–1 cm (rabbit) immediately underneath the outer capsule, parallel to the long axis of the spleen. Avoid perforation of the spleen (which is only a few millimeters thick). Afterward replace the spleen into the abdominal cavity and close the abdominal incisure with silk sutures. The skin can be closed with metal clips. Repeat at intervals of 3 weeks. Antibodies may already occur after a single intrasplenic immunization step (Spitz, 1986).

F. Intramuscular and Intradermal Injection and Injection into Footpads

These routes of immunization are more painful to the animal than subcutaneous, subscapular, and intravenous injection. Intradermal application of antigen should be avoided when using Freund's complete adjuvant because of ulceration of the skin within a few days. Granulomas formed in footpads may be painful for the animals. Therefore we do not recommend these routes as standard procedures. However, it should be mentioned that under certain conditions an injection into footpads, and intradermal application, of antigen were reported to be much more immunogenic than subcutaneous immunization into the skin of the back. The antigen injected into food pads is drained into popliteal and inguinal lymph nodes (can be removed for production of monoclonals antibodies in mice and rats; Mirza *et al.*, 1987). If footpads are injected, a board over half the area of the cage (when wire mesh is used in the cage) allows the rabbit to support its weight without hurting its feet.

VII. General Schedule for Immunization

A. Toxic Constituents of Antigen Solutions

All proteins or peptides to be used for immunization must be dissolved (suspended) in nontoxic buffers such as PBS, Tris-HCl, N-2-hydroxyethylpiperazine-N'-2-ethanesulfonic acid (HEPES), maleate, saline, and so on. Be sure that toxic compounds and ions have been removed or inactivated [e.g., azide, cacodylate, CN^-, SCN^-, 2-mercaptoethanol, PMSF, diisopropylfluorophosphate (DFP)]. Low concentrations of the following substances may be present when a total of 2 ml (rabbit) or 200 μl (mouse) of protein solution emulsified with Freund's adjuvant is used for local injections (subcutaneous or subscapular): Triton X-100 (0.5%), SDS (0.2%), urea (2 M), EDTA (10 mM), dithioerythritol/ dithiothreitol (DTE/DTT) (1 mM).

Intravenous injection requires removal of free EDTA (1 mM is tolerable) or chelation of EDTA by addition of an excess of $CaCl_2$. Total salt concentration should be 0.1–0.5 M. Free detergents not bound to proteins or urea are tolerable at the concentrations listed above. Toxic components must be removed by dialysis prior to intravenous administration. Another possibility for removal of toxic components is aluminium hydroxide precipitation of the antigen (see Section V).

B. Dosage

1. Proteins

The amount of native or SDS-denatured proteins, including fusion proteins, to be routinely used for subcutaneous, subscapular, intramuscular, or intraperitoneal routes of immunization is 200 μg (rabbit) or 10–20 μg (mice) per injection. If the protein is difficult to purify, amounts as small as 50 μg (rabbit) or 1 μg (mice) per injection may be tried. Adsorption of a protein to nitrocellulose and injection into the spleen is the method of choice if only minute amounts of protein (microgram levels in rabbits, below 1 μg in mice) are available. Highly conserved proteins such as tubulin may require greater quantities of protein for each immunization step (1 mg in rabbits; Osborn and Weber, 1982) than less conserved proteins (see Section IV). However, it should be kept in mind that large amounts of proteins used for immunization increase the probability of antibody production against contaminating components. Furthermore, high concentrations of immunogen may stimulate populations of B cells with low affinity for the immunogen and thus may evoke production of low-affinity antibodies. Finally, high protein concentrations are more likely to induce immunological tolerance than lower concentrations of antigen.

2. Peptides

The total amount of peptide (with a chain length of 10–30 amino acids) to be used for each injection is 500–1000 μg in rabbits and 50–100 μg in mice. The corresponding amount of carrier protein is about 1 mg of protein per milligram peptide when using glutaraldehyde or MBS as cross-linker and keyhole limpet hemocyanin (KLH) or chicken ovalbumin as carrier proteins. Peptides with chain lengths greater than 25 amino acids may be immunogenic by themselves and even smaller peptides may occasionally act as complete antigens (requiring the presence of both MHC II and T cell receptor epitopes). Glutaraldehyde cross-linking of peptides not coupled to carrier protein is recommended to further increase immunogenicity. If not immunogenic by themselves, couple peptides to carrier proteins and use them for immunization. A mixture of coupled and uncoupled (cross-linked) peptides may be used routinely for immunization.

C. Routine Schedule for Immunizing Rabbits

1. Take preimmune serum of the rabbit to be immunized (2–5 ml/rabbit) and store frozen. At this stage it is recommended that preimmune serum be screened for the existence of autoantibodies against the protein preparations or cellular and tissue structures to be studied.

2. Immunize with 200 μg of protein or with 500 μg of peptide coupled to 1 mg of carrier protein. Adsorb proteins (or coupled peptides) to bentonite and suspend in 1 ml of PBS (or any other physiological solution). Emulsify with 1 ml of Freund's complete adjuvant. The final volume of the antigen–adjuvant mixture will be about 2 ml.

3. Inject the antigen–adjuvant mixture under one shoulder blade (subscapular route; see Fig. 2).

4. Administer a booster injection 3 weeks later under the other scapula with the same amount and volume of antigen emulsified with Freund's incomplete adjuvant.

5. Ten days later take blood samples and assay for antibodies. If the antiserum can be used at a dilution of 1:30 in immunofluorescence or at 1:100 in enzyme-linked immunoblot or dot-blot assay save 10 ml of serum (one bleed of approximately 20 ml of blood). It is possible that during the following booster steps specificity and affinity of the antibody may change.

6. Three weeks after the first booster injection repeat the process with the same volume of antigen–adjuvant mixture as before, but use lower amounts of antigen (down to 100 μg of protein and 250 μg of coupled peptide).

7. Check the antibody titer 10 days later. Take a serum sample (e.g., 10 ml) if the titer is satisfactory.

8. Repeat the sixth step 3 and 6 weeks after the second booster and check antibody titers in between. Take serum samples if titers are satisfactory.

9. Regardless of the antibody titer obtained until this point (four booster injections), start with a series of intravenous injections. Adsorb 10–50 μg of protein or 50–100 μg of coupled peptide to bentonite as described above. Dilute to a final volume of 1–1.5 ml in saline. *Note:* Do not emulsify with Freund's adjuvant! Inject slowly into the marginal ear vein in a distal-to-proximal direction. Although circulating immunoglobulins may bind to the intravenously injected antigen the immune complexes may be even more immunogenic than the free antigens because immune complexes are known to be avidly taken up by Fc receptor-mediated endocytosis of macrophages.

10. Repeat intravenous injections for the next 4 days (total of five applications).

11. Seven days (4–10 days) after the last injection the antibody titers will reach peak values, which may rapidly fall during the following days and weeks. Two weeks after the last intravenous injection another series of two to five injections can be made (intravenous booster). Serum should be taken 4–10 days after the last of these intravenous applications.

If no significant antibody titer has been developed until this point, sacrifice the rabbit because there is little hope that this individual will ever start producing good antibodies.

Note: Intravenous application of antigen is a strong stimulant for production of precipitating antibodies. If the main purpose of immunization is to obtain precipitating antibodies (e.g., to be used for nephelometry) the two series of intravenous injections may be initiated 2 weeks after the first booster injection. This protocol is routinely used for commercial production of precipitating antibodies against μ chains (Dr. Münscher, Behring Company, Marburg, Germany, personal communication).

D. Routine Schedule for Immunizing Mice

1. Take small samples of preimmune serum (20–40 μl) and store frozen.

2. Adsorb 20 μg of protein or 50 μg of peptide (coupled to 50 μg of carrier protein) to bentonite and dilute with PBS to 250 μl. Emulsify with 300 μl of Freund's complete adjuvant. Inject into the peritoneal cavity (Fig. 3).

3. Repeat this step at intervals of 3 weeks. In mice the full booster potential may be reached after a 6-week interval, so that the booster intervals can be extended if the antigen is rare (Dr. Bernhardt, Behring Werke, Marburg, Germany, personal communication). The use of Freund's complete adjuvant stimulates development of ascites fluid in the peritoneal cavity. The antibody titers contained in ascites fluid may be similar to those in serum.

4. Assay serum samples (obtained by tail bleeds) 2 weeks after the third injection (second booster) and after each following injection.

5. If antibody titer is satisfactory (taking perhaps 6–9 months in mice) try to induce ascites by injection of 20–40 μl antigen solution (containing 10–20 μg of

protein adsorbed to bentonite) emulsified with a 10-fold excess volume of Freund's complete adjuvant (total volume of 200–400 μl).

6. Repeat weekly until ascites develops. Antigen should be added to every third injection. Normally ascites occurs after three to six injections. The mouse becomes noticeably large (abdominal swelling), with a rapid increase in weight. Tap the ascites fluid at the site shown in Fig. 3, using an 18- to 20-gauge needle attached to a 5- to 10-ml syringe. Centrifuge the ascites fluid at 10,000 g for 10 min, discard the pellet (debris, cells), and freeze the antibody-containing supernatant until use.

7. If ascites does not develop or if induction of ascites is not an acceptable procedure, final exsanguination is the method of choice (see below) to obtain about 2 ml of blood (1 ml of serum).

VIII. Collection of Blood

A. Rabbits

The rabbit should be restrained in a box. The box shown in Fig. 4 is a homemade version that meets all requirements. If a box is not available it is also possible to collect blood from the animal sitting free on the table or wrapped (rolled up) in a blanket with the head protruding over the free edge. It is important to keep rabbits quiet and calm. Two ways of bleeding are possible, as described in the following sections.

1. Puncture of Central Artery or Marginal Vein of the Ear

Insert an 18- to 20-gauge needle into the distal half of the central artery (Fig. 4) (distal-to-proximal direction). Let the blood drip into a conical centrifuge tube. It is also possible to draw the blood slowly into a syringe (5–10 ml) attached to the needle. This way of bleeding is most conveniently done if an assistant holds the ear and pulls it gently at its tip while the experimenter holds the needle (syringe) and/or the centrifuge tube for collecting blood.

The marginal vein may also be used for puncture. In this case the vein should be compressed at the base of the ear to block venous backflow (see Fig. 4).

2. Incision of Marginal Ear Vein

Set a transverse incision through the skin and the underlying marginal ear vein. The first incision should be made most proximally and the following ones 0.5 cm distal from the previous incisions (Fig. 4). The incision is made with a sharp scalpel blade while the ear is firmly held and slightly extended at its tip. Before setting the incision, the fur over the vein should be shaved and rubbed with petroleum jelly or a fat-based ointment to avoid soaking and early clotting of blood in the fur of the ear. After the cut has been made, place a centrifuge tube

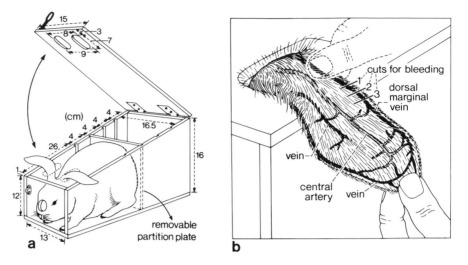

Fig. 4 (a) Rabbit box for bleeding and intravenous injection. The purpose of removable partition plates is to prevent the rabbit from retreating and kicking out (and potentially breaking its back) but not to compress the animal. Ears are slipped through holes in the lid. The box is constructed of nontransparent PVC. (b) Anatomy of the vascular system of the left ear of the rabbit. Sites for cuts and compression of the marginal ear vein are also indicated.

beneath the cut and block venous return by gently compressing the base of the ear, taking care not to compress the artery. Cycles of spontaneous peripheral vasoconstriction and dilation usually cause the blood flow to cease from time to time. If superficial clots have formed, wipe them away with a cotton swap or a piece of soft cleaning paper tissue.

When enough blood has been collected (do not take more than 10 ml/kg body wt at one time and not more than 20 ml/kg body wt in a month) stop the blood flow by releasing pressure from the base of the ear. Press cotton wool or paper tissue (soft cleaning paper) to the surface over the incision. Occasionally blood flow must be stopped by a few minutes of compression directly on the incision (in particular when having used local vasodilating compounds). Before setting the rabbit back in its hutch, gently clean the ear of blood clots but avoid reopening the wound.

3. Vasodilation

Often blood circulation in the ear is kept at low levels. Try to improve circulation by placing the ventral side of the ear in the palm of one hand and flapping the dorsal aspect of the ear with the other hand. Strong vasodilation can be obtained by wiping the dorsal side of the ear with Finalgon cream (contains nicotinic acid and extract of Spanish pepper; Thomae, Bieberach, Germany),

Roti-Histol (extract of orange peel; Roth, Karlsruhe, Germany), toluene, or xylene. The blood flow will be excellent but may be more difficult to stop. It is important to clean the ear thoroughly afterward with ethanol and soap and then to apply cream or petroleum jelly. Otherwise chronic inflammation and scaling will occur (seen particularly after use of Roti-Histol, toluene, or xylene).

B. Mice

The usual way of taking blood samples is to cut the tip of the tail and collect drops of blood in a tube. If the blood flow stops, gently compress the base of the tail between the thumb and the forefinger and "milk" the tail by base-to-tipward movements. Up to 150 μl of blood may be obtained by this procedure. It is important to keep the mouse and its tail warm during this procedure (an infrared lamp is useful, as is hanging the tail in warm water previous to cutting the tip). Causing vasodilation by rubbing the tail (except its tip) with Finalgon cream (see above) will increase blood yield. A convenient way to immobilize mice during this procedure is to fix the base of the tail to the bench with adhesive tape and to calm the mouse with a covering cardboard box, as shown in Fig. 5.

Another procedure for bleeding mice is to collect blood from the retroorbital plexus. This technique requires experience and should be used only if tail bleeding does not work (an excellent description is provided by Herbert, 1978).

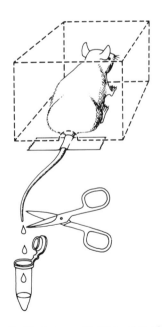

Fig. 5 Technique of tail bleeding of mice.

C. Exsanguination of Rabbits and Mice

1. Rabbits

To collect a large, final amount of blood first obtain as much blood as possible by bleeding from the ear artery or vein (e.g., 50 ml). Afterward the rabbit can be anesthesized with an overdose of pentobarbital sodium (Nembutal) (100 mg/kg) slowly injected into the marginal ear vein. When anesthesia is deep, hang the rabbit head down by holding its hind limbs (or ropes tied to them). Set a deep cut through the throat down to the spine. An assistant deflects the head at its ears and the flowing blood is collected with a funnel inserted in a blood bottle (large centrifuge tube). A total of 50–150 ml of blood can be expected by this method.

Another way to exsanguinate rabbits is to puncture the chambers of the heart through the left fourth intercostal space close to the sternum. A rather thick, 5-cm long needle (18 gauge) attached to a 50-ml syringe should be used. This method requires experience, and is more easily performed under visual control after opening the thorax. First draw blood from the left heart chamber and then from the right chamber. Blood circulation through the lungs is impaired due to shrinkage of the lungs when the thoracic cavity has been opened. Much blood will congest in the right heart and the large supplying body veins.

2. Mice

Anesthesize the mouse deeply with ether and puncture the heart through the fourth intercostal space at the sternal margin. Use a 20-gauge needle attached to a 2-ml syringe. If not experienced, open the chest and puncture the left chamber of the heart under visual control. If blood flow ceases, puncture the right chamber (see above). A total of 2 ml of blood can be expected by this method.

IX. Preparation, Storage, and Shipping of Serum

The blood collected will spontaneously coagulate within a period of 5–30 min at room temperature. Incubate closed blood bottles (centrifuge tubes) for 60 min at 37°C to allow complete coagulation. Clot retraction (shrinkage of the clot) will begin during this period. Try to separate the clot from the wall of the tube with a Pasteur pipette or spatula. Allow clot retraction to complete during the following 12–24 hr at 4°C. During this period the clear, slightly yellowish or pinkish serum becomes separated from the clotted blood. Carefully remove the bulk of serum fluid with a Pasteur pipette. Take care not to stir the clot. Afterward centrifuge at 10,000 g for 10 min at 4°C. The sediment consists of blood clot material. Remove the supernatant (serum) and discard the pellet. Often serum has a pink or even red color caused by hemolysis of erythrocytes or by aspiration of unclotted material from the sediment. In the latter case centrifuge again to remove clot particles.

Serum can be stored frozen (below $-20°C$) for many years without significant loss of activity. If stored at 4°C add 0.1% (w/v) sodium azide to prevent bacterial and fungal growth. If a sample (e.g., a 10-ml aliquot) is thawed, aliquot it in ten 1-ml portions to avoid repeated thawing and freezing of the same sample. If one of the 1-ml samples is thawed, aliquot it in ten 100-μl samples, and so on, thus reducing freeze-thaw cycles to about three to five, until final use of the antibody. Repeated thawing and freezing (>10 times) may cause gradual denaturation (precipitation) of immunoglobulins. Diluted antisera and affinity-purified IgGs are much more sensitive to freezing and thawing. Affinity-purified immunoglobulins should be stabilized and prevented from sticking to the wall of the tubes by addition of BSA (0.1%) and using siliconized tubes.

Nonfrozen serum samples can be sent through the mail without any significant loss of activity (up to 1 week is tolerable). In this case 0.1% azide should be added to prevent bacterial growth. The safest and most convenient way to ship serum samples is to lyophilize them and envelope the powder in sealed PVC pockets. Affinity-purified immunoglobulins should be sent frozen in dry ice.

X. Determination of Antibody Titer (Test of Reactivity)

To decide when to take large samples of serum from the immunized animal, the titer of antibody must be determined. The most frequently used methods for testing reactivity of the antibody against the native protein are enzyme-linked solid-phase assays and immunostaining of whole cells or tissue sections (see articles 3, 8, and 10–16).

Two Solid-Phase Assays for Determination of Antibody Titer

There are several methods to quantify antibody titers by solid-phase immunoassays. The underlying principle of the assays described below is to adsorb the antigen to a solid phase (i.e., to nitrocellulose or microtiter wells), which is then incubated with the antiserum (primary antibody). The amount of immunoglobulins captured by the solid phase-bound antigen can be determined indirectly by a secondary incubation step in which radiolabeled or enzyme-linked anti-immunoglobulins (secondary antibodies) are used. The amount of radioactivity or enzyme bound is an indirect measure of the total amount of primary immunoglobulin molecules captured by the antigen. Amounts of enzyme are measured (visualized) by the conversion of a colorless chromogen to a visible, colored product that can be inspected visually or quantified photometrically. If the purified antigen is available only in small amounts it is also possible to use an antigen-enriched protein fraction for adsorption to the solid phase. However, in this case the animal must have been immunized with the purified antigen (e.g., excised from SDS-polyacrylamide gels) and the specificity of the antibody should be further checked by immunoblots to be sure that the antibody titer measured does not result from binding to contaminants. Two convenient protocols for assaying antibody titers against peptides and proteins are described in the following sections.

1. Dot-Blot Method for Assaying Antibody Titers against Peptides

1. Couple peptide–antigen to a carrier protein that is different from the carrier protein used for immunization. Using a different carrier protein is important to discriminate between antibody titers against the peptide itself, the peptide–carrier interface, or the carrier. Instead of using peptides coupled to carriers it is also possible to use the uncoupled peptides for dot-blot assay. However, coupled peptides give better signals.

2. Prepare a solution containing 1 mg coupled peptide (or uncoupled peptide) per milliliter PBS. Place 1-μl drops on a strip of nitrocellulose. The distance between adjacent drops should be 1 cm. Let dry for 10 min, using a hair drier. Mark the site of each spot with a soft pencil at the margins of the strip.

3. Block the nitrocellulose afterward with a blocking solution containing 5% (w/v) low-fat milk powder in PBS or 1% gelatin in PBS (pH 7.4). Incubate at room temperature for 0.5–1 hr (with constant agitation). When using low-fat milk powder be sure that the peptide used for immunization was not coupled to bovine serum albumin, which may be present in milk powder.

4. Remove the blocking solution and rinse three times (5 min each) with PBS. Cut the strip in pieces, each containing a peptide spot. Incubate the pieces in separate trays with 0.5 ml of antiserum diluted 1:10, 1:100, 1:1000, and so on, with blocking solution. Incubate at room temperature for 1–2 hr.

5. Remove antibody and wash thoroughly with blocking solution (five times, 3 min each) at room temperature with agitation.

6. Visualize bound immunoglobulins with peroxidase-labeled second antibodies as described in article 6.

Note: In most cases detectable anti-peptide titers (1:10 to 1:100) will occur after the second booster injection. Continue boostering until titers of 1:500 or higher have developed. If the anti-peptide titer is above 1:100, it is worth obtaining serum from the animal.

2. Control Experiments

1. Add 20 μg of peptide coupled to carrier (or uncoupled peptide if the dot-blot assay is performed with uncoupled peptides) to 0.5 ml of the diluted antibody solution 15 min prior to incubation of the dots. Binding of antibody must be abolished or considerably reduced by this absorption.

2. Some peptide antibodies may be directed against protective side groups of the activated amino acids used for synthesis. In particular, protective groups of arginine may not be completely split off during synthesis. To check this possibility, perform dot-blot assays with unrelated peptides or, even better, with a nonsense peptide consisting of the same kind and number of amino acid residues but synthesized in a different order. Unrelated or nonsense peptides should not block antibody binding.

3. Direct Enzyme-Linked Immunosorbent Assay

1. Dissolve antigen in coating buffer (15 mM Na$_2$CO$_3$, 35 mM NaHCO$_3$, pH 9.5) at a concentration of 2 μg in 100 μl.

2. Pipette 50 μl of the antigen-coating buffer solution into each well of a 96-well polystyrene microtiter plate. Cover the plate and incubate at room temperature for at least 60 min. Longer incubation periods (e.g., overnight at 4°C) will increase the amount of bound protein. Approximately 100 ng of protein will bind to the wall of the well. All of the following steps are done at room temperature.

3. Remove the antigen solution (and all following solutions from wells) by inverting the plate and pushing it upside down onto a sheet of filter paper. Wash twice (5 min each) with Tris–Tween washing buffer (TTW; 10 mM Tris, 150 mM NaCl, 0.1% Tween-20, pH 7.5). In all washing steps the wells should be filled to the top (~350 μl).

4. Block unspecific protein binding sites with 3% (w/v) low-fat milk powder or 1% (w/v) gelatin in TTW. Fill the wells to the top. Incubate for 60 min.

5. Wash the wells three times (5 min each) with TTW.

6. Remove TTW and incubate with 50 μl of antiserum (diluted 1:10, 1:50, 1:100, etc., in TTW) for 60 min.

7. Wash the wells three times (5 min each) with TTW.

8. Remove the TTW and incubate with 50 μl of horseradish peroxidase-conjugated secondary antibody diluted 1:500 to 1:1000 in TTW for 30 min.

9. Wash the wells three times (5 min each) with TTW.

10. Remove the TTW and initiate the chromogenic enzyme reaction by addition of 50 μl per well of 1.5 mg o-phenylenediamine and 1 μl of 30% H$_2$O$_2$ (0.03%) dissolved in 1 ml of citrate–phosphate buffer (0.2 M Na$_2$HPO$_4$, 0.1 M citric acid, pH 5.0). Use only freshly prepared solution. Add H$_2$O$_2$ immediately prior to use.

11. Incubate for 15–30 min. Stop the reaction by addition of 50 μl of 0.5 M H$_2$SO$_4$ per well.

12. Read the plate at 492 nm.

XI. Test of Specificity

Specificity must be established by confrontation of the antibody with the bulk of cellular (tissue) proteins and demonstration that the antibody reacts selectively with the antigen or its homolog used for immunization. It must be emphasized that it is not sufficient to document ''specificity'' by demonstration of antibody binding to the purified antigen. This kind of experiment will simply show reactivity but not specificity. A combination of immunoblotting (SDS-denatured proteins) and immunoprecipitation (native proteins) would be ideal to establish specificity. Immunostaining may provide additional support for specificity if only those cellular (tissue) structures that are known to contain the antigen are labeled. Addition of an excess of the purified antigen to the antiserum must completely abolish immunoreactivity (absorption experiment). Moreover, it is recommended to further confirm results with samples of immunoglobulins affinity purified to the antigen (see article 5). If affinity-purified samples give results identical to those of the nonpurified serum samples, experiments may be further conducted with complete serum (immunoprecipitation and immunodiffusion often work better with complete serum than with affinity-purified samples).

Antibodies specific for a given peptide sequence (as assayed by dot-blot) may not recognize the native protein from which the sequence was taken. In our hands more than 50% of antisera specific for a peptide will not bind to the full protein. The most likely reason for this failure is that the sequence chosen adopts an abnormal conformation that does not occur in the native protein and may also not occur in the protein transferrerd to nitrocellulose by electroblotting. If the peptide-specific antibody reacts with either the native and/or the electroblotted protein, affinity purification to the protein (protein band) is recommended to remove all the immunoglobulins that react with nonsense conformations of the peptide used for immunization. Other epitopes recognized by peptide-specific antibodies may result from noncleaved protection groups or from intermolecular cross-links produced by the coupling reagents (e.g., glutaraldehyde, MBS). Such epitopes are most likely not present in the native protein, but may occur in glutaraldehyde-fixed tissues.

XII. Production and Selection of Isoform-Specific Immunoglobulins

The rapidly growing list of protein isoforms resulting from alternative splicing and gene duplications (gene families) demands the production of isoform-specific antibodies to unravel further the functional role of each isoform. In principle, there are four ways to obtain isoform-specific antibodies.

1. Raising monoclonal antibodies that selectively recognize a single isoform: This procedure is often time consuming and expensive but may be rewarded by excellent antibodies.

2. Raising antibodies against isoform-specific protein sequences obtained by peptide synthesis or purification of appropriate fusion proteins (see above).

3. Pharmacologically suppressing the immune response to conserved (immunodominant) stretches of proteins not containing isoform-specific sequence: This method has been successfully used in several cases, most impressively for the production of polyclonal antibodies against tubulin isoforms (Lewis *et al.*, 1987). The basic principle of this method is to first immunize rabbits with a protein preparation containing the conserved portions present in all tubulin isoforms. Immunization is done subcutaneously with 0.5 mg protein emulsified with Freund's complete adjuvant. On the following three days proliferating B cell clones are killed by intraperitoneal injection of the alkylating antitumor drug cyclophosphamide (daily dose of 65 mg per kilogram of body weight). This treatment is thought to destroy all proliferating B cells specific for the conserved portion of the antigen (e.g., tubulin). Two weeks later this procedure may be repeated (Matthew and Sandrock, 1987). Two weeks after the last injection of cyclophosphamide, the animal is immunized with the antigen (e.g., tubulin) isoform of interest (emulsified with Freund's complete adjuvant) in the hope of stimulating the remaining B cells to respond to the isoform-specific stretches of the sequence.

4. A fourth, particularly straightforward way is the affinity selection of isoform-specific immunoglobulin fractions.

Isoform-specific immunoglobulin fractions can be purified from polyclonal antisera raised against the whole sequence of a protein by using isoform-specific peptides or fusion proteins immobilized to solid phases. This straightforward strategy should always be considered before starting to immunize with isoform-specific peptides or fusion proteins. If a polyclonal antibody reacts with the peptide synthesized (or fusion protein) the fraction of immunoglobulins specific for this sequence can be isolated easily by affinity selection.

a. For affinity selection the fusion protein or peptide (or peptide coupled to carrier protein) can be directly bound to affinity column material such as AffiGel 10/15 (Bio-Rad, Richmond, CA) or CNBr-activated Sepharose. If the bound peptide (fusion protein) is still recognized by the antiserum, affinity purification will be possible.

b. Another affinity selection procedure is to use fusion proteins, or peptides coupled to an appropriate carrier protein (e.g., ovalbumin), and subject them to SDS-PAGE and subsequent electroblotting. If the antiserum still binds to the blotted band, the isoform-specific immunoglobulins can be selected by blot-affinity purification as described in article 5. We usually elute the bound immunoglobulins by a 10-min incubation of the nitrocellulose strips with PBS at

56°C (Drenckhahn and Franz, 1986). This method has two advantages as compared to low-pH elution of antibodies: first, there is no need for pH adjustment and further dilution of the eluted immunoglobulins. Second, the yield of active immunoglobulins is normally higher than that obtained by low-pH elution.

c. A third affinity selection procedure for immunoglobulins is simply to drop the fusion protein or the peptide coupled to a carrier (or even the uncoupled peptide) on nitrocellulose and treat the dots as described above for the dot-blot assay (Section X). The bound immunoglobulins can be eluted with PBS at 56°C (10 min).

XIII. Purification of Immunoglobulins

In most assays high-titer polyclonal antisera can be applied without the need of further purification. For the following purposes it may be necessary to purify immunoglobulins from the antiserum:

1. To reduce background staining often observed with total sera
2. To label IGs with biotin, fluorochromes, or other detection molecules
3. To produce anti-idiotypic antibodies
4. To purify Fab, Fc, and F(ab′)$_2$ fragments for special applications (e.g., for microinjection into cells)

Out of a number of different purification strategies described in the literature we found three methods particularly useful: ammonium sulfate precipitation, ion-exchange chromatography (using DEAE/CM cellulose), and adsorption to protein A beads (to isolate the IgG fraction).

A. Ammonium Sulfate Precipitation

Proteins are soluble in aqueous solution because their polar and ionic side groups form hydrogen bonds with water molecules. To precipitate a dissolved protein it is necessary to reduce the number of hydrogen bonds between the solvent (H$_2$O) and the protein. Addition of high concentrations of salts such as ammonium sulfate removes water molecules from proteins, thus decreasing the number of hydrogen bonds and increasing the tendency of the protein to precipitate. Different protein fractions will successively precipitate with increasing salt concentration. The majority of rabbit immunoglobulins can be precipitated with a 40% saturated ammonium sulfate solution. Mouse immunoglobulins precipitate at 45–50% saturated ammonium sulfate.

1. A saturated ammonium sulfate solution is made by adding 800 g of solid ammonium sulfate to 1000 ml of H$_2$O. Heat until the salt dissolves completely, quickly filter while still warm, and let cool (crystals formed during cooling

should remain in the bottle). At room temperature the saturated solution is 4.1 M (767 g/1000 ml). Adjust the pH to 7.4 with ammonium hydroxide.

2. Centrifuge the antiserum for 30 min at 3000 g, 4°C. Discard the pellet.

3. Transfer the supernatant to a beaker and add an equal volume of PBS at 4°C.

4. Under constant stirring with a magnetic stirrer slowly add an equal volume of 80% (rabbit) or 100% (mice) saturated ammonium sulfate solution (equilibrated to room temperature).

5. Let immunoglobulins precipitate at 4°C overnight. Alternatively let them precipitate for 30 min at room temperature; the yield at room temperature, however, is usually not as good as is obtained by overnight incubation at 4°C. Low-temperature precipitation (on ice) is necessary when the immunoglobulin concentration is around 1 mg/ml or below.

6. Centrifuge the precipitate (whitish color) for 30 min at 3000 g, 4°C, and remove the supernatant.

7. If entrapment of serum proteins (coprecipitation) occurs (i.e., as evidenced by pink or yellowish color of the precipitate) the pellet may be washed carefully with ice-cold 40% (rabbit) or 50% (mouse) ammonium sulfate. This step will remove the bulk of entrapped albumin, haptoglobin, hemoglobin, and some other serum components.

8. Resuspend the pellet in PBS (less than 0.5 vol of the starting serum volume) and dialyze against several liters of PBS (three changes) at 4°C overnight. Fix the dialysis bag to the upper margin of the beaker and do not stir. Due to its high density the 40% ammonium sulfate solution tends to sink to the bottom of the beaker.

9. If you wish to lyophilize the immunoglobulin fraction (to save storage space), dialyze against PBS diluted 1:10 with H_2O.

Ammonium sulfate precipitation will result in a partial purification of immunoglobulins. Typically, the purity of immunoglobulins is about 80% as judged by SDS-PAGE under reducing conditions. The predominant bands will be immunoglobulin light chains (20–25 kDa) and heavy chains (50–55 kDa). Bands at 150–180 and 100–110 kDa may result from incomplete reduction of disulfide bonds (full immunoglobulin molecules, heavy chain dimers) during sample preparation for SDS-PAGE.

B. Ion–Exchange Chromatography

For purification of immunoglobulins (preferably prepurified by ammonium sulfate precipitation) by ion-exchange chromatography a combination of different ion-exchange matrices is used.

1. Centrifuge the antibody solution dialyzed against PBS at 10,000 g for 30 min (4°C).

2. Pour a 1-cm thick column containing 6 ml DEAE-cellulose overlayered with 6 ml of CM-cellulose and equilibrated in PBS, pH 7.2. Apply 10 ml of the supernatant to the column.

3. Elute with PBS (pH 7.2) and collect fractions of 1 ml.

4. Monitor the separation of IgG by reading absorbance at 280 nm (absorbance of 1.35 corresponds to ~1 mg of IgG/ml).

C. Isolation of Immunoglobulin G by Adsorption to Protein A Beads

Protein A is a component of the cell wall of most strains of the bacterium *Staphylococcus aureus*. It is a polypeptide with a molecular mass of 42 kDa and contains four potential binding sites for the conserved region of the Fc domain of IgG molecules, two of which may be occupied at the same time per protein A molecule.

Isolation of IgG from a crude antibody solution by using a protein A bead column (protein A attached to an insoluble matrix such as agarose or Sepharose 4B) is an effective single-step procedure (Kessler, 1975, 1976).

Protein A bead columns are commercially available. Their IgG-binding capacity is in the range of 10–20 mg/ml of wet beads. Serum contains approximately 10 mg of IgG molecules/ml.

1. Take 2 ml of rabbit antiserum or ammonium sulfate-purified immunoglobulin fraction and add an equal volume of PBS, pH 7.5.

2. Apply the solution to a small column (e.g., Poly-Prep columns; Bio-Rad) containing 2.5 ml protein A beads equilibrated with PBS.

3. Wash with 10 column volumes of PBS.

4. Elute with 100 mM sodium citrate (pH 2.8) and collect fractions of 0.5 ml.

5. Immediately after elution the low pH of the fractions should be adjusted to pH 7.5 by careful addition of drops of 1 M NaOH (or 2 M Tris) with vigorous agitation.

6. Determine the IgG concentration by reading the absorbance at 280 nm (1 mg of IgG/ml, ~1.35). The degree of purity should be checked SDS-PAGE. Only heavy chains (50–55 kDa) or light chains (20–25 kDa) should be seen under reducing conditions.

Note: The affinity of protein A for IgG varies considerably, depending on the animal species; variation may occur even among different IgG subclasses of the same species. In general, protein A purification is the method of choice for purifying rabbit IgGs. However, in one case we observed that a rabbit antiserum contained a fraction of high-affinity Igs, specific for the brush border protein villin, that did not bind protein A. Only IgGs of lower affinity were bound. Therefore we recommend saving the antiserum fraction not bound by the column and checking it for antibody titer.

D. Selection of Immunoglobulin Fractions for Immunocytochemistry

High-titer antibodies can normally be used without further affinity purification for immunostaining at both the light and electron microscope levels. However, affinity purification of antisera to the antigen will usually decrease background staining significantly (see also Section XI). Affinity purification is particularly important for immunogold labeling of ultrathin sections of frozen or resin-embedded tissues (Drenckhahn and Dermietzel, 1988). In our hands, the Ig fraction obtained by blot-affinity purification gives much better immunogold labeling of ultrathin sections of aldehyde-fixed and plastic-embedded tissues than do Ig fractions obtained by affinity purification to the native antigen. For example, Igs purified to the native band 3-anion exchanger of erythrocytes (coupled to Sepharose) were excellent in immunostaining of unfixed tissue sections and blood smears but resulted in poor labeling of plastic sections. However, when the same antiserum was affinity-purified on nitrocellulose blots of SDS slab gels, an Ig fraction was selected that bound with high affinity and low background to the anion exchanger in plastic sections of aldehyde-fixed tissue (Drenckhahn and Merte, 1987).

E. Removal of Cross-Reactive Immunoglobulin Fractions

Depending on the reason for cross-reactivity of an antiserum with unrelated proteins and cellular structures (see Section II) there are several ways to remove the cross-reactive immunoglobulins.

1. The most common reason for cross-reactivity is the generation of immunoglobulins against contaminants of the antigen preparation. Affinity purification of the antiserum to the antigen preparation used for immunization will often lower the proportion of contaminating immunoglobulins to the level of the minor contaminant in the antigen preparation. If the antigen of interest (or a fragment of it) is available in pure form, affinity purification will eliminate all contaminating immunoglobulins. Contaminating immunoglobulins may also be removed by absorption of the antiserum with an excess of the contaminant or protein and cell fractions containing the contaminant.

2. If cross-reactivity is due to shared epitopes of otherwise different proteins (gene products), absorb the antiserum with an excess of the cross-reactive protein or cell and tissue fraction containing it. Another possibility to eliminate cross-reactivity is to affinity purify the antibody to proteolytic fragments or synthetic peptide sequences of the immunogen or to use appropriate fusion proteins for purification, as outlined in section XII. If the given protein fragment, fusion protein, or peptide contains antibody-binding sites not shared by the cross-reactive protein (trial-and-error experiment) it is possible to obtain a clean, non-cross-reactive immunoglobulin fraction. This procedure can also be used to separate isoform-specific immunoglobulin fractions from nonisoform specific polyclonal antisera (see Section XII).

3. The third class of cross-reacting immunoglobulins consists of autoantibodies. If the antigen or the cellular structure against which autoantibodies have developed can be purified, use the same strategy for purification as described above for contaminating immunoglobulins. In any case, affinity purification of the antiserum to the immunogen will effectively remove autoantibodies. In some cases it may be of interest to identify the autoantigen and perhaps to discover a new protein (many examples of this are to be found in the literature e.g., Beckerle, 1986; Earnshaw *et al.*, 1986; Reimer *et al.*, 1987; Amagi *et al.*, 1991).

Acknowledgments

We are grateful to the following colleagues for methodological suggestions and for critical reading of the manuscript: Dr. D. Bernhard and Dr. Münscher (Behring Company, Marburg), Dr. W.-G. Forssmann (University of Hannover), Dr. Murphy and K. Wallis (Johns Hopkins Medical School, Baltimore), Dr. K. Radsak (University of Marburg), K. Denzer, Dr. D. Höfer, Dr. A. Kollert-Jöns, Dr. P. Kugler, and Dr. B. Püschel (University of Würzburg).

Supported by grants from the Deutsche Forschungsgemeinschaft to Detlev Drenckhahn (SFB 176, SFB 286).

References

Allison, A. C., and Byars, N. E. (1986). *J. Immunol. Methods* **95,** 157–168.

Alving, C. R. (1991). *J. Immunol. Methods* **14,** 1–13.

Amagai, M., Klaus-Kovtun, V., and Stanley, J. R. (1991). *Cell* **67,** 869–877.

Arnon, R., Shapira, M., and Jacob, C. O. (1983). *J. Immunol.* **61,** 261–273.

Beckerle, M. C. (1986). *J. Cell Biol.* **103,** 1679–1687.

Benjamin, C. D. (1984). *Annu. Rev. Immunol.* **2,** 67–101.

Berzowsky, J. A., and Berkower, I. J. (1984). *In* "Fundamental Immunology" (W. E. Paul, ed.), pp. 595–644. Raven, New York.

Bessler, W. G., and Hauschildt, S. (1987). *Forum Mikrobiol.* **4,** 106–111.

Carpenter, R. R., and Barsales, P. B. (1967). *J. Immunol.* **98,** 844–853.

Conolly, J. A., Kalnins, V. I., and Barber, B. H. (1979). *Nature (London)* **282,** 511–513.

Dalsgaard, K. (1987). *Vet. Immunol. Immunopathol.* **17,** 145–152.

Darnell, J. E., Lodish, H. F., and Baltimore, D. (1990). "Molecular Cell Biology." Sci. Am. Books, New York.

Davies, D. R., and Metzger, H. (1983). *Annu. Rev. Immunol.* **1,** 87–117.

Davis, D., Davis, A., and Gregoriadis, G. (1987). *Immunol. Lett.* **14,** 341–348.

Djabali, K., Portier, M.-M., Gros, F., Blobel, G., and Georgatos, S. D. (1991). *Cell* **64,** 109–121.

Drenckhahn, D., and Dermietzel, R. J. (1988). *J. Cell Biol.* **107,** 1037–1048.

Drenckhahn, D., and Franz, H. (1986). *J. Cell Biol.* **102,** 1943–1952.

Drenckhahn, D., and Merte, C. (1987). *Eur. J. Cell Biol.* **45,** 107–115.

Eidelman, O., Schlegel, R., Tralka, T. S., and Blumenthal, R. (1984). *J. Biol. Chem.* **259,** 4622–4628.

Earnshaw, W. C., Bordwell, B. J., Marino, C., and Rothfield, N. (1986). *J. Clin. Invest.* **77,** 426–430.

Enders, B., Hundt, E., Bernhardt, D., Schorlemmer, H.-U., Weinmann, E., and Küpper, H. A. (1990). "Vaccines 90," pp. 29–45. Cold Spring Harbor Lab. Press, Cold Spring Harbor, New York.

Fitzgerald, T. J. (1991). *Vaccine* **9,** 266–272.

Freund, J., and McDermott, K. (1942). *Proc. Soc. Exp. Biol. Med.* **49,** 548–553.

Gallily, R., and Garvey, J. S. (1968). *J. Immunol.* **101,** 924–929.

Gaulton, G. N., and Greene, M. I. (1986). *Annu. Rev. Immunol.* **4,** 253–280.

Getzoff, E. D., Geysen, H. M., Rodda, S. J., Alexander, H., Tainer, J. A., and Lerner, R. A. (1987). *Science* **235**, 1191–1196.

Geysen, H. M., Tainer, J. A., Rodda, S. J., Mason, T. J., Alexander, H., Getzoff, E. D., and Lerner, R. A. (1987). *Science* **235**, 1184–1190.

Glenney, I. R., and Glenney, P. (1983). *Cell* **34**, 503–512.

Goodman, J. W. (1987). *In* "Basic and Clinical Immunology" (D. P. Stites, J. D. Stobo, and J. V. Wells, eds.), pp. 20–36. Appleton & Lange, Norwalk, Connecticut.

Gröschel-Stewart, U., and Drenckhahn, D. (1982). *Collagen Relat. Res.* **2**, 381–463.

Grohmann, U., Romeni, L., Binaglia, L., Fioretti, M. C., and Pucetti, P. (1991). *J. Immunol. Methods* **137**, 9–16.

Gurvich, A. E., and Korukova, A. (1986). *J. Immunol. Methods* **87**, 161–167.

Harlow, E. D., and Lane, D. (1988). "Antibodies, a Laboratory Manual." Cold Spring Harbor Lab. Press, Cold Spring Harbor, New York.

Herbert, W. J. (1978). *In* "Handbook of Experimental Immunology" (D. M. Weir, ed.), Vol. 3, pp. A4–A4.29. Blackwell, Oxford.

Hopp, T. P. (1986). *J. Immunol. Methods* **88**, 1–18.

Hunter, R. L., Bennett, B., Howerton, D., Buynitzky, S., and Check, I. J. (1989). *In* "Immunological Adjuvants and Vaccines" (G. Gregoriadis, A. C. Allison, and G. Poste, eds.), pp. 133–144. Plenum, New York.

Jerne, N. K. (1974). *Ann. Immunol. (Inst. Pasteur)* **125C**, 373–389.

Jöns, T., and Drenckhahn, D. (1992). *EMBO J.* **11**, 2863–2867.

Johnson, A. G., Gaines, S., and Landy, M. (1956). *J. Exp. Med.* **103**, 225–246.

Kessler, S. W. (1975). *J. Immunol.* **115**, 1617–1623.

Kessler, S. W. (1976). *J. Immunol.* **117**, 1482–1490.

Knudsen, K. A. (1985). *Anal. Biochem.* **47**, 285–288.

Köhler, H., Kaveri, S., Kieber-Emmons, T., Morrow, W. J. W., Müller, S., and Raychaudhuri, S. (1989). (J. Langone, ed.), Methods in Enzymology, *In* "Antibodies, Antigens, and Molecular Mimicry" Vol. 178, pp. 3–35. Academic Press, San Diego.

Kordeli, E., and Bennett, V. (1991). *J. Cell Biol.* **114**, 1243–1259.

Kyte, J., and Doolittle, R. F. (1982). *J. Mol. Biol.* **157**, 105–132.

Larson, D. M., Fujiwara, K., Alexander, R. W., and Gimbrone, M. A. (1984). *Lab. Invest.* **50**, 401–407.

Lee, C., Levin, A., and Branton, D. (1987). *Anal. Biochem.* **166**, 308–312.

Lewis, S. A., Gu, W., and Cowan, N. J. (1987). *Cell* **49**, 539–548.

Marston, A. O. (1986). *Biochem. J.* **240**, 1–12.

Matthew, W. D., and Sandrock, A. W. (1987). *Immunol. Methods* **100**, 73–82.

McLean, G. W., Owsianka, A. M., Subak-Sharpe, J. H., and Marsden, H. S. (1991). *Immunol. Methods* **137**, 149–157.

Meyer, K.-H., Behringer, D. M., and Veh, J. W. (1991). *J. Histochem. Cytochem.* **6**, 749–760.

Mirza, I. H., Wilkin, T. J., Cantarini, M., and Moore, K. (1987). *J. Immunol. Methods* **105**, 235–243.

Mishell, R. I., and Shiigi, M. (1980). "Selected Methods in Cellular Immunology." Freeman, San Francisco.

Nigg, I. A. (1988). *Int. Rev. Cytol.* **110**, 27–92.

Nilsson, B. O., Svalander, P. C., and Larsson, A. (1987). *J. Immunol. Methods* **99**, 67–75.

Osborn, M., and Weber, K. (1982). *Methods Cell Biol.* **24**, 97–132.

Osborn, M., Franke, W. W., and Weber, K. (1977). *Proc. Natl. Acad. Sci. U.S.A.* **74**, 2490–2494.

Peters, J. H., and Baumgarten, H. (1992). "Monoclonal Antibodies" Springer-Verlag, Berlin.

Potter, E. V., and Stollerman, G. H. (1961). *J. Immunol.* **87**, 110–113.

Qureshi, N., Takayama, K., and Ribi, E. (1982). *J. Biol. Chem.* **257**, 11808–11815.

Reading, C. L. (1982). *J. Immunol. Methods* **53**, 261–291.

Reimer, G., Raska, I., Tan, E. M., and Scheer, U. (1987). *Virchows Arch B.* **54**, 131–143.

Sege, K., and Peterson, P. A. (1978). *Proc. Natl. Acad. Sci. U.S.A.* **75**, 2443–2447.

Spitz, M. (1986). *In* "Immunochemical Techniques, Part I: Hybridoma Technology and Monoclonal Antibodies" (J. Langone and H. Van Vunakis, eds.), Methods in Enzymology, Vol. 121, pp. 33–41.

Tam, J. P., and Lu, Y.-A. (1989). *Proc. Natl. Acad. Sci. U.S.A.* **86,** 9084–9088.

van Rooijen, N., and van Nieuwmegen, R. (1983). *In* "Immunochemical Techniques, Part F: Conventional Antibodies, Fc Receptors, and Cytotoxicity" (J. Langone and H. Van Vunakis, eds.), Methods in Enzymology, Vol. 93, pp. 83–95. Academic Press, San Diego.

Vanselow, B. A. (1987). *Vet. Bull.* **57,** 881–896.

Waksman, B. H. (1979). *Springer Semin. Immunopathol.* **2,** 5–33.

Weir, D. M., ed. (1978). "Handbook of Experimental Immunology," Vol. 3. Blackwell, Oxford.

CHAPTER 3

Making Monoclonal Antibodies

David J. Asai and John K. Wilder

Department of Biological Sciences
Purdue University
West Lafayette, Indiana 47907

I. Introduction

The vertebrate immune system presents the investigator with the means to "custom order" probes specific for virtually any biological molecule (reviewed in Kindt and Capra, 1984). Monoclonal antibody technology provides a powerful advantage over conventional serology in that the former enables the investigator to obtain an infinite quantity of antibody molecules of absolute homogeneity. Monoclonal antibody technology and polyclonal serology are governed by the same set of rules prescribed by the vertebrate immune response. Thus monoclonal antibodies are not necessarily any more specific or of higher affinity than the corresponding polyclonal sera. The point of departure between the two methods is that each monoclonal antibody represents a single component of the overall polyclonal response. Therefore, monoclonal antibodies are most useful when the investigator wishes to concentrate her/his attention on a single epitope at a time. Generally, conventional polyclonal antisera emphasize similarities among homologous antigens, whereas monoclonal antibodies can amplify discrete differences among antigens.

Since the initial report (Köhler and Milstein, 1975), monoclonal antibodies have been used in an almost infinitely broad spectrum of basic and applied biological research. At the same time, methods for obtaining hybridomas and for assaying antibodies have evolved significantly.

Since 1980, our laboratory has performed over 70 successful fusions, mostly directed against cytoskeletal proteins. In this article we present, in what we hope to be useful detail, our current procedures. We are indebted to many people who have helped refine these methods, especially Rick Harmon, Natalie Stein, and Jeff Frelinger (this group is otherwise affectionately known as the Natalie Stein School of Monoclonal Antibodies) who, when they were at the University of Southern California, taught us their protocol (Harmon *et al.,* 1982).

This article assumes that the reader has the experience and equipment required for mammalian cell culture and is aimed at the investigator who will use monoclonal antibodies in cell biological applications. And, of course, this chapter is neither the first nor the last word on monoclonal antibody procedures. There are many excellent resources (see, e.g., Goding, 1983; Campbell, 1984; Harlow and Lane, 1988).

II. First Principles

A. Monoclonal or Polyclonal?

Making monoclonal antibodies is laborious and expensive. Therefore it is important to consider carefully the application of the antibody prior to initiating the effort. For example, most applications involving the identification, quantification, or intracellular localization of an antigen can usually be accomplished with specific polyclonal antisera. Indeed, because the polyclonal antiserum usually reacts with several epitopes, the detection of the antigen may be much more reliable with a polyclonal reagent because the reaction does not depend on the presence of a single epitope. A polyclonal antiserum is preferred when it is used to identify an antigen or a fragment of the antigen in a complex mixture derived from a source different from that which the antigen originally was purified; for example, screening a cDNA library expressed in bacteria (Obar and Holzbaur, article 18 in this volume). Polyclonal antisera can be elicited that are highly specific for the antigen of interest (Drenckhahn *et al.*, article 2 in this volume). If necessary, affinity methods may be applied to purify antibodies from the polyclonal antiserum (see, e.g., Fuller *et al.*, 1975; see also Tang, article 5 in this volume).

Monoclonal antibodies should be favorably considered when it is of interest to aim an antibody at a specific epitope on the antigen. There are several illustrations of the exquisite specificity of monoclonal antibodies. For example, antibodies have been directed exclusively against a particular isotype in a family of proteins (see, e.g., Banerjee *et al.*, 1988) or to specific posttranslational modifications (e.g., acetylated tubulin, Piperno and Fuller, 1985; phosphorylated MAP 1B, Asai *et al.*, 1985, Sato *et al.*, 1986). Monoclonal antibodies have also been used to identify new components of a partially isolated cellular structure (e.g., mitotic phosphoproteins, Davis *et al.*, 1983). Another application in which monoclonal antibodies provide useful specificity is the inhibition of an activity by the binding of the monoclonal antibody near the active site of an enzyme (e.g., certain monoclonal antibodies to kinesin inhibit its motor function whereas others do not, Ingold *et al.*, 1988).

B. Hybridoma Selection in HAT[1]

A hybridoma is the result of the physical fusion between a B lymphocyte and a mouse myeloma cell. The B lymphocyte, stimulated *in vivo* prior to the fusion to secrete a specific antibody, is immortalized by its union with the tumor cell. Because the myelomas, fused or unfused, have an unlimited life span, it is necessary to distinguish between the lymphocyte–myeloma hybridomas and all other cells in the culture. This differentiation takes advantage of the purine synthetic pathways and the HAT selection is targeted at the enzyme hypoxanthine–guanine phosphoribosyltransferase, or HGPRT (Littlefield, 1964).

The HGPRT is essential to the purine salvage synthetic pathway. An HGPRT$^-$ cell must rely on its *de novo* pathway, synthesizing purines from nonpurine materials. 8-Azaguanine, an analog of guanine, enters the metabolic system via the salvage pathway. If a cell incorporates azaguanine in its DNA, it

---

[1] Adapted from Lerner, E. A. (1980). "The Yellow Brick Road to the Happy Hybridoma." Yale University.

cannot replicate the DNA and the cell will not proliferate. A cell that is HGPRT$^-$, therefore lacking the salvage pathway, is resistant to 8-azaguanine.

Selection in HAT (hypoxanthine, aminopterin, thymidine) allows only HGPRT$^+$ cells to survive. The antimetabolite aminopterin has two effects on the cell: it blocks *do novo* purine synthesis and it prevents the methylation of deoxyuridylate to form thymidylate. Hypoxanthine serves as a salvageable purine if aminopterin blocks *de novo* synthesis. And the thymidine bypasses the requirement for the methylation reaction, blocked by aminopterin, that leads to thymidylate. Thus, HAT selects for HGPRT$^+$ cells because the aminopterin will kill HGPRT$^-$ cells that rely solely on the *de novo* pathway.

The myeloma cells are selected in 8-azaguanine prior to the fusion to ensure that they are HGPRT$^-$. Myelomas by themselves will therefore not survive in HAT-containing medium. Only the hybridomas in which the B lymphocyte contributes the HGPRT activity will survive and proliferate in HAT (the unfused B cells and the B cell–B cell fusomas will also survive in HAT but they will not proliferate and will eventually die).

C. Myeloma Cell Line

Since the initial description of monoclonal antibodies by Köhler and Milstein (1975), the myeloma fusion partner has been refined. We utilize the M5 myeloma that was derived from SP2/0Ag14 (Schulman *et al.*, 1978). The SP2 cell (itself a hybridoma) does not synthesize either the light chain or heavy chain of immunoglobulin, ensuring that all of the antibody obtained from the hybridoma will be homogeneous. The M5 line was adapted to grow in horse serum and was provided by Joseph Davie.

D. Prior to Beginning Procedure

A successful fusion requires significant quantities of time, effort, attention, and supplies. It is imperative to prepare completely before beginning. A checklist with some of the important things to consider is as follows.

1. What will the antibodies be used for and is it a significant advantage to have monoclonal antibodies instead of conventional polyclonal antisera? See Section II,A (above) and article 1 by Golub (in this volume).

2. Is there an adequate stockpile of antigen that will be used for the immunizations and for the binding assays? It is a good idea to pool all of the antigen prior to starting the immunization and to check the purity of the antigen pool. In certain cases it may be desirable to immunize with a slightly heterogeneous mixture of proteins and to screen with a cleaner preparation.

3. Is the assay [usually an enzyme-linked immunosorbent assay (ELISA) or some other solid-phase binding assay] appropriate for the application of the antibody? Does it work reliably? Soon after fusion, the several hundred culture

wells will have to be assayed for antibody. The assay must be rapid (i.e., 1 day) because it will be necessary to decide which cultures to feed and split and which ones to allow to overgrow and die. The assay must be reliable because it will serve as the basis for identifying which cultures to keep and which to discard. The assay must be able to handle large numbers of samples because several hundred cultures will need to be tested. For these reasons, binding assays that are based on 96-well microtiter plate templates are highly desirable. Other assays (e.g., immunofluorescence microscopy, Western blotting, and enzyme inhibition), although occasionally necessary, are less desirable because they are more tedious and less amenable to rapid sampling of hundreds of samples. The investigator must decide on the kind of information that is most relevant to the application of the antibodies, tempered with the practical requirements of speed, accuracy, and large numbers. For example, the investigator may plan to utilize the monoclonal antibody to study an antigen in its native conformation and, therefore, the binding assay in this case should use undenatured antigen.

It is critical that the assay be working prior to fusion. Often an antibody (perhaps a polyclonal antiserum) is already available to the same antigen and can be used to establish the assay. We routinely obtain some serum from the immunized mouse and use this serum in an ELISA immediately prior to fusion in order to test the serum for antibody activity and to establish that the binding assay is working well. If an antibody to the antigen of interest is not available, then another antibody and its antigen should be used to mimic the assay conditions.

4. Is enough time available for fusion? The immunization process usually requires 6–8 weeks. After fusion, the selection, cloning, and expansion of the hybridomas requires 8–12 weeks. Nurturing of the hybridomas requires daily attention.

III. Overview of Protocol

A brief overview of the hybridoma protocol follows. The details of the various steps are presented in Section IV, and are ordered and numbered according to the organization of this section.

A. Before Fusion

1. The mouse is immunized, a process requiring 6–12 weeks. When the mouse is producing the antibodies of interest, it is sacrificed for the fusion.

2. Feeder medium is used throughout the initial stages of cell culture. Prior to fusion, adequate quantities of normal medium and feeder medium must be made and tested for sterility. For a typical fusion, feeder medium from six to eight spleens is required.

3. The M5 myelomas are selected in 8-azaguanine 2 weeks prior to the fusion. After selection, the M5 cells are expanded so that they are in exponential growth phase at the time of the fusion.

4. The assay is working reliably. Adequate antigen has been stockpiled for all of the anticipated binding assays.

B. Fusion

5. The immunized mouse is sacrificed. Its spleen is removed aseptically. The spleen cells are dissociated, washed, and counted. The M5 cells are harvested and counted. The splenocytes and the M5 cells are mixed together and fused with polyethylene glycol. The fusomas are plated into 96-well dishes.

C. Selection of Hybridomas

6. Aminopterin is added either on the day of the fusion or the next day, depending on specific requirements (discussed in Section IV,F).

D. After Fusion

7. The hybridomas are fed every 5 days. Usually some wells are ready to be assayed for antibody 10 days after fusion.

8. The wells are assayed when they begin to overgrow. Typically the first assay is spread over a 10-day period because of variability in hybridoma growth rates and in the number of colonies per well.

9. The positive wells (i.e., cells secreting antibody of interest) are cloned by limiting dilution. Single clones are identified visually. Single clones are assayed.

10. Positive clones are expanded in culture. The cells are weaned from feeder medium during this step.

11. The expanded single clones are frozen.

12. Positive clones can also be injected into mice to produce ascites serum, which contains a high concentration of antibody.

IV. Hybridoma Protocols

A. Immunization

We stimulate the immune system of the living mouse to respond positively to the antigen before performing the fusion.

1. Mice

The myeloma M5 cells were derived from a BALB/c mouse and we use the same strain to avoid potential cytotoxic reactions in culture. Female BALB/c mice, aged 8–10 weeks, are used for the immunizations. Normally, we immunize two mice simultaneously, check their sera for the appropriate antibody, and then use the mouse that is responding better.

2. Antigen

Purified protein ($M_r > 20K$) is injected either as a solution, or as a crushed slice from a polyacrylamide gel, or as a strip from a nitrocellulose blot. If the antigen is in solution, it is injected in a total volume of 0.5 ml/injection. We aim for 10–20 μg of protein in the initial injection and 5–10 μg of protein in the booster injections. Larger quantities are used if the antigen is particularly weak in eliciting an immune response or if the antigen is a mixture of proteins. To produce antibodies against a specific sequence, synthetic peptides are made with an amino-terminal cysteine through which the peptide is coupled to the carrier protein with the SMCC [sulfosuccinimidy 4-(N-maleimidomethyl) cyclohexane-1-carboxylate] reaction (Pierce Chemicals, Rockford, IL). We divide the peptide into two portions and couple one portion to bovine serum albumin and the other portion to thyroglobulin. We use the peptide–thyroglobulin conjugate as the immunogen and the peptide–albumin in the binding assays. In this way, antibodies to the carrier are not detected in the binding assay.

The first injection is with antigen (usually dissolved in phosphate-buffered saline) emulsified in an equal volume of Freund's complete adjuvant. Booster injections are with antigen emulsified in Freund's incomplete adjuvant. Emulsification is conveniently accomplished by repeated back-and-forth passage of the mixture between a "double-ended" 20-gauge steel needle constructed from two hypodermic needles soldered tip to tip. All injections are administered subcutaneously on the back of the mouse. The skin on the back of the mouse is loose and it is easy to "skin pop" the antigen with a syringe. The resulting lump persists for several weeks and marks the site of the injection; subsequent injections are placed in new sites.

3. Schedule of Injections

Soluble antigen is administered according to the following schedule: first injection (complete adjuvant, 10–20 μg of protein, 0.5-ml total volume); rest 2 weeks; booster 1 (incomplete adjuvant, 5–10 μg of protein, 0.5 ml); rest 1 week; booster 2 (same as booster 1); rest 1 week; booster 3 (same as booster 1); rest 1 week. At this point, a few drops of blood are obtained from the tail vein [the mouse is warmed by a light bulb in a box for several minutes; approximately

1 cm of its tail is amputated with a razor blade; the blood is collected in a microfuge tube with 20 μl of citrate–saline solution (0.136 M sodium citrate, 0.145 M sodium chloride); the red blood cells are removed by centrifugation and the serum is drawn]. The serum is diluted 250- to 500-fold and applied to a Western blot. If the mouse is producing antibodies, it is ready for the final boost. If it is not responding, additional booster injections are administered according to the same schedule.

The final boost is 5–10 μg of protein in 0.5 ml in the absence of adjuvant. Approximately 100 μl is injected with a 27-gauge needle into the tail vein (raised with ethanol). The remainder is injected intraperitoneally into the mouse. The mouse is sacrificed 4 days later and its spleen cells used in the fusion.

4. Variations

If the antigen is purified from a polyacrylamide gel, it either can be eluted from the polyacrylamide, in which case the soluble antigen is injected according to the schedule above, or stained bands can be excised, crushed, and injected as polyacrylamide. In the latter case, approximately the same quantity of protein per injection as is recommended for soluble proteins is injected. Adjuvants are not necessary if polyacrylamide is used. The final booster should be performed with soluble (eluted) protein. Proteins can also be blotted to nitrocellulose, the appropriate portion of the blot excised with a razor blade, and the nitrocellulose dissolved in dimethylsulfoxide (DMSO) and injected with adjuvant.

A convenient alternative is to blot the protein to nitrocellulose, excise the band of nitrocellulose that contains the protein, and insert the nitrocellulose strip under the skin on the back of the mouse. The nitrocellulose carrier serves as a depot that ensures the long-term exposure of the mouse to the antigen. We try to immunize with approximately 10 μg of protein on the strip of nitrocellulose. The nitrocellulose is folded up and inserted through a small incision made between the shoulder blades. The mouse is anesthetized with the inhalant methoxyfluorane during the surgery. The incision is closed with a sterile wound clip. A single booster strip of nitrocellulose is applied approximately 1 month later. The serum is tested and, when the mouse is producing antibodies of interest, it is sacrificed for the fusion.

B. Media

Serum-free (SF) medium: 1× Dulbecco's modified Eagle's medium (DMEM) (4500 mg of glucose/liter, with glutamine, without sodium pyruvate, without sodium bicarbonate), 20 mM N-2-hydroxyethylpiperazine-N'-2-ethanesulfonic acid (HEPES), 2 mM glutamine, 1× nonessential amino acids, 5×10^{-5} M 2-mercaptoethanol, penicillin–streptomycin (100 IU penicillin/ml, 100 μg streptomycin/ml), 3.7 g sodium bicarbonate/liter; filter sterilized through sterile 0.2-μm pore size membranes

Horse serum (HS) medium: SF medium plus 10% added horse serum. The serum is heated to 56°C for 45 min to inactivate the complement

Fetal calf serum (FCS) medium: SF medium plus 5% heat-inactivated fetal calf serum plus 5% heat-inactivated horse serum

Feeder medium: Spleens from nonimmunized 8- to 12-week-old BALB/c mice are obtained aseptically and dissociated into HS medium. Dissociation is conveniently accomplished by scraping the spleen through a sterile screen. The cells from one spleen are mixed with 10 ml of HS medium and placed in the CO_2 incubator for 4 days. At the end of the incubation, the medium is centrifuged to remove the cells. The supernatant is 10× feeder medium and can be stored at −20°C for 2 months.

50× HT (per 100-ml final volume): 19.4 mg thymidine and 68.1 mg hypoxanthine, dissolved in warm sterile water; filter sterilize and store in 1-ml aliquots at −20°C

Aminopterin (final concentration, ~$1.3 \times 10^{-6}M$)

"100×" aminopterin (670-4010AD; GIBCO, Grand Island, NY) is used as a 33× stock and stored at 4°C

"50×" aminopterin (A5159; Sigma, St. Louis, MO) is diluted into one-half the prescribed volume and this is used as a 33× stock and stored at 4°C

8-Azaguanine (1000×): 200 mg of azaguanine dissolved in 10 ml of water (solubilize with NaOH); filter sterilize and store in aliquots at −20°C

C. M5 Myeloma Cells

The M5 cells are split 1 : 20 into HS medium containing 20 μg azaguanine/ml. After 4–6 days,the cells are again split 1 : 20 into fresh azaguanine-containing HS medium and grown for another 4 days. After selection, the M5 cells are split 1 : 5 and passaged once in HS medium and grown to confluency (4–6 days) in a single T25 flask. The contents of the T25 flask are then split into three T75 flasks and expanded in HS medium for 2 days prior to fusion. Cells should be in the exponential growth phase at the time of fusion.

D. ELISA Procedure

1. Use Nunc-immuno plates, Polysorp F96 (6106L75; Thomas Scientific, Swedesboro, NJ) or Corning (Corning, NY) ELISA plates (Cat. No. 25801).

2. Dilute antigen in a carbonate buffer (0.01 M Na_2CO_3, 0.035 M $NaHCO_3$, pH 9.6) to a final concentration of 50–200 ng of protein/50 μl. Pipette 50 μl/well, incubate overnight at 4°C, or for 5 hr at 37°C.

3. Dump out the antigen. Whack the plate dry. Block each well with 200 μl of 2.5% bovine serum albumin (BSA) in phosphate- or Tris-buffered saline. Block for 1–2 hr at 37°C.

4. Dump out the BSA. Whack the plate dry. Wash each well once with phosphate-buffered saline (PBS)–0.05% Tween-20. Whack the plate dry.

5. Add 50 μl of antibody. For the first assay (about 10 days after fusion), 100 μl of the culture supernatant is drawn and half of this is applied directly to the ELISA well. For the second assay (after cloning), only 50 μl of the culture supernatant is drawn, and it is possible to dilute this 1:1 with PBS before applying it to the ELISA well. Make sure to include a negative control (e.g., culture supernatant from M5 cells or unused medium) in a desirable configuration (e.g., upper left corner). Incubate at 4°C for two or more hours.

6. Dump out the antibody. Whack the plate dry. Wash each well three times with PBS–Tween. Whack the plate dry.

7. Add 50 μl of diluted secondary antibody. For alkaline phosphatase-conjugated secondary antibody (anti-mouse IgG or anti-mouse IgG + IgM), dilute 1:1000 in PBS–0.5% BSA. Incubate at 4°C for 1–2 hr.

8. Discard the secondary antibody. Dry the plate on a paper towel. Wash each well three times with PBS–Tween. Dry the plate again.

9. Add 200 μl of freshly made developing solution per well [Substrate buffer: 48.5 ml diethanolamine, 400 ml water, 0.2 g NaN_3; adjust pH to 9.8 with HCl and add water to 500 mL; store in the dark at 4°C. Developing solution: dissolve p-nitrophenyl phosphate tablets (5 mg/tablet; 104, Sigma) to a final concentration of 1 mg in 1 ml of substrate buffer]. Allow color development at room temperature for 5 min to 1 hr.

10. Optional: Stop the reaction with 50 μl/well of 3 M NaOH.

11. Read on plate reader at 405 nm.

E. Fusion

Prior to begining fusion, go through the following checklist:

Healthy M5 cells, recently selected in azaguanine, 5–10 × 10^7 cells
Feeder medium, 200 ml, supplemented with 1× HT (or 1× HAT)
SF medium, 100 ml
Ten 96-well flat-bottomed microtiter dishes
Sterile pipettes, 1 and 10 ml
Sterile plastic 50- and 15-ml conical bottomed (screw-capped) tubes
Sterile syringe (3 ml), 25-gauge needle
Sterile syringe (1 ml), 18-gauge needle
Two small sterile petri dishes, covered
Tabletop centrifuge to pellet cells
Polyethylene glycol, M_r 1500 (PEG 1500; Cat. No. 783641; Boehringer
    Mannheim, Indianapolis, IN), stored at 4°C; bring to room temperature

Inverted phase-contrast microscope

Hemacytometer

Trypan blue

Four disposable glass tubes to dilute and mix cells for counting in trypan blue

600-ml beaker three-quarters filled with warm (37–40°C) water

Clock with a second hand

Sterile pipette tips

Multichannel pipetter

Sterile petri dishes or troughs from which to dispense the cells

Marker pen

Sterile Pasteur pipettes

WARNING: PRACTICE THE FUSION PROCEDURE BEFOREHAND

1. It is generally a good idea to obtain immune serum from the mouse immediately before sacrifice. The mouse is anesthesized with methoxyfluorane and bled by puncturing the retroorbital sinus with half of a 100-μl glass capillary pipette. The blood is collected into sodium citrate to inhibit coagulation (see Section IV,A,3). The blood is collected into a 1.5-ml microfuge tube and the serum obtained by centrifuging out the cells. It should be possible to obtain 0.5 ml of blood. Mouse antibodies are sensitive to repeated freezing and thawing so the serum should be aliquoted into small volumes and frozen once. The immune serum is used as a positive control during the hybridoma screening assays.

2. Sacrifice the mouse by cervical dislocation (the mouse was last injected intraperitoneally and intravenously 4 days prior to fusion). Aseptically (using ethanol-sterilized dissecting tools and cleaning the site of incision with ethanol) remove the enlarged spleen (left side of the abdomen) and place it into a small petri dish with a few milliliters of sterile SF medium. The spleen should be rapidly removed in one piece and freed of any adhering tissue. *Note: All handling of the mouse should occur in a room separate from the culture room.* Only the petri dish containing the spleen should enter the culture room.

3. Transfer the spleen from the petri dish in which it was transported from the surgery room to a fresh petri dish containing SF medium and place the new dish in the laminar flow hood. *Note: The dish that transported the spleen from the surgery room to the culture room should never enter the hood.* Dissociate the spleen by repeatedly "injecting" the spleen with SF medium through the 25-gauge needle and teasing the spleen apart with forceps. Transfer the cells to a sterile 15-ml tube. Allow debris to settle for a few minutes. Transfer the suspension of cells (leaving behind the debris at the bottom of the tube) to a fresh sterile 15-ml tube. Centrifuge. Resuspend the splenocytes in 10 ml of SF medium and count them in trypan blue on the hemacytometer. A typical spleen from a hyperimmunized animal will yield $2-3 \times 10^8$ lymphocytes.

4. Harvest the M5 myeloma cells. Wash them in SF medium. Resuspend them in SF medium. Count them with the hemacytometer.

5. Aim for a ratio of three splenocytes to one M5 cell. If aminopterin is added on day 2 (see Section IV,F), use 1.2×10^8 splenocytes and 4×10^7 M5 cells plated into ten 96-well dishes. This should yield one to three colonies of hybridomas per well. If aminopterin is added on the day of the fusion (i.e., day 1), increasing by a factor of 5 the concentrations of both the splenocytes and the M5 cells and preserving the 3 : 1 ratio will result in one to three colonies per well.

6. Pipette the appropriate number of splenocytes and M5 cells into the same 50-ml tube, mix, and centrifuge. Remove all of the supernatant. Loosen the pellet by tapping the tube on the floor of the hood.

7. The fusion is accomplished by the controlled addition of PEG followed by SF medium with constant agitation at 37°C. Move the beaker of warm water into the hood. Hold the tube of cells in one hand and gently swirl it in the water (avoid splashing the water into the tube!). All of the components will be added by pipetting with the other hand while continuously swirling the tube.

8. Using the syringe with the 18-gauge needle, add 1 ml of PEG dropwise over a 45-sec interval. *It is extremely important to maintain a steady rate of dripping throughout all of these addition steps.*

9. Incubate (swirling in the water) the cells and the PEG for 1 min. If the pellet does not completely suspend, tap it gently, or use the tip of a pipette to loosen the pellet.

10. Add 1 ml of SF medium, steadily dropwise, over 30 sec.

11. Incubate for 1 min.

12. Add 2 ml of SF medium, steadily dropwise, over 1 min.

13. Incubate for 2 min.

14. Add 4 ml of SF medium, steadily dropwise, over 2 min.

15. Incubate for 4 min.

16. Add 8 ml of SF medium, steadily dropwise, over 3 min.

17. Centrifuge at 500 g for 5 min. Discard the supernatant. Resuspend the pellet in 175 ml of 1× (HS) feeder medium. (If HAT is to be added on day 1, include 1× HAT, and suspend the cells in 200 ml of feeder medium plus HAT.)

18. Plate the hybridomas into the wells of ten 96-well (full area) flat-bottomed microtiter dishes. If HAT has been added, plate 200 μl/well. (If the HAT is to be added on day 2, plate 175 μl/well and, on day 2, gently add 25 μl of HAT in feeder medium to each well.)

19. Grow the hybridomas in a leveled 37°C incubator, in high humidity, and in 8% CO_2.

F. HAT Selection

We aim for one to three hybridoma colonies per microtiter well because this is a convenient number in the cloning work that occurs later. If the investigator wishes to limit the period of time during which the initial hybridomas are assayed, then a larger number of colonies per well is preferred because they will tend to overgrow simultaneously and rapidly. However, if the number of colonies per well is greater than a few, there is an increased likelihood that the particular antibody activity that is sought will be "swamped out" by faster growing, uninteresting cells and the interesting cells may be lost.

It is our experience that different growth patterns occur, depending on when the aminopterin is added to the hybridomas. If aminopterin is added immediately there are few survivors, presumably because the fusion procedure initially weakens the cells. For this reason, we increase by fivefold the numbers of cells per well if aminopterin is added on day 1. On the other hand, if aminopterin is added on day 2, there is a greater survival among the fusomas, but many of them are fragile. Because there is a greater proportion of survivors, adding aminopterin on day 2 increases the probability of recovering a rare antibody activity. The drawback, however, is that up to 50% of the colonies are highly fragile and begin to "ghost" (i.e., die off; the dying cells lose their contrast when viewed with phase-contrast microscopy) after several days. Therefore they require extremely careful and individualized attention (see Section IV,M, below). To generalize: when aminopterin is added on day 1, there are fewer survivors and only the hardier fusomas survive; when aminopterin is added on day 2, there are many more survivors (five times as many), but approximately half of these remain fragile.

Finally, growth rates are different depending on when the aminopterin is added. If aminopterin is added immediately, the cells tend to grow more slowly and are ready to be assayed about 15 days after the fusion. If aminopterin is withheld until day 2, the cells appear to enjoy a "jump start" and grow faster, ready to be assayed about 10 days after the fusion.

The investigator should consider these factors and experiment with them. If the antigen elicits a strong and diverse antibody response, it is likely that it will be easy to find and clone the interesting hybridomas and HAT selection should begin immediately. On the other hand, if it appears that the antibody activity that is sought is rare, the investigator should consider adding aminopterin on the second day.

G. Feeding Hybridomas

Prior to the first feeding (3–5 days after the fusion), the wells are examined by inverted microscopy in order to obtain an estimate of the number of colonies per well. (*Note:* These are faster growing colonies.)

The hybridomas are fed every 5 days with feeder medium supplemented with $1 \times$ HT. The first feeding also includes aminopterin. The upper 90 μl of the old medium are removed and replaced with 100 μl of fresh medium, gently pipetted along the side of the well (to avoid disrupting the colonies). A multichannel micropipetter expedites these procedures.

H. Assaying Hybridomas

Usually by the tenth day after fusion, some of the wells are ready to be assayed. Although it is possible simply to assay all of the wells at the same time, it is preferable to assay wells only as they begin to overgrow (yellowing of the medium, usually visible over a light box). The well that is to be assayed is identified, 100 μl of the medium removed under sterile conditions (enough for two 50-μl assays), and the well then fed with feeder medium supplemented with HT. Because each well is assayed at a different time, the initial screening of the ca. 960 wells is spread over several days. We find it practical to screen the fusion in groups of 200–300 wells, performing an ELISA every other day over a 1-week period. The investigator needs to keep careful records so that wells are not confused during this process. The ELISA results determine which wells are saved and which are discarded (allowed to overgrow without further feedings). It is critical that the assay results are obtained promptly because the hybridomas at this stage will begin to die if they are not immediately split and fed.

I. Cloning Hybridomas

The wells that are identified as producing the antibody of interest are cloned by limiting dilution. Often there are more positive wells than is practical to clone at one time. Our strategy is to select a few fast-growing colonies and a few slower growing colonies and to clone these (cloning more than 10 initial wells at once is technically challenging). The other colonies that are producing antibody of interest can be frozen in 10% DMSO, 50% serum (final concentrations). These are stable at −70°C for a few months. It is possible to expand the colonies in the wells of 24-well dishes and to freeze the cells in the dish, making sure that the cover is sealed with Parafilm to the bottom.

The contents of a positive well are gently suspended in its medium (~200-μl total volume). Approximately three-fourths of the cells are expanded as a polyclonal culture in the well of a 24-well plate (1-ml total volume) in feeder medium supplemented with 0.5× HT. These are expanded into a T25 flask and frozen. The other one-fourth of the cells is suspended in 5 ml of cloning medium (feeder medium made in 5% HS and 5%FCS, supplemented with 1× HT). From this suspension, 0.5 ml is transferred to a 15-ml tube containing 9.5 ml of cloning medium, and the rest is plated into a T25 flask. The cells in the tube are mixed gently, then 1 ml is transferred to a second tube containing 9 ml of cloning medium. This 1 : 9 dilution is repeated twice more so that four tubes of serially diluted cells are obtained. The contents of each tube are then plated, 100 μl/well, into 96-well *half-area, flat-bottomed* microtiter dishes and grown in a leveled incubator (avoid jarring the incubator during the growth of single clones). The clones are not fed during the entire two or more weeks required for cloning.

Five days after setting up the cloning plates, the cloning plates are carefully removed from the incubator and viewed with an inverted microscope. Each well is examined and wells containing clearly single clones are marked and noted. Five days after the initial mapping, the wells are reexamined in order to ensure that there remain single clones (and that a slower growing second clone has not materialized). It is important to identify rigorously the single clones. If there is doubt about whether a clone is derived from a single cell, avoid it.

Usually 12–14 days after setting up the cloning plates, the faster growing clones are ready to be assayed (the medium should be turning yellow and the diameter of the clone should be one-quarter to one-third the diameter of the half-area well). Medium from the well is removed at the meniscus with a sterile pipette tip, taking 55–60 μl for the ELISA (the supernatant can be diluted 1 : 1 in PBS in order to perform duplicate assays). Positive clones are expanded in culture (see Section IV,J).

J. Expanding Clones in Culture

It is necessary to expand clones slowly; that is, slowly moving them to gradually larger volumes. This is especially the case for the hybridomas from which the aminopterin was withheld until day 2. Occasionally, a clone will not grow well when diluted. In this case, it is important to have a "back-up" culture so that the dilution may be repeated. The goal is to expand the clones in order to have enough cells to freeze and/or inject into primed mice for the production of ascites serum, and to wean the hybridomas from the feeder medium.

When the diameter of the clone is at least one-quarter the diameter of the half-area well, the cells are suspended and transferred to the well of a 24-well plate in 0.5 ml of feeder medium made in 5% HS and 5% FCS, and supplemented with 1× HT. The back-up is the original half-area well on the cloning dish—100 μl of the contents from the larger well are added back to the emptied half-area well. When the larger well begins to turn yellow and the cells near confluency (3–6 days after transfer), 0.5 ml of the same medium is added to the well and another 0.5 ml of HS medium (no feeder medium, no FCS, no HT) is added. When the well is again beginning to overgrow, 1 ml of the spent medium is removed (this can be used in a Western blot), and the cells are transferred to the well of a six-well plate in 1.5 ml of feeder–5% HS–5% FCS–1× HT and 1.5 ml of HS medium. A back-up culture is created by returning 1 ml to the well of the 24-well plate. Finally, when the largest well is approaching confluency, the cells are transferred to a T25 flask in 5 ml of HS medium. The well is reseeded with 2 ml of this suspension. When the flask begins to overgrow, another 5 ml of HS medium is added.

K. Freezing Hybridomas

Single clones are frozen after they have been expanded in culture. In addition, it is a good idea to expand the polyclonal cultures (identified after the first assay and prior to being cloned) and freeze these as well. Occasionally it is necessary to recover cells from frozen polyclonal cultures.

The contents of one confluent T25 flask are frozen in two 1-ml vials in 50% HS medium, 40% added HS, 10% DMSO. Best results are obtained when the cells are cooled gradually. The cells can be placed in a covered styrofoam box (at room temperature) and frozen at −70°C for 2 days. Alternatively, more sophisticated and efficient freezing systems [e.g., molded plastic racks insulated with isopropyl alcohol; e.g., Cat. No. 5100-0001 (Nalgene Labware, Rochester, NY)] can be used. After the cells are equilibrated at −70°C, they are transferred to liquid nitrogen.

Survival of the frozen cells is enhanced if they are thawed rapidly, washed in SF medium, and initially grown in a small volume of feeder medium–5% HS–5% FCS.

L. Growing Ascites Serum

Ascitic growth of the hybridomas yields serum with a high titer of antibody (usually >1000 times the concentration of antibody in the culture supernatant). BALB/c mice are primed with 0.5 ml (intraperitoneal injection) of pristane (2,6,10,14-tetramethylpentadecane) 3–5 days prior to the injection of the cells. The expanded clones (at least 1×10^6 cells/mouse) are injected intraperitoneally into the primed mice. The mice are monitored frequently. They are ready (usually 7–14 days after injection of the cells) to be tapped when their bellies are noticeably (twice normal) swollen. They are usually lethargic and may be cold to the touch. Mice are tapped by the insertion of a sterile 18-gauge hypodermic needle into the peritoneum and the yellow-red liquid is collected into a sterile 50-cm^3 tube. Mice may be tapped once or twice more (on consecutive days). The mouse should be sacrificed before the tumor kills the mouse.

The fluid (up to 5 ml/mouse per tapping) is clotted overnight in the refrigerator and the serum obtained by centrifuging. The serum is aliquoted into small volumes and stored frozen.

An alternative to ascitic growth of the hybridomas is the use of large-scale bioreactors (often based on hollow-fiber technology). When the hybridomas are growing well in a bioreactor, they quickly deplete the circulating medium. Therefore it is critical that the medium be monitored regularly (usually by measuring glucose levels) and replaced when necessary.

M. Extraordinary Efforts

Soon after fusion it is necessary to treat each well individually. Some grow faster than others. Some are hardier than others. Especially if the antibody sought is rare, individualized and continuous care is required. Particularly frustrating is the "ghosting" of hybridomas. This is especially prevalent among hybridomas from a fusion in which the aminopterin was added on day 2.

"Ghosting" is the death of a colony so that only faint outlines of the cells are visible. Some ghosting colonies can be saved if they are recognized early enough. The usual treatment involves enriching the medium (see below).

A few variations can be used to help a clone survive. Feeder cells can be used in place of feeder medium. One spleen from an unimmunized mouse is suspended (final concentration) in 100 ml of medium and fed to the hybridomas. Feeder cell medium usually stimulates growth slightly better than feeder medium. Some investigators prefer to add thymocytes as a feeder layer. Ghosting cultures can sometimes be saved by replacing the medium more frequently than every 5 days. Finally, we find that fetal calf serum supports growth better than horse serum. However, because it appears to take some time for the cells to adapt to FCS, we find that 5% FCS and 5% HS is better than placing the cells directly into 10% FCS medium.

V. Perspective

The immune system is programmed to respond to a large number of antigens. Sophisticated strategies can now be employed to harness the immune response (e.g., *in vitro* immunization of lymphocytes, epitope selection of a recombinant cDNA expression library). Despite the power of these new technologies, conventional antisera and monoclonal antibodies are often found to be satisfactory and more accessible to the majority of cell and molecular biologists.

The difference in cost and effort between conventional antisera and monoclonal antibodies is significant. Therefore, the single most important consideration for the investigator is an honest appraisal of what the antibody he/she seeks will be used for. In many instances a specific, high-titer polyclonal antiserum will be adequate. When it is desirable to direct an antibody to a particular epitope, the appropriate reagent may be a monoclonal antibody. Once the decision to pursue monoclonal antibodies has been made, the procedure is reasonably straightforward but time consuming.

Acknowledgments

We thank the many people who have advised us during the refinement of our protocols. We thank the members of the Asailum and our collaborators who have helped to make our hybridoma work worthwhile. This laboratory has been supported by grants from the National Science Foundation and from a Cancer Center Core Grant from the National Cancer Institute awarded to the Purdue Cancer Center.

References

Asai, D. J., Thompson, W. C., Wilson, L., Dresden, C. F., Schulman, H., and Purich, D. L. (1985). Microtubule-associated proteins (MAPs): A monoclonal antibody to MAP1 decorates microtubules *in vitro* but stains stress fibers and not microtubules *in vivo*. *Proc. Natl. Acad. Sci. U.S.A.* **82**, 1434–1438.

Banerjee, A., Roach, M. C., Wall, K. A., Lopata, M. A., Cleveland, D. W., and Luduena, R. F. (1988). A monoclonal antibody against the Type II isotype of β-tubulin. *J. Biol. Chem.* **263,** 3029-3034.

Campbell, A. M. (1984). "Monoclonal Antibody Technology." Elsevier, Amsterdam.

Davis, F. M., Tsao, T. Y., Fowler, S. K., and Rao, P. N. (1983). Monoclonal antibodies to mitotic cells. *Proc. Natl. Acad. Sci. U.S.A.* **80,** 2926-2930.

Fuller, G. M., Brinkley, B. R., and Boughter, J. M. (1975). Immunofluorescence of mitotic spindles by using monospecific antibody against bovine brain tubulin. *Science* **187,** 948-950.

Goding, J. W. (1983). "Monoclonal Antibodies: Principles and Practice." Academic Press, New York.

Harlow, E., and Lane, D. (1988). "Antibodies: A Laboratory Manual." Cold Spring Harbor Lab. Press, Cold Spring Harbor, New York.

Harmon, R. C., Shigekawa, B., Stein, N. M., and Frelinger, J. A. (1982). Problems associated with determining the specificity of monoclonal antibodies to major histocompatibility complex antigens. *In* "Hybridomas in Cancer Diagnosis and Treatment" (H. F. Oettgen and M. S. Mitchell, eds.), pp. 21–30. Raven, New York.

Ingold, A. L., Cohn, S. A., and Scholey, J. M. (1988). Inhibition of kinesin-driven microtubule motility by monoclonal antibodies to kinesin heavy chains. *J. Cell Biol.* **107,** 2657–2667.

Kindt, T. J., and Capra, J. D. (1984). "The Antibody Enigma." Plenum, New York.

Köhler, G., and Milstein, C. (1975). Continuous cultures of fused cells secreting antibody of predefined specificity. *Nature (London)* **256,** 495–497.

Littlefield, J. W. (1964). Selection of hybrids from matings of fibroblasts *in vitro* and their presumed recombinants. *Science* **145,** 709–710.

Piperno, G., and Fuller, M. T. (1985). Monoclonal antibodies specific for an acetylated form of α-tubulin recognize the antigen in cilia and flagella from a variety of organisms. *J. Cell Biol.* **101,** 2085–2094.

Sato, C., Nishizawa, K., Nakayama, T., Nose, K., Takasaki, Y., Hirose, S., and Nakamura, H. (1986). Intranuclear appearance of the phosphorylated form of cytoskeleton-associated 350-kDa proteins in U1-ribonucleoprotein regions after growth stimulation of fibroblasts. *Proc. Natl. Acad. Sci. U.S.A.* **83,** 7287–7291.

Schulman, M., Wilde, C. D., and Köhler, G. (1978). A better cell line for making hybridomas secreting specific antibodies. *Nature (London)* **276,** 269–270.

CHAPTER 4

Purification of Mouse Antibodies and Fab Fragments

Thomas J. Smith

Department of Biological Sciences
Purdue University
West Lafayette, Indiana 47907

I. Introduction

Immunoglobulins are large proteins ($\sim$150 kDa) that are composed of two heavy chains ($\sim$50 kDa each) and two light chains ($\sim$25 kDa each). The basic architecture of an antibody is shown in Fig. 1. Two "arms" of the antibody, called Fabs, bind antigen. The rest of the antibody is a highly conserved region

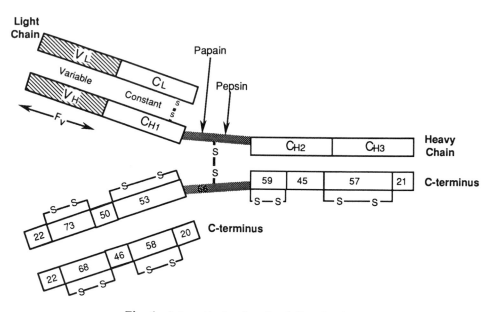

Fig. 1 Schematic drawing of an IgG molecule.

called the Fc. Between the Fab arms and the Fc is a highly flexible region called the hinge. As shown in Fig. 1, this region is highly sensitive to proteolysis. There are several types of heavy chains expressed in mammals, but IgG and IgM are predominantly found in the serum. In most cases, the B cells switch from IgM to IgG production during the maturation process. Over time monoclonal antibodies have become an essential tool in biochemical and structural studies. Therefore the purification of antibodies and fragments of antibodies have become essential procedures in many laboratories. Fortunately, antibodies have several unique properties that can be exploited for purification.

II. Purification of Murine IgG Monoclonal Antibodies

There are several fast, preparatory ways to isolate murine IgG monoclonal antibodies. When determining which method should be used, one must consider the physical properties of the antibody. One such determinant is the isotype of the antibody. As will be discussed, different isotypes of murine antibodies bind to the various columns with different affinities. Another important antibody property is solubility. Some antibodies are not very soluble in low ionic strength buffers and therefore cannot be isolated by ion-exchange chromatography. However, this property can be used to produce pure material.

A. Affinity Chromatography

Protein A is a 42,000-Da, cell membrane protein from *Staphylococcus aureus*. This bacterial protein binds to the Fc region of most mouse and human but not rat immunoglobulins (Langone, 1978). Protein G is a cell wall protein from β-hemolytic streptococci of the C and G strains and binds well to all isotypes of mouse and human antibodies but weakly to rat antibodies (Björck and Kronvall, 1984; Åkerström and Björck, 1986). The most common commercially available form of protein G is one from which the albumin-binding domain has been removed. The basic procedures for protein A and protein G affinity chromatography are the same. The sample containing antibody is loaded onto the column at neutral pH, the unbound material is washed from the column matrix with neutral pH buffers, and the antibody is eluted with low-pH buffers. The acidic eluant is then quickly neutralized to prevent denaturation of the antibody.

1. Protein A Chromatography

Each murine isotype has a different affinity for protein A. This property can be utilized when the sample is contaminated with other isotypes.

The following are examples of washing and elution conditions for each of the different isotypes, and should be refined for each hybridoma. For example, we have often found it advantageous to perform an initial precipitation of the antibody with ammonium sulfate (60% final saturation). This allows one to work with smaller volumes and to remove much of the contaminating albumin proteins. We have also found that it may be necessary to use lower pH buffers than suggested in order to elute the antibody in reasonable volumes and in a timely way (Fig. 2).

Materials

Protein A–Sepharose 4B (Pharmacia Fine Chemicals, Piscataway, NJ)
Sodium phosphate buffer, pH 7.0–8.0 (0.1 M)
Acetate buffer, pH 2.0–2.5 (0.1 M)
Tris-HCl, pH 7.6 (2 M)
Sodium citrate buffers, at pH 6.0, 5.5, 4.5, and 3.5 (0.1 M)

a. (Optional) Precipitation of Antibody from Ascites Fluid with Ammonium Sulfate.

The antibody is precipitated from ascites fluid with a 60% ammonium sulfate solution (final concentration). One can obtain this concentration of ammonium sulfate by using either a stock solution of saturated ammonium sulfate or by adding dry ammonium sulfate to the ascites solution (approximately 349 g/liter final volume). The solution is stirred for several hours or overnight at 4°C. The precipitate is collected by a 10-min centrifugation at 10,000 g. The pellets are solubilized and dialyzed against the sodium phosphate buffer. This initial pre-

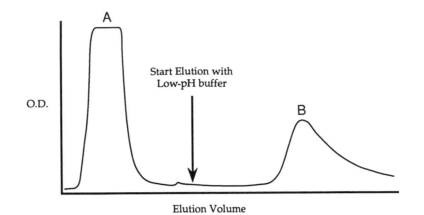

Fig. 2 Elution profile of ascites fluid from a protein G column. Ascites fluid was loaded onto the column and washed with 0.1 M sodium phosphate buffer (pH 7). Peak A, unbound material. The antibody is then eluted with 0.05 M sodium acetate buffer (pH 2.0) (peak B).

cipitation step also allows efficient removal of the lipid material that is often found in the ascites fluid.

b. Removal of Lipids and Large Aggregates from Antibody Solution.

The antibody solution is clarified with a 10-min centrifugation at 10,000 g. The lipids form an oily layer on top of the tube and can be discarded. The antibody solution can be further clarified by filtration (e.g., with a 0.4-μm pore size syringe filter).

c. Preparation of Column for Loading.

The column should first be washed with the acetate buffer to remove any previously bound antibody material. Once the eluant from the column proves to be significantly free of protein (an $OD_{280\ nm}$ of less than 0.05), then the column is equilibrated with the phosphate buffer. In our laboratory, we use the fast protein liquid chromatography (FPLC; Pharmacia Fine Chemicals, Piscataway, NJ) system to pump buffers through standard columns of protein A–agarose beads (bed volumes from 1 to 5 ml), to monitor the elution of proteins from the column, and to collect the resulting peaks. This procedure can also be done on a small scale without the FPLC system by using columns made from Pasteur pipettes (glass wool in the bottom) and by measuring the OD_{280} of each fraction.

d. Loading Sample and Washing Unbound Material from Column.

Aliquots of the sample are then put onto the column and allowed to bind to the matrix. We let the sample pass slowly over the column in order to obtain maximum binding. Gravity usually pushes the sample into the matrix slowly enough. The amount of sample loaded is dependent on the bed volume and antibody concentration. The wash eluant is monitored by enzyme-linked immunosorbent assay (ELISA) to determine how much antibody has escaped the first pass and the loading volume is adjusted accordingly. The binding capacity information is usually included with the column material.

e. Elution of IgG.

In all of the following cases, the antibody fractions are collected as 1-ml fractions in tubes containing 100 μl of 2.0 M Tris buffer to neutralize the acetate elution buffer (Ey *et al.*, 1978). For large-scale preparations, we often collect the antibody peak directly into a large flask containing the concentrated Tris buffer.

1. IgG_1 does not bind well to most protein A columns. If other isotypes are present in small amounts then IgG_1 is even further excluded from the column. This makes it impractical to use protein A affinity chromatography to purify large amounts of IgG_1. However, it is possible to use the protein A column to isolate small amounts of this isotype. A pH 6.0, 0.1 M sodium citrate buffer is sufficient to elute IgG_1.

2. IgG_{2a} is eluted from the column with a 0.1 M sodium citrate buffer, pH 4.5.

3. IgG_{2b} is eluted from the column with a 0.1 M sodium citrate buffer, pH 3.5.

4. IgG_3 elutes at pH 4.5 together with IgG_{2a}. However, the two isotypes can be separated from each other if a long, thin protein A column is used in conjunction with a shallow pH gradient (pH 5.0–4.4). IgG_{2a} elutes at pH 5.0 and IgG_3 elutes at pH 4.5 (Seppällä *et al.*, 1981).

The samples of purified antibody are then dialyzed against the appropriate buffers or stored as an ammonium sulfate precipitate.

2. Protein G Chromatography

The procedure is the same as that for protein A except that antibodies often bind tighter to protein G than protein A. We used acetate buffers at pH 2.0 to elute one IgG_{2a} antibody. This low pH did not affect antibody activity.

B. Ion-Exchange Chromatography

Ion-exchange chromatography can also be employed to purify antibodies from ascites fluid. This method is useful for purifying those antibodies that do not bind well to protein A or protein G (e.g., IgG_1) or those that might be particularly sensitive to the low pH used during the elution process. It has been reported that such high-resolution techniques as Mono-Q chromatography can, using the FPLC system, also separate IgG_1 from IgG_3 isotypes (Hill *et al.*, 1982; Clezardin *et al.*, 1985). We have also had similar results (see Fig. 3). The following section discusses the protocol using ion-exchange chromatography with the FPLC system. Similar, albeit not as good, results can be obtained by standard chromatography, using such resins as Fast Flow DEAE (Pharmacia Fine Chemicals, Piscataway, NJ).

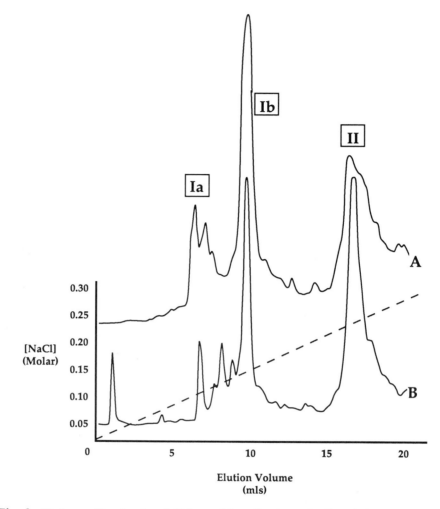

Fig. 3 Elution profile of ascites fluid from a Mono-Q column. Profile A is from the ammonium sulfate precipitate of ascites fluid, and profile B is from whole ascites fluid. The buffer used was 10 mM Tris-HCl, the salt gradient was approximately 0.1 M NaCl/7 ml, and the pressure was approximately 1 MPa. The group of peaks labeled Ia were found to be other isotypes (endogenous) of murine antibody, peak Ib was the monoclonal antibody MAb-17, and peak II was found to be albumin.

Materials

Mono-Q and FPLC systems (Pharmacia/LKB)
Tris-HCl, pH 7.8 (20 mM)
Tris-HCl, pH 7.8 (20 mM) and 1 M NaCl

1. Precipitation of Antibody from Ascites Fluid with Ammonium Sulfate

This is the same procedure as found in the previous section. While this step is not absolutely required, it helps to keep the column cleaner by removing much of the other serum proteins that precipitate and block the column. If unfractionated ascites fluid is used, expect a precipitate to form over time, especially if the sample is kept at room temperature.

The pellet from this step is then solubilized and dialyzed against the 20 mM Tris (no NaCl) buffer.

2. Examining Dialyzed Sample

It is important to examine any precipitate that may form while dialyzing against this low ionic strength buffer. With all of the antibodies that we have worked with, there is always some precipitation at this step. In most cases a small amount of antibody is irreversibly denatured due to the high salt concentration. Occasionally, the antibody precipitates because the salt concentration was too high. This latter situation can easily be tested by isolating the precipitate and diluting it in Tris buffer to see if it resolubilizes. In about one-quarter of the cases a large amount of the antibody precipitated. Often this precipitate resolubilizes with the addition of high NaCl concentrations. Again, it is important to perform this step carefully so that all of the antibody is not lost.

3. Loading Sample and Eluting with an NaCl Gradient

A 0.5-ml portion of this solution is then loaded onto a 1-ml Pharmacia Mono-Q column and eluted with a linear salt gradient of 0.1 M NaCl per 7 ml of eluant containing 10 mM Tris, pH 7.8. The flow rate should be 1 ml/min, with a pressure of about 1 MPa. The peaks are collected with Pharmacia FRAC-100. Examples of Mono-Q elution profiles of ascites fluid and ammonium sulfate-precipitated antibody are shown in Fig. 3. Note that the ammonium sulfate-precipitated material had a much higher proportion of antibody to albumin than did the ascites fluid.

4. Examining Samples with ELISA, SDS-PAGE, and Pooling

Because each antibody runs differently on the Mono-Q column, we routinely perform ELISA and sodium dodecyl sulfate-polyacrylamide gel electrophoresis (SDS-PAGE) on each of the fractions for every new antibody. An example of one of the purification results is shown in Fig. 4.

Several notes should be made about the FPLC columns. Material that shows any cloudiness should not be loaded onto the column. Such material causes the back pressure to increase significantly with just a few runs of sample. Check the

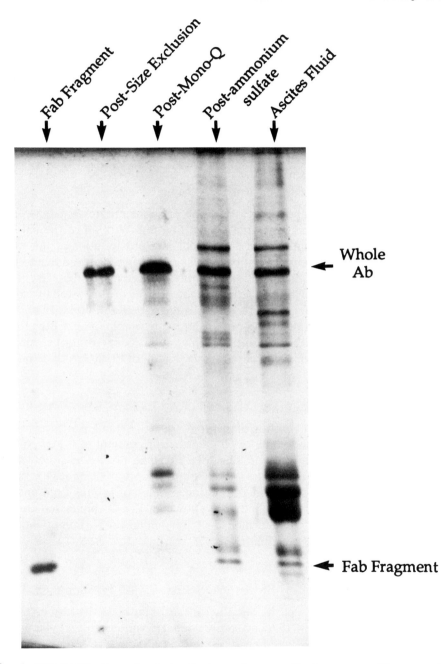

Fig. 4 SDS-PAGE of samples during the ion-exchange purification process. The ammonium sulfate precipitation step removes much of the albumin. The Mono-Q step removes most of the rest of the contaminants. It should be noted that slower gradients removed the rest of the albumin and allowed the size-exclusion chromatography step to be eliminated. Also shown here is purified Fab.

manual for cleaning instructions. We have found that injecting 1 M NaOH and letting it incubate for 30 min will remove most of the obstructing material. Sometimes proteins bind to the beads through hydrophobic interactions. This binding is promoted by salting-out effects caused by NaCl. In these cases it is often beneficial to use salts with higher chaotropic character such as $LiClO_4$ instead of NaCl.

C. Other Methods

In the next section, an example of an antibody that could not be purified by one of the above techniques will be discussed. This particular antibody is called MAb-23 and is a monoclonal antibody against human rhinovirus 14 (NIm-IA epitope). This monoclonal was developed in the laboratory of R. Rueckert (University of Wisconsin).

Properties of This Antibody
1. Does not bind well to protein G column
2. Precipitates in 20 mM Tris, pH 7.8

Purification Protocol

1. Antibody was precipitated with ammonium sulfate as previously described.

2. Precipitate was dialyzed against 20 mM Tris buffer, pH 7.8. At this step, a large amount of precipitate formed. The precipitate was found to be mostly antibody.

3. Precipitate was collected with a 10-min centrifugation at 10,000 g. Precipitate was resolubilized in 0.2 M phosphate buffer, pH 7.2. At this point the antibody was approximately 95% pure. However SDS-PAGE analysis showed that some contaminants were still in solution.

4. Antibody was chromatographed with a standard column containing S-300 Sephacryl beads, using the same phosphate buffer as above. This resulted in a single broad peak. Fractions of this broad peak were examined by SDS-PAGE (Fig. 4). By comparing the reducing versus nonreducing sides of the gel, it was easy to determine which samples contained immunoglobulins. From the gel, it appeared that the first two samples probably contained IgM (from the high molecular weight of the nonreduced species and the complex pattern in the presence of 2-mercaptoethanol). The last sample also contained some contaminants, but these are of low molecular weight and are probably albumins. The other fractions were pooled and found to react strongly with antibody HRV14. This particular antibody was found to be isotype IgG_3.

III. Generation of Fab and Fab' Fragments

Once purified antibody is obtained, it is often desirable to generate the antigen-binding fragments Fab, F(ab')$_2$, or Fab'. Fab fragments are created by papain proteolysis at the hinge region to yield two 45,000-Da fragments. If antibodies are cleaved with pepsin, then most of the Fc region is digested away, leaving the Fab arms attached via the disulfide linkages in the hinge region. These F(ab)'$_2$ fragments can be separated under mild reducing conditions to yield Fab' fragments. Iodoacetamide can be used to block the free cysteines. Therefore Fab' fragments differ from Fab fragments in that they have longer heavy chain carboxyl termini than Fab fragments and there are free cysteines in the Fab' fragments.

A. Papain Digestion

Many use variations of the conditions first described by Porter (1959). It is advisable to use small quantities of antibody and to test various parameters in order to optimize digestion conditions. Some antibodies are highly sensitive to papain whereas others are not. If the Fab fragments are to be used for crystallization experiments, it is often not enough to cleave the antibody. It may be necessary to optimize for a more homogeneous population of Fabs with respect to pI.

Porter's Conditions
1. IgG in a solution containing

> Phosphate (10 mM), pH 7.3
> NaCl (0.15 M)
> Ethylenediaminetetraacetic acid (EDTA) (1 mM)
> 2-Mercaptoethanol (5 mM)
> Papain (enzyme-to-antibody ratio, 1:100)

2. Incubate for 1 hr at 37°C or for an optimal time.
3. Stop the reaction with 10 mM (final concentration) iodoacetamide. Let the iodoacetamide react for 15 min at 17°C.

To optimize conditions, the time of digestion, the amount of 2-mercaptoethanol (with a concomitant increase in the iodoacetamide concentration), and the enzyme-to-antibody ratio can all be varied. It is best to start with different enzyme-to-antibody ratios and to vary the time of digestion. In our laboratory, an IgG$_3$ class antibody was completely cleaved in less than 4 hr with an enzyme-

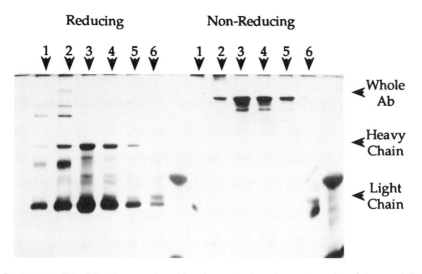

Fig. 5 SDS-PAGE of fractions produced by size-exclusion chromatography of the precipitating antibody under reducing and nonreducing conditions. The band patterns of lanes 1 and 2 are typical of IgM (contamination), whereas lanes 3–5 are examples of purified IgG.

to-antibody ratio of 1:500. However, an IgG_{2a} antibody required 16 hr with an enzyme-to-antibody ratio of 1:100.

Small samples are digested and then examined by SDS-PAGE. Often, because of the 2-mercaptoethanol present during digestion, the disulfide bonds between the antibody chains are broken. Because of this, it is uncommon to see only 150,000-Da bands in nonreducing SDS-PAGE gels even if the antibody is undigested. Therefore the samples should be reduced and the presence of a 50,000-Da band representing the heavy chain can be used to follow the extent of digestion. Fully digested, reduced antibody should be composed of chains approximately 24,000 Da in size. Once approximate conditions are found for cleavage, small pilot digestions are performed. The amount of antibody to be digested is increased to about 1 mg and only two conditions are varied. The Fab fragments are then purified with the FPLC system (see section IV) and the Fab fragments examined for ionic purity. It is common to see Fab fragments with two to four different pI values after digestion. By looking at samples digested under slightly different conditions, different pI species can be selected. Once the conditions are refined, then the amount of digested material is increased and the resulting Fab fragments are purified. It should be noted that ionically pure samples of Fab are often necessary for crystallographic studies, but such purity is not critical for other experiments.

One potential problem with this procedure occurs when the purified antibody is not very soluble. By increasing the salt (NaCl) concentration in the reaction mixture, the antibodies often are solubilized. In some cases we have found that even high salt concentrations are insufficient for solubilization. For these antibodies we used the cloudy, precipitated solution of antibody and found that, once the Fc portion was removed by the papain, the resulting fragments were highly soluble. The problem with this method is that, unless the antibody concentration was measured by methods other than OD_{280}, it may be difficult to reproduce the digestion conditions.

B. Digestion with Pepsin

Pepsin, like papain, is fairly nonspecific in its cleavage of protein. However, unlike papain, pepsin is active at low pH and is denatured at neutral or high pH. This allows for rapid inactivation of the enzyme after digestion, by merely raising the pH of the reaction mixture.

Murine IgG subclasses show marked differences in their sensitivities to pepsin cleavage (Lamoy, 1986; Parham, 1983). The general pattern of sensitivity to proteolysis is as follows:

$$IgG_2b >> IgG_3 > IgG_{2a} > IgG_1$$

The following protocol (Lamoy, 1986; Parham, 1983) can be used to find the optimum conditions for a particular antibody.

Materials
Immunoglobulin (2–10 mg/ml) in 0.1 M sodium acetate at pH 7.0
Acetic acid (2 M)
Tris-HCl (2 M), pH 8.0

Method

1. Add the acetic acid solution, dropwise, to the immunoglobulin solution until the pH is 4.2 (IgG$_1$ or IgG$_{2a}$) or 4.5 (IgG$_3$).
2. Warm the solution to 37°C.
3. Add 1 mg of pepsin for each 33 mg of antibody.
4. Stop digestion with a 1:40 volume of Tris-HCl.

As with the papain digestion, the time of digestion needs to be determined for each different antibody.

IV. Fab Purification

A. Protein A/G Purification

There are several ways to purify Fab fragments from the reaction mixture. One common way is to pass the digested material over a protein A or G column.

This column is run in a fashion similar to that previously described, except that the Fc fragments and whole antibodies will bind to the column whereas the Fab fragments will not. The problems with this method are that small protein fragments may also pass through the column unbound and some isotypes do not bind well to the matrix.

B. Ion-Exchange Chromatography

Another way to purify Fab fragments is by ion-exchange chromatography. Fab fragments often have alkaline pI values whereas intact antibodies are more neutral. Therefore, most Fab fragments do not bind to DEAE or Mono-Q columns, but Fc fragments and whole antibodies do. Mono-Q chromatography may be performed in the following way.

Materials
Tris-HCl (20 mM), pH 7.8
Tris-HCl (20 mM), pH 7.8, containing 1 M NaCl
FPLC system with a Mono-Q column

Method

1. Dialyze the digested sample against 20 mM Tris-HCl buffer, pH 7.8 (no salt).

2. A 0.5-ml portion of this solution is then loaded onto a 1-ml Pharmacia Mono-Q column and eluted with a linear salt gradient of 0.1 M NaCl (per 7 ml of eluant) containing 20 mM Tris, pH 7.8. The flow rate is 1 ml/min, and the pressure should be 1 MPa. The peaks are collected with Pharmacia FRAC-100. Fab fragments elute in the void volume. The next peak to elute is intact antibody, followed by Fc fragments. To obtain better ionic purity at this stage, the pH can be raised. Proteins bind to the positively charged column if the pH is at or higher than the pI. At higher pH values Fab fragments with lower pI values will be retained slightly longer than the higher pI forms. It is not necessary to use the FPLC system for this procedure, but better separation and faster runs will be obtained if it is. Fast Flow DEAE is the next best choice of matrix material.

(Fab′)$_2$ fragments are much like Fab fragments in that they do not bind well to Mono-Q or DEAE columns. Therefore (Fab′)$_2$ fragments can be purified in the same way as Fab fragments. In addition, (Fab′)$_2$ fragments can be purified by size-exclusion chromatography because their molecular weight is twice that of the resulting Fc fragment.

V. Further Purification of Fab Fragments

If further purification of the Fab fragments is necessary, a chromatofocusing column can be used. Chromatofocusing separates proteins according to pI. The general principle behind chromatofocusing is that the column matrix is composed of tertiary and quaternary amines (amines that can and cannot be deprotonated, respectively). The protein sample is loaded onto the column at a pH high enough that the protein has a net negative charge and can bind. A lower pH solution of ampholytes is then passed over the column and the protons from the ampholytes are exchanged to the column matrix and to the protein. Because the solution of ampholytes is composed of a mixture of molecules with many different pI values, this proton exchange causes a pH gradient within the column and the eluting buffer. Eventually the protein is unable to remain bound to the matrix, and elutes from the column. This occurs at approximately the pI of the protein. Again, a column produced for the FPLC system (Mono-P) or a standard column with chromatofocusing material from Pharmacia can be used.

Materials

Buffer A: 20 mM diethanolamine, pH 9.5

Buffer B: 1:10 dilution of Polybuffer 96 (Pharmacia) at pH 6.0 (acidified with HCl)

Mono-P (Pharmacia) column: Flow rate, 0.5 ml/min

Procedure

1. Make sure the column is clean. Injections of 1 M NaOH and/or 1 M HCl are usually sufficient.

2. Inject an aliquot of 1 M NaOH and monitor the pH. The pH of the eluant will first become high and then slowly equilibrate to the pH of buffer A. This step removes all residual ampholytes and equilibrates the tertiary amines to that pH (removes protons).

3. Inject an aliquot of the Fab solution. Approximately 1 ml of sample, which has been dialyzed against buffer A, is loaded onto the column.

4. After the sample has been loaded, switch to 100% buffer B. The column will now generate the pH gradient.

5. Collect samples with FRAC-100 (Pharmacia) and examine the peaks with isoelectric focusing gels. Because the peaks are already in ampholytes, they can go directly to the isoelectric focusing (IEF) gels.

Several things that can be done to improve the resolution. We have found that decreasing the concentration of ampholytes to 1:20 greatly improves the resolution. This is because the concentration of hydronium ions is less in the ampholyte solution and therefore takes longer to exchange with the tertiary amines in the column. Sometimes the resolution may be improved by the addition of small amounts of salts (e.g., 10 mM NaCl).

Shown in Fig. 6 is the elution profile of one sample of a Fab fragment.

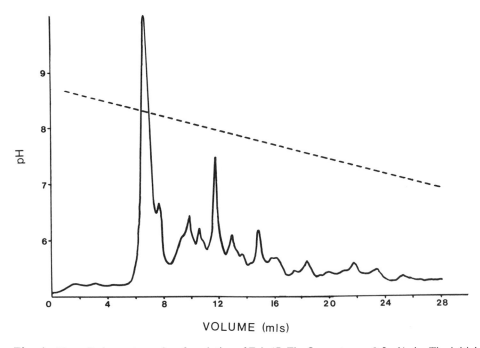

Fig. 6 Mono-P chromatography of a solution of Fab-17. The flow rate was 0.5 ml/min. The initial buffer was 20 mM monoethanolamine, pH 9.0. After the sample was loaded onto the column, the buffer was switched to a 1:20 dilution of Polybuffer 96 at pH 6.0. The dashed line represents the elution pH.

VI. Crystallization of Fab Fragments

Our group has crystallized Fab fragments by using basic principles of protein crystallization. Perhaps most important to crystallization is the purity of the sample. When proteins pack into ordered crystalline arrays, any contamination of the sample can cause the crystallization process to become poisoned. The worst kinds of contaminants are those that are chemically similar to the crystallizing species. This is a problem with Fab fragments because papain and pepsin cleavages are nonspecific. There are always several different ionic species of Fab after papain cleavage. Such heterogeneity has been shown to affect crystallization greatly, but can be easily remedied by one of the methods described above. In our experience, if the Fab sample was pure crystallization was rarely a problem.

The initial conditions for crystallizing our Fab fragments were obtained from the literature. From the literature, it is clear that different Fab fragments crystallize under similar conditions. We were able to obtain small crystals of two different Fab fragments by the vapor diffusion method (Fig. 7) and the following search conditions.

1. Polyethylene glycol 8000 as precipitant
   a. Fab solution: Fab (4–10 mg/ml) dialyzed against low ionic strength buffer (e.g., 10 mM Tris, pH 7.5)
   b. Reservoir:
      Sodium phosphate buffer (0.1 M), pH 6.0–9.0
      Polyethylene glycol 8000 (7–20%)
      Sodium azide (1 mM)
2. Ammonium sulfate as precipitate
   a. Fab solution: Same as above.
   b. Reservoir:
      Sodium phosphate buffer (0.1 M), pH 6.0–9.0
      Ammonium sulfate (final concentration, 20–50%)
      Sodium azide (1 mM)

One milliliter of the reservoir solution was placed into the well. Silicone grease is applied to the rim of each well. Five microliters of this solution is

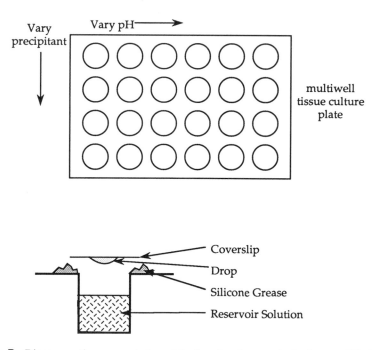

Fig. 7 Diagrammatic representation of the hanging drop method of vapor diffusion.

placed onto a plastic coverslip with 5 μl of Fab solution. The coverslip is inverted and placed onto the well, ensuring that the silicone forms a tight seal. Place in a quiet area. Crystals often form within 2 weeks (Fig. 8).

Notes about Crystallization

The Fab sample was dialyzed against low ionic strength buffer so that the buffer of the reservoir overwhelmed the weak buffer of the Fab solution. In this way, the same solution of Fab can be used for many different conditions.

The pI values of the Fab fragments were determined before crystallization by using isoelectropoint focusing. As the pH of the reservoir approached the pI of the Fab, the crystals formed faster and required less precipitant. For example, Fab-17 (pI of 8.5) crystals formed overnight at pH 7.2, but took over more than 11 days to grow at pH 6.5. However, the fast crystal growth at higher pH values yielded small, unusable crystals. The best crystals were obtained around pH 6.5.

When the drops were composed of 10 μl of the Fab solution and 5 μl of the reservoir solution, the crystals grew to larger dimensions than when 5 μl of

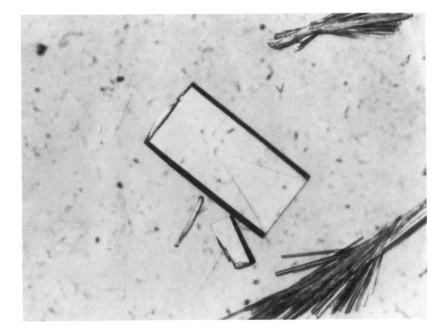

Fig. 8 Crystals of Fab-17. This is one of three different crystal forms of Fab-17. These crystals were grown with 40% ammonium sulfate as precipitate and 0.1 M sodium phosphate buffer (pH 6.5).

protein solution was mixed with 5 µl of reservoir solution. This is probably not merely due to the increased amount of total protein, but because of slower crystal growth due to the increased equilibration time.

In the case of Fab-17, the method of antibody purification may have an effect on crystallization. It was found that antibody purified with the Mono-Q method yielded different Fab fragments than did antibody purified by protein G affinity chromatography. These two samples of Fab crystallized into different crystal forms. It is possible that the low pH needed to elute the antibodies from the column somehow affected the antibodies so that their cleavage with papain was different than when the antibody was subjected only to the high salt of ion exchange.

It should also be noted that there may be solubility problems with the Fab fragment. Centricon-10 (Amicon, Inc., Beverly, MA) concentrators are used to concentrate Fab-17. However, at the refrigerated temperatures of the concentrator this Fab fragment precipitates. If the concentrator is warmed to room temperature, then the Fab goes back into solution immediately.

VII. Summary

Although monoclonal antibodies can have different properties, it should still be possible to purify any antibody to homogeneity. Often, by utilizing these differences, the purification procedure can be improved. All of the methods discussed are fairly simple and fast. With 1 week of work, over 50 mg of purified antibody can be realized. These methods can also yield material of sufficient quality for structural studies.

References

Åkerström, B., and Björck, L. (1986). A physiochemical study of protein G, a molecule with unique immunoglobulin G-binding properties. *J. Biol. Chem.* **261,** 10240–10247.

Björck, L., and Kronvall, G. (1984). Purification and some properties of streptococcal protein G, a novel IgG-binding reagent. *J. Immunol.* **133,** 969–974.

Clezardin, P., MacGregor, J. L., Manach, M., Boukercle, H., and Dechavanne, M. (1985). One-step procedure for the rapid isolation of mouse monoclonal antibodies and their antigen binding fragments by fast protein ligand chromatography on a Mono-Q anion exchange column. *J. Chromatogr.* **319,** 67–77.

Ey, P. L., Prowse, S. J., and Jenkin, C. R. (1978). Isolation of pure IgG_1, IgG_{2a} and IgG_{2b} immunoglobulin from mouse serum using protein A-sepharose. *Immunochemistry* **15,** 429–436.

Hill, E. A., Penny, R. I., and Anderton, B. A. (1982). *Int. Symp. Chromatogr., 14th, London* Abstr. C101.

Lamoy, E. (1986). Preparation of $F(ab')_2$ fragments from mouse IgG of various subclasses. *In* "Immunochemical Techniques, Part I: Hybridoma Technology and Monoclonal Antibodies" (J. Langone and H. Van Vunakis, eds.), Methods in Enzymology, Vol. 121, pp. 652–663. Academic Press, Orlando, Florida.

Langone, J. J. (1978). [^{125}I]Protein A: A tracer for general use in immunoassay. *J. Immunol. Methods* **24,** 269–285.

Parham, P. (1983). On the fragmentation of monoclonal IgG$_1$, IgG$_{2a}$, and IgG$_{2b}$ from BALB/c mice. *J. Immunol.* **131,** 2895.

Porter, R. R. (1959). The hydrolysis of rabbit γ-globulin and antibodies with crystalline papain. *Biochem. J.* **73,** 119–126.

Seppällä, I., Sarvas, H., Pèterfy, F., and Mäkelä, O. (1981). Occurrence and specificity of antibodies against group-specific polysaccharides in beta-hemolytic streptococcal infections. *Scand. J. Immunol.* **14,** 335.

CHAPTER 5

Blot–Affinity Purification of Antibodies

Wen-Jing Y. Tang

Kewalo Marine Laboratory
Pacific Biomedical Research Center
University of Hawaii
Honolulu, Hawaii 96813

I. Introduction

Blot-affinity purification (BAP) was developed by Olmsted (1981), and has been widely used. It utilizes the superb resolving power of sodium dodecyl sulfate-polyacrylamide gel electrophoresis (SDS-PAGE) to separate proteins in a mixture and the subsequent electrophoretic transfer of the individual bands from the gel to a diazotized paper blot. The separated, immobilized protein bands are then used as antigens to isolate their corresponding antibodies from the heterogeneous antiserum. This method is much simpler than column affinity purification. In addition, it offers a solution to obtain pure antigen, and thus its specific antibody, when a particular protein cannot be purified with conventional methods such as affinity column chromatography or high-performance liquid chromatography (HPLC). Smith and Fisher (1984) did an extensive study of this method with nitrocellulose (NC) as the solid support. With the inclusion of the detergent Tween-20 in all the buffers, greater than 80% of the antibodies could be recovered from the blot. They also found that the binding of the antigen onto NC was irreversible even after being boiled in SDS; thus, the blot could be used for repeated antibody isolation.

We have used BAP to isolate antibodies specific to the heavy chains of dynein of the sea urchin sperm flagella. Using our experience as a case example, this article describes in detail the isolation of these antibodies, the problems we encountered in this pursuit, how the purified antibodies had helped us in achieving our goals, and what their limitations were. For this purpose, a brief description of our system is presented below.

Dynein is a microtubule-based ATPase. It hydrolyses ATP to ADP and the energy released can be coupled to microtubule-mediated cell motility. In sea urchin sperm flagella, the outer arm dynein can be solubilized from the axoneme with a 0.6 M NaCl extraction medium. This 21S form of dynein has a complex structure: it contains two heavy chains (HCs), α and β (>400 kDa[1]), three intermediate chains (ICs; 80–120 kDa), and several light chains (LCs; 15–25 kDa) with a combined molecular weight of about 1.3×10^6. The two HCs are similar in size. When analyzed by SDS-PAGE, they comigrate in most gel systems. There is a small difference in their electrophoretic mobility when a 3–6% gradient gel of the Laemmli system is used, with α-HC being the slower migrator (Bell *et al.*, 1979). The two HCs are also similar in that they both possess ATPase activities and, in the presence of Mg·ATP and vanadate, near-ultraviolet (UV) irradiation causes them to cleave (V1 cleavage): specifically, each results in two V1-cleaved fragments, HUV1 (~228 kDa) and LUV1 (~200 kDa) (Lee-Eiford *et al.*, 1986). If dialyzed against a low-salt medium, the 21S dynein particle is dissociated into two subfractions that can be separated by zonal centrifugation (Tang *et al.*, 1982). β-Heavy chain and the three ICs co-sediment around 9 to 10S (β subfraction) whereas α-HC aggregates and together with several minor polypeptides sediments over the range of 12 to 30S (α subfraction). The two HCs are also different in their response to trypsin. In a low-salt medium, limited trypsin digestion of β-HC yields discrete polypeptides of 195, 130, and 110 kDa, whereas the digestion of α-HC gives rise to a multitude of faint bands when analyzed by SDS-gel electrophoresis (Ow *et al.*, 1987).

To help us understand the structure and function of the dynein ATPase, we have raised rabbit polyclonal antibodies against SDS-gel-purified dynein HCs isolated from the sperm flagella of the sea urchin *Tripneustes gratilla* and used

---

[1] The correct molecular weights of the sea urchin sperm flagellar dynein β-HC and its tryptic and V1 cleavage fragments are now known (Gibbons *et al.*, 1991). However, to avoid confusion, the molecular weights used here follow the apparent molecular weights used in the original papers cited.

BAP to isolate antibodies specific to either α-HC, β-HC, or V1-cleaved dynein fragments. Test blots on which trypsin-digested or V1-cleaved dynein had been immobilized were used to verify the specificity of the isolated antibodies.

II. SDS-Polyacrylamide Gel Electrophoresis and Blotting

For protein separation, the Dreyfuss system (Dreyfuss *et al.*, 1984), which can resolve polypeptides of widely varying molecular weights, is used. For example, a 12.5% gel can resolve polypeptides from ~5 to 250 kDa and a 6.5% gel covers a range from less than 20 kDa to >500 kDa.

A. Reagents

Running gel stock solution [acrylamide–bisacrylamide (33.5–0.3%)]: Dissolve 33.5 g of acrylamide [Serva (Heidelberg, Germany), twice recrystallized] and 0.3 g of N,N'-bis-methylene acrylamide (Serva) in water to a final volume of 100 ml. Filter through Whatman (Clifton, NJ) No. 3 filter paper and store in a dark bottle at 4°C

Running gel buffer (1 M Tris, pH 9.1): Mix 48.4 g of Tris base (Trizma; Sigma, St. Louis, MO), 60 ml of 1 N HCl, and water to a final volume of 400 ml

Ammonium persulfate (APS), 3%: Dissolve 0.3 g of APS (Serva) in water to a final volume of 10 ml. Aliquot and freeze

Sodium dodecyl sulfate (SDS), 10%: Dissolve 10 g of SDS (Sigma) in water to a final volume of 100 ml. Store at room temperature

N,N,N',N'-Tetramethylenediamine (TEMED; Sigma)

Stacking gel stock solution [acrylamide–bisacrylamide (30–0.44%)]: Dissolve 30 g of acrylamide and 0.44 g of bisacrylamide in water to a final volume of 100 ml. Filter and store in a dark bottle at 4°C

Stacking gel buffer (0.5 M Tris, pH 6.8): Mix 6.06 g of Tris, 45 ml of 1 N HCl, and 1 ml of 1% bromphenol blue. Add water to a final volume of 100 ml

Tank buffer: Dissolve 6 g of Tris, 28.8 g of glycine, and 2 g of SDS (Sigma) in water to a final volume of 2 liters

Running gel (6.5%), 100 ml: Mix 39.0 ml of water, 38.0 ml of running gel buffer, 19.5 ml of running gel stock, 1.0 ml of 10% SDS, and 50 μl of TEMED. Add 2.5 ml of 3% APS to start polymerization

Stacking gel (4%), 10 ml: Mix 6.1 ml of water, 2.5 ml of stacking gel buffer, 1.3 ml of stacking gel stock, 0.1 ml of 10% SDS, and 10 μl of TEMED. Add 0.1 ml of 3% APS to start polymerization

Sample buffer (5×), pH 6.8: A 10-ml stock contains 0.61 g of Tris, 4.5 ml of 1 N HCl, 0.5 g of SDS, 5 ml of glycerol, and a few grains of bromphenol blue. Store at 4°C. Before mixing with samples, take an aliquot out and add 1/20 vol of 100% 2-mercaptoethanol

Transfer buffer: 1 liter of solution contains 6 g of Tris base, 28.8 g of glycine, 0.1 g of SDS (ultrapure; BDH, Poole, England), and 200 ml methanol

Ponceau S staining solution: Dissolve 0.2 g of Ponceau S in 100 ml of 3% trichloroacetic acid. The solution is reusable

B. Preparation of Nitrocellulose-Immobilized Antigen

1. Prepare four preparative, discontinuous Dreyfuss slab gels (6.5% running gel and 4% stacking gel), 1.5 mm in thickness. (Protein transfer is less efficient if 3-mm gels are used.) A blank (without teeth) comb is inserted into each stacking gel.

2. Three different dynein preparations are used as gel samples:

α subfraction: Contains mainly α-HC and some minor proteins

β subfraction: Contains mainly β-HC and the three ICs

V1-cleaved dynein: Contains both HUV1 and LUV1 from α- and β-HCs

Each sample is mixed with a one-fourth volume of the $5\times$ sample buffer, heated in a boiling water bath for 3 min, and chilled on ice. For the α or β subfraction, 250 μg is loaded onto each gel so that four gels have a total loading of $\sim$1 mg. For V1-cleaved dynein, 500 μg is loaded per gel so that HUV1 and LUV1 each has $\sim$1 mg in four gels. Protein concentration is based on the Bradford assay (Bradford, 1976), using bovine serum albumin (BSA) as the standard.

3. Electrophorese at 150 V for 3 hr. Gels are immersed in the lower tank buffer of a Protean II apparatus (Bio-Rad, Richmond, CA), which acts as a heat sink so that cooling is not necessary.

4. Cut and briefly stain a vertical strip from the slab gel to locate the relevant bands. Then horizontally cut the regions containing the relevant protein bands and equilibrate them with the transfer buffer.

5. Set up the blotting system. With a Trans-blot cell (Bio-Rad), blotting onto NC (BA 83; Schleicher & Schuell, Keene, NH) is done at 25 V for 15 hr or at 60 V for 6 hr at $\sim$15°C.

6. Air dry the blots and stain with Ponceau S for about 5 min at room temperature. Rinse briefly, air dry again, and store the blots between two sheets of filter paper at 4°C.

C. Preparation of Test Blots

1. Prepare 1.0-mm thick, 6.5% running and 4% stacking mini-Dreyfuss gels. Combs with 15 teeth are used in the stacking gels.

2. Two different dynein preparations are used as test samples:

Trypsin-digested dynein: Contains the trypsin-digested products of both α- and β-HCs

V1-cleaved dynein

Load 2 μg into each well. The presence of bromphenol blue in the stacking gel makes the wells easily visible. Gels are run at 120 V for 1 hr with the Bio-Rad mini-Protean dual slab cell.

3. Equilibrate the gels with transfer buffer and blot the minigels onto NC (mini-trans-blot cell; Bio-Rad) at 60 V for 1 hr.

4. Air dry the blots, stain with Ponceau S, and mark the lanes with a soft pencil. Store blots at 4°C.

III. Antibody Adsorption, Elution, and Storage

A. Reagents

Tris-buffered saline with Tween-20 (TBST) (10 mM Tris, pH 8.0–150 mM NaCl–0.05% Tween-20): Mix 1.21 g of Tris base, 8.4 g of NaCl, 6 ml of 1 N HCl, water, and 0.5 ml of Tween-20 [polyoxyethylenesorbitan monolaurate (Bio-Rad), enzyme immunoassay (EIA) purity] to a final volume of 1 liter

Blocking solution: (5% nonfat powdered milk in TBST): Add 5 g of Carnation nonfat powdered milk to 100 ml TBST; mix well

Glycine elution buffer, pH 2.8: (0.1 M glycine–0.5 M NaCl–0.05% Tween-20): Mix 0.75 g of glycine, 2.9 g of NaCl, 3 ml of 1 N HCl, water, and 50 μl of Tween-20 to a final volume of 100 ml

Tris buffer (1 M), pH 8.1: Mix 12.1 g of Tris base, 40 ml of 1 N HCl, and water to a final volume of 100 ml

B. Procedure

1. In a small container with a cover, block the NC-bound antigen strips with 50 ml of 5% milk (Johnson *et al.*, 1984) in TBST for 30 min at room temperature with mild agitation to saturate the nonspecific binding sites on the NC blot. Rinse with TBST three times briefly to remove milk.

2. For antibody adsorption, incubate the blocked NC-bound antigen strips with 5 ml of 100-fold-diluted antiserum [50 μl of antiserum in 5 ml of TBST with 50 μl of 10% BSA, (fraction V) and 50 μl of 5% NaN$_3$] overnight at room temperature with mild agitation.

3. Transfer the diluted serum to a test tube and save for future use. Rinse the NC strips three times with 10 ml of TBST for 5 min with mild agitation to remove unbound antibodies.

4. Elute antibody from the washed NC strips with 2 ml of glycine elution buffer for 3 min over ice with gentle mixing. Immediately transfer the eluted antibody to a test tube containing 0.3 ml 1 M Tris, pH 8.1, for neutralization. Repeat this elution–neutralization once. Rinse the NC blot with 2 ml of TBST. Combine the two eluates and the TBST rinse. Add 0.75 ml of 10% BSA and 75 μl of 5% NaN$_3$ and store at 4°C.

5. Rinse the NC-bound antigen strips with more TBST until the pH becomes neutral. The strips are now ready for another round of adsorption and elution. For readsorption, replenish the used, diluted antiserum with 5 μl of the original antiserum and go through steps 2 to 5 again.

IV. Quality Evaluation of the Eluted Antibodies

A. Reagents

TBST

Milk (5%) in TBST

Secondary antibody: Anti-rabbit IgG, alkaline phosphatase conjugated (Promega, Madison, WI)

Alkaline phosphatase assay buffer (AP buffer) (100 mM Tris, pH 9.5–100 mM NaCl–5 mM MgCl$_2$): Mix 2.4 g of Tris base, 1.16 g of NaCl, 1 ml of 1 N HCl, and water to 200 ml

p-Nitroblue tetrazolium chloride (NBT): 50 mg/ml in 70% dimethylformamide

5-Bromo-4-chloro-3-indolyl phosphate, p-toluidine salt (BCIP): 50 mg/ml in dimethylformamide (the sodium salt is not soluble in this solvent)

Color development solution: To 15 ml AP buffer, add 100 μl of NBT and mix well; then add 50 μl of BCIP and mix again. This solution should be made shortly before needed

B. Procedure

1. Cut the required number of strips from the test blot, prewet with TBST, put each into a slot of a homemade Lucite tray. Add 2 ml of 5% milk in TBST to block the strips at room temperature for 30 min. Rinse with TBST briefly three times to remove milk.

2. Dilute a required amount of the eluted antibody with TBST and add 2 ml to each test strip. Incubate at room temperature for 3 hr (the dilution factor should be determined for each antibody).

3. Rinse with 2 ml of TBST three times, 5 min each time, to remove the unbound antibodies.

4. Dilute the secondary antibody with TBST according to the recommendation of the supplier. Add 1–2 ml to each test strip and incubate at room temperature for 1 hr.

5. Rinse the test strips three times with 2 ml of TBST, and a fourth time with AP buffer.

6. Prepare color development solution. Add 1–2 ml to each test strip. A purplish color should appear gradually. Stop the reaction by rinsing with water.

C. Results

By following the procedures described above, we have isolated antibodies against α-HC (anti-α), β-HC (anti-β), HUV1 (anti-HUV1) and LUV1 (anti-LUV1). Both anti-α and anti-β stain mainly the HC region of the undigested dynein. Because it is difficult to differentiate between α-HC and β-HC on the blot, test strips of trypsin-digested dynein are used for the evaluation of their respective antibodies. Figure 1a shows the general protein staining pattern of the strip by amido black. Figure 1b shows that anti-β strongly stained the 195- and 130-kDa bands and faintly stained the 110-kDa and several other bands of unknown origin. The pattern of staining by anti-α is different. Many closely spaced bands appeared (Fig. 1c), including the 195- and 130-kDa band regions. To test whether this staining of the 195- and the 130-kDa bands was due to anti-β contamination, anti-α was diluted three times with TBST to reduce the salt concentration and the NC-bound β-HC strips were added to it for negative adsorption. After overnight incubation, this anti-α was tested again with the same test procedure. Comparison of Fig. 1c with Fig. 1d indicates that, after negative adsorption, staining of the 195- and 130-kDa regions by anti-α decreased significantly whereas the other bands were basically unaffected. Figure 2a and b are the amido black staining results of dynein before and after V1 cleavage, whereas Fig. 2c and d show the test results of the antibodies isolated from V1-cleaved dynein fragments. Anti-HUV1 stained predominantly HUV1, as expected; however, anti-LUV1 stained both HUV1 and LUV1 with nearly equal intensity.

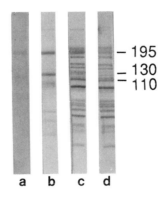

Fig. 1 Test strips of trypsin-digested dynein were treated with (a) amido black, (b) anti-β, (c) anti-α before negative adsorption, and (d) anti-α after negative adsorption. Following the primary antibody incubation, (b), (c), and (d) were further incubated with the alkaline phosphatase-conjugated secondary antibody and stained with NBT and BCIP.

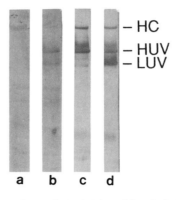

Fig. 2 Amido black was used to stain test strips of dynein before (a) and after (b) V1 cleavage. For antibody staining, strips immobilized with V1-cleaved dynein were treated with either anti-HUV1 (c) or anti-LUV1 (d) followed by secondary antibody incubation and NBT–BCIP staining.

V. Discussion

In our earlier studies, the α- and β-HCs of flagellar outer arm dynein were often resolved when analyzed on 3–6% gels of the Laemmli system. In theory, it should be possible to separate them by SDS-PAGE and subsequently blot them for direct affinity purification. However, for reasons unknown to us, gel resolution of the two heavy chains has been inconsistent. Therefore dynein must be dialyzed against low salt and separated into the α and β subfractions by sedimenting in a sucrose gradient before relatively pure antigens of α-HC and β-HC can be obtained by SDS-PAGE and blotting.

To blot the high molecular weight polypeptides of dynein, the double-strength transfer buffer (compared to that of Towbin *et al.,* 1979) containing SDS developed by Otter *et al.* (1987) was used. However, the high field strength blotting condition they recommended was not suitable for our system. The prescribed high voltage (84 V) caused portions of the dynein HCs to pass through the first sheet of NC whereas some still remained on the gel. It seems that high field strength pushed the protein molecules through the membrane pores before they could bind to the membrane. We found the efficiency of blotting the high molecular weight polypeptides in the preparative gels at low voltage (24 V) to be much better.

Among the four blot-purified antibodies, anti-α (after negative adsorption), anti-β, and anti-HUV1 all demonstrated expected, specific staining patterns on the test blots. However, we do not have any explanation for the unusual test result of anti-LUV1. It is unlikely that the substantial staining of HUV1 by anti-LUV1 was due to the contamination of anti-HUV1 because the mobility of the two V1 fragments on SDS-polyacrylamide gels was quite different. Blotting and subsequent excision of bands should result in little cross-contamination between the two fragments. If HUV1 and LUV1 share some common epitopes, then why does anti-HUV1 stain the LUV1 region only faintly?

We found the stability of the anti-β remarkable. By storing it undiluted at 4°C in the presence of sodium azide, it has been used repeatedly for more than a year to positively identify clones carrying fragments of the dynein β-chain gene. The stability of anti-α and anti-HUV1 has not been checked in this way.

Nitrocellulose has a high binding capacity (80 $\mu g/cm^2$) for proteins (Bers and Garfin, 1985). Similar to Smith and Fisher (1984), we also found the binding of β-HC and HUV1 to the NC blot to be stable (the α-HC blot was not extensively tested). Repeated adsorption and elution (more than 10 times) did not weaken the antibody-binding strength of these blots as long as the antiserum used for adsorption was being replenished.

Besides NC, the Immobilon membrane from Millipore (Bedford, MA) has been used quite commonly. It has strong physical strength and its protein blot can be used directly for microsequence analysis. We have compared the antibodies purified from the NC and the Immobilon blot and found that, at least with our antigens, antibodies purified from the NC blots showed fewer nonspecific staining bands than those purified from the Immobilon membranes.

Although low pH is commonly used for antibody elution from blots, not all antibodies can be eluted this way. When antibodies bind to their antigens with high affinity, sometimes even 8 M urea or 4 M guanidine-HCl cannot elute these antibodies from the blot. Under such condition, 3 M potassium thiocyanate– 0.5 M NH$_4$OH can be used (Earnshaw and Rothfield, 1985). However, it is necessary to remove the thiocyanate right after elution by passing the antibody through a Sephadex G-25 spin column.

We have used purified anti-β and anti-HUV1 to screen a λgt11 cDNA expression library made from the mRNA of sea urchin embryos to isolate clones carrying cDNA sequences encoding portions of dynein β-HC. From about 2 million clones, 64 were immunopositive toward both anti-β and anti-HUV1. Further analysis and the eventual sequencing of the complete β-HC (Gibbons *et al.*, 1991) indicated that only 1 of the 64 positive clones contained a piece of the real dynein β-HC gene. This sequence information also suggested that the putative, partial β-HC sequence obtained earlier with four independent antibody applications (Foltz and Asai, 1990) was probably not dynein. Therefore it is important to bear in mind that information obtained from antibody studies can be used only as supporting rather than conclusive evidence. However, in the absence of other information, antibodies can be a useful tool, as proved by the success in identifying the clone containing a fragment of the dynein β-HC gene.

References

Bell, C. W., Fronk, E., and Gibbons, I. R. (1979). Polypeptide subunits of dynein 1 from sea urchin sperm flagella. *J. Supramol. Struct.* **11,** 311–317.

Bers, G., and Garfin, D. (1985). Protein and nucleic acid blotting and immunobiochemical detection. *BioTechniques* **3,** 276–288.

Bradford, M. M. (1976). A rapid and sensitive method for the quantitation of microgram quantities of protein utilizing the principle of protein-dye binding. *Anal. Biochem.* **72,** 248–254.

Dreyfuss, G., Adam, S. A., and Choi, Y. D. (1984). Physical change in cytoplasmic messenger ribonucleoproteins in cells treated with inhibitors of mRNA transcription. *Mol. Cell. Biol.* **4,** 415–423.

Earnshaw, W. C., and Rothfield, N. F. (1985). Identification of a family of human centromere proteins using autoimmune sera from patients with scleroderma. *Chromosoma* **91,** 313–321.

Foltz, K. R., and Asai, D. J. (1990). Molecular cloning and expression of sea urchin embryonic ciliary dynein β heavy chain. *Cell Motil. Cytoskeleton* **16,** 33–46.

Gibbons, I. R., Asai, D. J., Ching, N. S., Dolecki, G. G., Mocz, G., Phillipson, C. A., Ren, H., Tang, W.-J. Y., and Gibbons, B. H. (1991). A PCR procedure to determine the sequence of large polypeptides by rapid walking through a cDNA library. *Proc. Natl. Acad. Sci. U.S.A.* **88,** 8563–8567.

Johnson, D. A., Gautsch, J. W., Sportsman, J. R., and Elder, J. H. (1984). Improved technique utilizing nonfat dry milk for analysis of proteins and nucleic acids transferred to nitrocellulose. *Gene Anal. Tech.* **1,** 3–8.

Lee-Eiford, A., Ow, R. A., and Gibbons, I. R. (1986). Specific cleavage of dynein heavy chains by ultraviolet irradiation in the presence of ATP and vanadate. *J. Biol. Chem.* **261,** 2337–2342.

Olmsted, J. B. (1981). Affinity purification of antibodies from diazotized paper blots of heterogeneous protein samples. *J. Biol. Chem.* **256,** 11955–11957.

Otter, T., King, S. M., and Witman, G. B. (1987). A two-step procedure for efficient electro-transfer of both high-molecular-weight (>400,000) and low-molecular-weight (<20,000) proteins. *Anal. Biochem.* **162,** 370–377.

Ow, R. A., Tang, W.-J. Y., Mocz, G., and Gibbons, I. R. (1987). Tryptic digestion of dynein 1 in low salt medium. *J. Biol. Chem.* **262,** 3409–3414.

Smith, D. E., and Fisher, P. A. (1984). Identification, developmental regulation and response to heat shock of two antigenically related forms of a major nuclear envelope protein in *Drosophila* embryos: Application of an improved method for affinity purification of antibodies using polypeptides immobilized on nitrocellulose blots. *J. Cell Biol.* **99,** 20–28.

Tang, W.-J. Y., Bell, C. W., Sale, W. S., and Gibbons, I. R. (1982). Structure of the dynein-1 outer arm in sea urchin sperm flagella. 1. Analysis by separation of subunits. *J. Biol. Chem.* **257,** 508–515.

Towbin, H., Staehelin, T., and Gordon, J. (1979). Electrophoretic transfer of proteins from polyacrylamide gels to nitrocellulose sheets: Procedure and some applications. *Proc. Natl. Acad. Sci. U.S.A.* **76,** 4350–4354.

CHAPTER 6

Immunoblotting

Joann J. Otto

Department of Biological Sciences
Purdue University
West Lafayette, Indiana 47907

I. Introduction

Immunoblotting or Western blotting is a procedure in which a replica of a separating gel is produced by transferring proteins from a gel to a membrane such as nitrocellulose. This replica, or blot, is subsequently incubated or stained with antibodies. Because the replica of the gel is concentrated on the suface of a sheet of membrane, the antibody incubation and wash times are significantly shorter than if a gel itself is stained with antibody. Immunoblotting provides information about an antigen; for example, the subunit molecular weight of the antigen(s) can be determined from its mobility in a sodium dodecyl sulfate (SDS) gel that is blotted. A blot also can give information about the purity of antibodies; if more than one band on a blot binds antibody, antibody contamination is a distinct possibility (see Section IV,C,3).

Most types of gels, both one and two dimensional, can be blotted provided they are strong enough to withstand physical manipulation. The procedures in this article are specifically designed for blotting one-dimensional SDS gels onto nitrocellulose. However, other types of membranes can also be used with these procedures. For example, the basic semidry electrophoretic procedure described here works for blotting onto polyvinylidene difluoride (PVDF) membrane as well (Matsudaira, 1987).

The basic steps in immunoblotting are as follows:

1. Separate the proteins of interest on a one- or two-dimensional gel.
2. Blot the gel onto nitrocellulose or other membrane.
3. Block the membrane with protein to prevent nonspecific antibody binding.
4. Incubate the membrane with primary antibody and allow it to bind to the antigen.
5. Wash the membrane to remove nonspecifically bound antibody.
6. Incubate the membrane with a secondary labeled reagent to bind the primary antibody.
7. Wash the membrane to remove nonspecifically bound labeled reagent.
8. Develop or expose the secondary reagent to locate the position of the primary antibody.

In this article, three different methods of blotting (step 2) one-dimensional SDS gels (Laemmli, 1970) and a basic method for steps 3–8 are presented.

II. Semidry and Wet Blotting Gels

Gels can be blotted either by submersed (wet) methods or by buffer-soaked filter paper (semidry) methods. Transfer can be accomplished either by passive diffusion or electrophoresis. In the wet electrophoretic method, the gel–nitrocellulose sandwich is placed in an electrophoretic chamber filled with transfer solution; some of these chambers require large volumes of solution (e.g., 3–4 liters) whereas many newer versions for minigels require less (e.g., 0.5 liter). In the wet passive method, at least 1–2 liters of solution is used. The wet method may be superior for high molecular weight proteins although I have not seen a quantitative measure of this. The semidry methods require little buffer (50–200 ml) and work well for a variety of proteins.

A. Semidry Blotting

1. Equipment and Reagents

Semidry blotting is done electrophoretically. Two pieces of equipment are required: a semidry blotter and a power supply. Semidry blotters can be purchased from several commercial companies [e.g., Bio-Rad (Richmond, CA), Hoefer (San Francisco, CA), and Pharmacia (Piscataway, NJ)] or be homemade. We have made our own from Plexiglas, graphite plates (available from Ultra Carbon Corp., Bay City, MI), and platinum wire (Fig. 1). One critical feature is sufficiently thick (at least 1 cm) Plexiglas so that the apparatus will not warp. Larger diameter blotters tend to warp such that sufficient contact between the graphite plates and the gel sandwich (see below) is difficult to achieve. Some commercial models use platinum and stainless steel plates rather than graphite to avoid the orange staining that sometimes occurs with graphite plates. However, staining is minimal with graphite plates if they are washed immediately after each use.

Power supplies can be purchased from many commercial companies. The key feature of the power supply is the ability to hold low constant current (0.8–1 mA/cm^2).

The transfer or blotting solution that is used most frequently consists of the following:

Semidry blotting solution:
0.039 M glycine, 0.048 M Tris base, 20% methanol, 0.0375–0.1% SDS

The concentration of SDS in the transfer solution can be varied. For most proteins, 0.0375% is sufficient. High molecular weight proteins or ones that are difficult to transfer may require higher concentrations; however, higher concentrations of SDS may inhibit protein binding to nitrocellulose. The concentration of methanol can also be varied. Although methanol inhibits swelling of the gel and enhances binding of protein to nitrocellulose, it has the negative feature of fixing protein in the gel and thereby reducing transfer.

2. Assembly of Gel Sandwich on Blotter

The materials required for the gel sandwich are the gel, nitrocellulose [available in either 0.2- or 0.45-μm pore sizes from several companies, such as Schleicher & Schuell (Keene, NH), Millipore (Bedford, MA), or Bio-Rad], and heavy filter paper [e.g., Whatman (Clifton, NJ) 3MM]. The gel sandwich consists of the gel adjacent to the nitrocellulose with both enclosed by heavy filter paper. The paper and nitrocellulose are saturated with transfer solution to carry the electrical current.

To assemble the sandwich (diagrammed in Fig. 2), first cut four to six pieces of heavy filter paper so that each is the same size and all are slightly larger than the

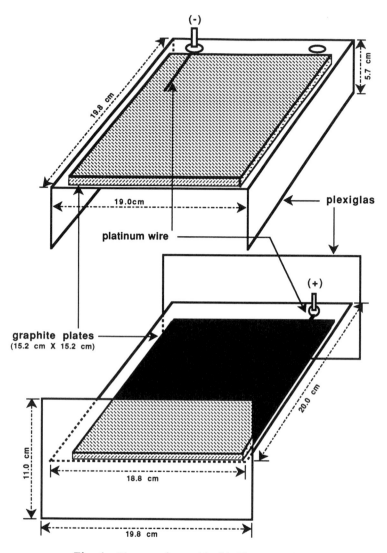

Fig. 1 Diagram of a semidry blotting apparatus.

gel to be blotted. Calculate the area of a piece of the filter paper so that later the amount of current required for blotting may be detemined. Cut a piece of nitrocellulose the same size as the gel to be blotted. Take care not to handle it directly and not to break it. (Dry nitrocellulose is extremely fragile, and fingers will contaminate it with protein.) Disassemble the gel apparatus and trim the gel so that only the part of interest is blotted. Usually it is best to remove the

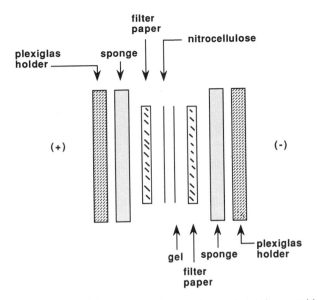

Fig. 2 Diagram of a gel sandwich. For wet blotting, the sandwich is assembled in the order shown. For semidry blotting, the Plexiglas holder and sponges are not used.

stacking gel because it is quite sticky; however, if the antigen does not enter the separating gel, leave the stack on. Wet the plate where the sandwich will sit. Wetting a lint-free wipe and smearing it on the plate is sufficient; do not leave bubbles. Wet the nitrocellulose and place it on top of the filter paper so that the nitrocellulose is placed on the anode (+) side of the blotter relative to the gel. Wet the gel in the transfer buffer and carefully lay it on the nitrocellulose. Try to make the top of the gel abut the top of the nitrocellulose or at least lie parallel to it so that it is easier to determine where lanes are, approximate molecular weights, and so on. Gently, with either fingers or a glass rod, remove any bubbles that are between the gel and nitrocellulose. Now wet and place the last pieces of filter paper on the gel. Roll a glass rod over the sandwich to remove any bubbles. Dry any excess buffer that has squeezed out around the sandwich so that all current runs through the sandwich. Carefully place the cathode (−) plate on top of the sandwich, connect the electrodes, and set current to 0.8–1 mA/cm^2 of filter paper. The higher amperage is used for proteins that are difficult to transfer. Transfer for 75 min for "average" proteins, less for small or easily transferred proteins, or up to 2 hr for proteins that are difficult to transfer. If a homemade blotting apparatus is used, weights on top of the blotter may be required to maintain even contact between the gel sandwich and the graphite plates.

B. Wet Blotting

1. Electrophoretic Blotting

a. Equipment and Reagents

For electrophoretic wet blotting as originally described by Towbin et al. (1979), a transfer chamber and appropriate power supply are required. Transfer chambers for either regular or minigels and power supplies can be purchased from companies such as Bio-Rad, Hoefer, and Pharmacia. A key feature for the power supply is the capacity to hold constant voltage from 30 to 200 V at low power. The transfer solution used most commonly is as follows:

Wet transfer solution: 0.05 M Tris base, 0.384 M glycine, 0.01% SDS, 20% methanol

b. Assembly of Gel Sandwich

The assembly of the gel sandwich is essentially identical to semidry blotting except that a plastic holder and sponges are used to press the nitrocellulose and gel together (see Fig. 2). The sandwich thus consists of (1) a sponge, (2) two or three pieces of heavy filter paper, (3) nitrocellulose, (4) the gel, (5) filter paper, and (6) sponge. This is either assembled on, or placed in, the plastic or Plexiglas holder that comes with the transfer chamber. It is best to assemble this sandwich in a dish filled with transfer buffer to avoid trapping bubbles; however, as for semidry blotting, a glass rod can be rolled over the package to remove bubbles. As with semidry blotting, the nitrocellulose is placed on the anode (+) side of the gel.

2. Passive Wet Blotting

In the absence of an appropriate power supply, blotting can be accomplished by diffusion. In this case, sponges and filter paper that are saturated with transfer buffer surround the gel and nitrocellulose. Pieces of Plexiglas with holes drilled in them hold the package together. A gel sandwich is assembled in a dish filled with transfer buffer as follows: a sheet of Plexiglas is placed in the dish and a piece of sponge that is saturated with transfer buffer is placed on it, followed by two or three sheets of heavy filter paper. Then a piece of nitrocellulose is placed on top of the filter paper and the gel placed on it. If replicate blots are desired, another piece of nitrocellulose can be added on the other side of the gel. Filter paper (two or three sheets) is added. At this point it is critical to remove bubbles. The easiest way to do this is to roll a glass rod over the package. Then another piece of sponge and Plexiglas are added. The entire sandwich is held together with rubber bands. It is left in the dish filled with transfer solution at room temperature for 1–2 days to allow diffusion to occur. Weights can be placed on top of the sandwich to press the nitrocellulose and gel together. A transfer solution that works well for many proteins is as follows:

Passive wet transfer solution: 0.15 M NaCl, 0.002 M disodium ethyelene-diaminetetraacetic cid (EDTA), 0.01 M NaN$_3$, 0.007 M 2-mercaptoethanol, 0.01 M Tris-HCl (pH 7.5)

III. Staining Nitrocellulose with Antibodies

A. Blocking Nonspecific Binding Sites on Nitrocellulose

Nitrocellulose and other such membranes have a high affinity for proteins (and usually nucleic acids). Therefore the entire paper must be coated with protein; otherwise the primary antibody will bind everywhere nonspecifically. Thus, before antibody incubation, the paper is incubated with a blocking solution. The solution that we find most effective is as follows:

Blocking solution: 3% bovine serum albumin (BSA), 0.25% gelatin, 0.15 M NaCl, 15 mM Tris-HCl (pH 7.4)

To prepare the blocking solution, heat either water or Tris–saline solution and completely dissolve the gelatin, cool the solution, and then add the BSA. Depending on the grade of BSA used, the pH can change substantially so be certain to recheck the pH if BSA and gelatin are simply added to Tris–saline. This solution can be reused at least 10 times; store it in the freezer between uses. Blocking solution is also used to dilute antibodies.

Others use blocking solutions containing nonfat dried milk or normal serum from the type of animal used to produce the primary antibody. Our experience with milk is minimal, but in our hands it does not block nonspecific binding as well as the solution with BSA–gelatin. We find that normal serum works reasonably well as a blocking agent; however, one runs the risk of the serum containing an autoimmune or other antibody that can confound results. In addition, protein A or G cannot be used as the secondary labeled reagent if serum is used to block, because the protein A or G will bind all over the blot.

With the BSA–gelatin blocking solution, the minimal time for blocking is 1 hr at 37°C. Alternatively, the blot can be left overnight in blocking solution at 4°C and then warmed to room temperature the next morning.

B. Primary Antibody Incubation

The primary antibody is diluted in blocking solution and incubated for 2–4 hr at room temperature with agitation. The dilution of antibody depends on its quality. As a starting concentration for most polyclonal antibodies, we use a 1/100 dilution of serum or 0.1 mg of IgG/ml. Usually, for monoclonal antibodies the working dilution can be from 1/1000 to 1/50 of supernatant or 0.01–0.2 mg of IgG/ml.

Agitate during the primary antibody incubation in order to conserve primary antibody and to expose the paper to fresh solution continuously. For single narrow strips of nitrocellulose, we place the strips in test tubes sealed with Parafilm and rock them on an aliquot mixer (e.g., Ames/Miles Hema-Tek aliquot mixer (Scientific Products, McGaw Park, IL)). For a 12 × 75 mm test tube, 0.4 ml of diluted antibody is usually sufficient to cover the nitrocellulose. Larger pieces of nitrocellulose are sealed into plastic food storage bags (e.g., Seal-A-Meal (Daisy Corp., Industrial Airport, KS)) and rocked on the aliquot mixer or on a shaking platform.

C. Washing the Blot

To remove nonspecifically bound antibody the nitrocellulose must be washed with a Tris–saline solution with and without detergent. The solutions are as follows:

Tris–saline solution: 0.15 M NaCl, 15 mM Tris-HCl (pH 7.4)
Tris–saline with Tween: 0.05% Tween-20 in Tris–saline solution

The washing is done either directly in the test tube that the nitrocellulose strip is in or in a dish such as a plastic food storage container. Five washes, 5 to 10 min in duration, are done in the following order: one Tris–saline solution wash, two Tris–saline solution with Tween washes, and two Tris–saline solution washes. Do not wash blots that have been incubated with different antibodies together because some antibodies can "jump" from one blot to another during washing (Hammarback and Vallee, 1990).

D. Incubation with Secondary Labeled Reagent

To detect the location of the primary antibody on the nitrocellulose, a labeled secondary reagent is used. Here one has many choices. If the primary antibody is from a species that binds protein A or G (see Appendix Table II), a labeled form of one of these proteins is convenient. Protein A or G conjugated with horse-radish peroxidase, alkaline phosphatase, or ^{125}I can be purchased from many companies. The advantages of enzyme labeled protein A or G are its long shelf life and the absence of radioactive waste. In addition, we find that peroxidase-labeled protein A can be reused 5–10 times (freeze at −20°C between uses) so it is quite economical. (The protein A that we use is from Kirkegaard & Perry, Inc. (Gaithersburg, MD), and is diluted 1/200 in blocking solution.) For some types of experiments, iodinated protein A or G is required.

Alternatively, if the primary antibody is from a species that does not bind protein A or G, a labeled secondary antibody directed against the species of the first can be used. Again, a variety of these are available from many commercial firms, and they are labeled in the same manner as protein A.

The protein A or G or secondary antibody incubation is from 2 to 4 hr at room temperature with agitation, as for the primary antibody incubation. Usually, the manufacturer suggests a range of working dilutions for the secondary reagent. It is important to do a control with secondary reagent alone to be certain that it does not recognize any protein on the blot. This control will also allow determination of whether the reagent concentration is too high. If it is, the entire blot may stain nonspecifically (See Section IV,C,2).

E. Washing the Blot

The blot is washed with Tris–saline with and without Tween-20, just as after the primary antibody incubation. Then a final wash at higher ionic strength is done with Tris–buffered saline (TBS):

Tris-buffered saline (TBS): 0.5 M NaCl, 20 mM Tris-HCl (pH 7.4)

This wash is for 5–10 min at room temperature with agitation.

F. Detection of Labeled Reagent

1. Iodinated Reagent

If the secondary reagent is labeled with iodine, then the nitrocellulose is simply dried between heavy filter paper. After the paper has dried, it is mounted on a stiff piece of paper or thin cardboard with cellophane tape. After covering the paper with a sheet of plastic wrap (e.g., Saran Wrap) to prevent the nitrocellulose from sticking to the film, the paper is exposed to X-ray film either wih or without an intensifying screen at $-70°C$. The exposure time varies with the strength of the signal. One advantage of using ^{125}I is that multiple exposures of the blot can be made.

2. Peroxidase-Labeled Reagent

If the secondary reagent is labeled with peroxidase, the substrate that is most convenient to use is made as follows immediately before use: (1) dissolve 30 mg of chloronaphthol in 10 ml of ice-cold methanol. Keep on ice in the dark until needed. Although chloronaphthol is not a known carcinogen or toxin, it is probably wise to use gloves for this step; (2) mix 30 μl of cold 30% H_2O_2 in 50 ml of TBS right before needed. Mix solutions 1 and 2 immediately before use. Pour the TBS off the nitrocellulose and add the substrate. Incubate on a shaking platform and watch carefully. When bands are dark blue, stop the reaction by removing the substrate and adding distilled H_2O. The water should be changed three times and then the blot can be dried between filter paper. It is relatively easy to overdevelop peroxidase-labeled secondary reagents. This is seen as a uniform blue background on the paper. Thus the reaction must be watched carefully. The blot should be stored protected from light to inhibit fading of the reaction product; sheet protectors lined with black paper are convenient holders for these blots.

An alternative substrate that has often been used for peroxidase is diaminobenzidine. Diaminobenzidine is a suspected carcinogen and has no obvious advantages over chloronaphthol for immunoblotting (it is necessary for immunostaining of cells and sections because diaminobenzidine is insoluble in the solutions used for these procedures).

Note that sodium azide inhibits peroxidase; be certain not to use any solution with azide when using peroxidase-labeled reagents.

3. Phosphatase-Labeled Reagents

If the secondary reagent is labeled with phosphatase, the following solutions are needed: (1) 0.1 M NaCl, 0.01 M MgSO$_4$, 0.1 M Tris-HCl (pH 9.5), (2) 50 mg of nitroblue tetrazolium/ml of 70% dimethylformamide, and (3) 50 mg of 5-bromo-4-chloro-3-indoyl phosphate (BCIP)/ml of 100% dimethylformamide. Store all these stock solutions in the refrigerator. To make the working solution, add 100 μl of stock 2 and 50 μl of stock 3 to 15 ml of solution 1. Pour off TBS and add the phosphatase developing solution at room temperature. Watch the blot for the appearance of purple bands. Stop the reaction with several changes of distilled water.

IV. Miscellaneous Information

A. Controls for Antibody Specificity

There are two critical controls that must be done to be certain that the bands visualized on the blot are in fact due to the primary antibody reacting with the antigen. Preimmune serum should be assayed for the presence of antibodies that react with proteins in the antigen preparation. The same is true for the secondary reagent. In addition, the presence of excess antigen in the primary antibody incubation should inhibit or eliminate antibody binding to the blot. If the final purification of the antigen was by one-dimensional gel electrophoresis (i.e., it was cut out of a gel or blot of a gel), then immunoblotting alone is an insufficient assay of antibody purity since potential antigen contaminants of the same molecular weight could be present at the same position in the blot or gel.

B. Staining All Proteins on the Blot

It is often convenient to stain the blot with a general protein stain, for example, to detect the efficiency of protein transfer or to detect molecular weight standards. There are several ways to do this. For a transient stain, Ponceau S is used (Salinovich and Montelaro, 1986). The blot can be stained with Ponceau S, the positions of markers or lanes marked with a pencil, then the Ponceau stain removed and the blot placed in blocking solution to prepare for antibody staining.

Ponceau S solution: 10 ml 2% Ponceau S concentrate (Cat. No. P 7767; Sigma), 90 ml deionized water

After blotting, place the nitrocellulose in the Ponceau S solution for 15–20 min, then destain with deionized water. The stain can be completely removed by continued washing with water or Tris–saline solution. The blot should be rinsed in Tris–saline solution before being placed in blocking solution. The Ponceau stain can be reused if it is stored in a tightly capped container. One problem we have had with Ponceau is that it occasionally does not stain. This seems to correlate with the quality of water used to dilute the concentrate. Deionized or reverse osmosis water appear to work best.

For a permanent stain, either Ponceau S or amido black can be used. For permanent Ponceau-stained blots, dry the stained blot once the background stain is washed away. The recipe for amido black stain is as follows:

Amido black stain: 0.1% amido black 10B (naphthol blue black or Buffalo black NBR), 5% methanol, 10% acetic acid

Stain 5 min, then destain in water or an ethanol:water:acetic acid (5 parts:5 parts:1 part) destaining solution. This will give a permanent record of the blot.

C. Troubleshooting

1. *The blot is blank*. There are many possible causes of blank blots:

a. The blot sandwich has been assembled incorrectly. For electrophoretic transfers, be certain that the nitrocellulose is on the anode (+) side of the gel.

b. The concentration of primary antibody is too low; try higher concentrations.

c. The concentration of secondary reagent is too low; try higher concentrations.

d. The substrate for an enzyme-labeled secondary reagent is not good. Test by placing 10–20 μl of secondary reagent in a test tube with 25–50 μl of the substrate. Does the substrate form the appropriate color? If not, prepare new substrate. For the peroxidase reaction, the most unstable reagent is H_2O_2. In fact, if in the middle of trying to develop a blot no bands have appeared by 10 min, either add completely fresh substrate or fresher H_2O_2. (The expiration date of the H_2O_2 should be on the bottle.)

e. The antigen is not present on the paper or it is present in such low amounts that it cannot be detected. Small antigens may pass through the nitrocellulose under the conditions used. This can be tested by placing a second sheet of nitrocellulose behind the first. If a positive reaction is seen on it, then the antigen is not binding to the first sheet of nitrocellulose. Remedy this problem by transferring for a shorter time or, if you are using 0.45-μm pore size nitrocellulose, switch to 0.20 μm to trap the antigen. Large antigens may not leave the gel under the conditions used. Stain the blotted gel with Coomassie blue to be certain that transfer was achieved or stain the nitrocellulose with Ponceau S or amido black to determine if proteins of the appropriate molecular mass were transferred. If little transfer of proteins of the expected size of the antigen has occurred, either increase the transfer time and/or the SDS concentration in the blotting solution or decrease the methanol concentration in the transfer solution. If the antigen is present in small amounts, either increase the amount loaded on the gel or enrich for the antigen by immunoprecipitation.

f. Only denatured antigen is recognized by the antibody. Supplement the blocking buffer and primary antibody solution with 0.1% SDS, 0.5% Nonidet P-40, and 0.5% Tween 20. This will keep the proteins on the blot denatured and allow the antibody to recognize the antigen. Although the proteins were denatured in the SDS gel, some renaturing can occur during the blotting and blocking steps.

2. *Reaction product covers the entire blot:* There are two likely causes of reaction product covering the entire blot. One cause is insufficient blocking; increase or change the proteins present in the blocking solution or increase the time for blocking. The second cause could be the use of too high a concentration of primary or secondary antibody; this usually appears in a banding pattern on the blot (see the next section).

3. *Antibody binds multiple bands on blot:* There are several potential causes for multiple bands:

a. The primary antibody is contaminated. Test by using alternative methods to characterize the antibody, for example, immunoprecipitation or immunodiffusion.

b. The primary or secondary antibody is being used at too high a concentration. Dilute the antibody and retest.

c. The antigen has been proteolyzed. Prepare fresh antigen in the presence of protease inhibitors. Some antigens are extremely sensitive to proteolysis. For these, add protease inhibitors to the gel electrophoresis sample buffer, heat it, add it to the antigen sample, and reheat the sample to minimize proteolysis.

d. The primary antibody recognizes an epitope shared by several proteins. This sometimes occurs with monoclonal antibodies. To determine if the bands that bind antibody share an epitope, affinity purify the antibody from each band (see article 5) and test them on a blot of the sample. If the antibody continues to react with all the bands, this is strong evidence that they are related; to be certain, however, the peptides (or epitopes) need to be sequenced.

e. Antibodies can jump from one blot to another if they are washed together (Hammaback and Vallee, 1990). If one stained band is common on multiple blots that have been washed together, a "jumping" antibody should be suspected; wash the blots separately.

V. Perspectives

The major new development in immunoblotting is the production of chemiluminescent substrates for both horseradish peroxidase and phosphatase-labeled secondary reagents (Bronstein et al., 1989). These substrates emit light that is detected by film. Thus, just as for ^{125}I-labeled secondary reagents, multiple exposures of the blot can be made. Chemical enhancers are available to increase the emission of light. Immunoblotting with these substrates in the presence of enhancers should provide a higher level of sensitivity compared to colorimetric substrates.

Acknowledgments

I thank Richard Heil-Chapdelaine and Julia Wulfkuhle for their helpful comments on the manuscript, R. Heil-Chapdelaine and Alonzo LaGrone for drawing the figures, and the NSF (MCD-9012165) and the American Cancer Society (CD-108) for supporting this research.

References

Bronstein, I., Edwards, B., and Voyta, J. C. (1989). Dioxetanes: Novel chemiluminescent enzyme substrates. Applications to immunoassays. *J. Biolumin. Chemilumin.* **4,** 99–111.

Hammarback, J. A., and Vallee, R. B. (1990). Antibody exchange immunochemistry. *J. Biol. Chem.* **265,** 12763–12766.

Laemmli, U. K. (1970). Cleavage of structural proteins during the assembly of the head of bacteriophage T2. *Nature (London)* **227,** 680–685.

Matsudaira, P. (1987). Sequence from picomole quantities of proteins electroblotted onto polyvinylidene difluoride membranes. *J. Biol. Chem.* **262,** 10035–10038.

Salinovich, O., and Montelaro, R. C. (1986). Reversible staining and peptide mapping of proteins transferred to nitrocellulose after separation by sodium dodecyl sulfate-polyacrylamide gel electrophoresis. *Anal. Biochem.* **156,** 341–347.

Towbin, H. Staehelin, T., and Gordon, J. (1979). Electrophoretic transfer of proteins from polyacrylamide gels to nitrocellulose sheets: Procedure and some applications. *Proc. Natl. Acad. Sci. U.S.A.* **76,** 4350–4354.

CHAPTER 7

Immunoprecipitation Methods

Joann J. Otto and Seung-won Lee

Department of Biological Sciences
Purdue University
West Lafayette, Indiana 47907

I. Introduction

Immunoprecipitation methods are used in several major techniques for research in cell biology. These include identifying the molecular mass of an antigen, characterizing the specificity of antibodies, identifying molecules associated with antigens, quantifying amounts of antigen with radioimmunoassays, and determining binding constants. This article focuses on the first three techniques, each of which uses gel electrophoresis to analyze the results. Many of the same steps are used for radioimmunoassays (Chard, 1982; Van Vunakis, 1980; Parker, 1990) and binding constant determination (Bennett and Stenbuck, 1980).

Although it is possible to obtain a precipitate solely with a primary, polyclonal antibody, the relative concentration of antigen and antibody must be precisely correct. For a given amount of antigen, too little antibody will yield no cross-linked antigen. With too much antibody, each antigen will have antibody bound and, again, no cross-links will form. The discovery that protein A, a component of the cell wall of the bacterium *Staphylococcus aureus,* specifically binds the Fc portion of IgG of many species (Kessler, 1975) has greatly simplified the task of immunoprecipitating antigens. More recently, another bacterial wall component, protein G, has been shown to bind the Fc portion of additional classes of IgG (see Appendix Table II). With these reagents, an excess of primary antibody can be used so that quantitative immunoprecipitation is possible.

The basic steps for immunoprecipitation include the following:

1. Solubilizing the antigen preparation
2. Clearing the preparation of any insoluble material and molecules that bind nonspecifically to protein A
3. Incubating with primary antibody
4. Precipitating the primary antibody with protein A or G attached to a matrix
5. Washing the immunoprecipitate
6. Resolving the immunoprecipitate on electrophoretic gels

This article presents a protocol for immunoprecipitation as well as variations that may be necessary under certain conditions. The work in our laboratory concerns cytoskeletal proteins, many of which are insoluble under physiological conditions. Therefore we often use extreme measures to solubilize antigens, which may not be necessary for others. We also often wish to separate the soluble and insoluble (including the cytoskeleton) cell fractions, and recipes for solutions needed to do this are included.

II. Reagents and Solutions

A. Antibodies

The primary antibody is directed against the antigen. For most kinds of analysis, the use of purified immunoglobulin is required rather than crude serum. We use either chromatography on Affi-Gel blue (Bio-Rad Laboratories, Richmond, CA) or protein A–Sepharose (Pharmacia, Piscataway, NJ) or ammonium sulfate (40% saturation) precipitation to purify the immunoglobulins. For antibodies that are not monospecific, affinity purification will be required. High-affinity antibodies are generally lost during affinity purification; it is therefore best to avoid this procedure if possible. Control antibodies are usually the preimmune immunoglobulins from the same animal used to produce the primary antibody or a nonimmune immunoglobulin from the same species. If the primary antibody is not in a class bound by protein A or G, a bridge antibody that is directed against the species of the primary antibody and that binds protein A or G is required.

B. Secondary Reagents

Protein A or G Sepharose or agarose [available from many companies, e.g., Pharmacia, Sigma (St. Louis, MO) or Pierce (Rockford, IL)] or formalin-fixed *S. aureus* [also widely available commercially; e.g., as Pansorbin (Calbiochem, La'Jolla, CA)]. Alternatively, a secondary antibody directed against the immunoglobulin of the primary antibody can be coupled to a matrix such as Sepharose or agarose (see Section III,E for a strategy by which to choose the most appropriate).

C. Cell Permeabilization Solution

Cell permeabilization solution (Ben-Zeev *et al.*, 1979) is used to solubilize the cytosolic pool of proteins in vertebrate cells, but it leaves the majority of the cytoskeleton and certain organelles, such as the nucleus, reasonably intact. (To solubilize the majority of cellular molecules, use lysis solution.)

Triton X-100 (0.5%, w/v)
NaCl (50 mM)
MgCl$_2$ (3 mM)
Sucrose (300 mM)
N-2-Hydroxyethylpiperazine-N'-2-ethane sulfonic acid (HEPES), pH 6.9 (10 mM)

D. Lysis Solution

Lysis solution solubilizes the majority of cellular proteins while allowing the antigen–antibody interaction to occur. However, most other protein–protein interactions are disrupted, and therefore it cannot be used to detect these. Thus it is most useful for characterizing the mass of the antigen and for quantification by radioimmunoassay.

Triton X-100 (1.0%, w/v)
Sodium dodecyl sulfate (SDS) (0.2%, w/v)

Sodium deoxycholate (0.5%, w/v)
NaCl (150 mM)
MgCl$_2$ (1 mM)
Ethylene glycol-bis (β-aminoethyl ether)-N,N,N',N'-tetraacetic acid (EGTA) (1 mM)
2-Mercaptoethanol (10 mM)
Tris-HCl, pH 7.4 (15 mM)

If the antigen is known or suspected to be proteolyzed easily, protease inhibitors should be included in the cell permeabilization or lysis solution. We routinely include phenylmethylsulfonyl fluoride (PMSF), aprotinin, leupeptin, and pepstatin A.

E. Wash Solution

Usually, the solution used to solubilize the antigen (i.e., permeabilization or lysis solution) is used to wash the immunoprecipitate; however, if the resulting immunoprecipitate contains numerous proteins, the concentration of NaCl can be increased to disrupt nonspecific binding. For many antibodies, up to 0.5 M NaCl can be added. Note that NaCl is used rather than KCl because potassium forms a precipitate in SDS, making it difficult to obtain useful SDS gels.

F. Sample Buffer

To elute the antigen–antibody complex from protein A or G, a solution with either at least 1% SDS or 8 M urea is required. Generally, the sample buffer for electrophoresis includes this. For example, the sample buffer for Laemmli (1970) gels is as follows:

Tris-HCl, pH 6.8 (0.0625 M)
SDS (2%)
Glycerin (10%)
2-Mercaptoethanol (5%)

III. Protocol for Immunoprecipitation

A. Sample Preparation

If the starting material is vertebrate cells or tissues, they can be solubilized with either cell permeabilization or lysis solution. If both the soluble and cytoskeletal fractions are to be analyzed, cells are first lysed with cell permeabilization solution, and the soluble material removed and saved. The remaining cytoskeletal fraction is washed with cell permeabilization solution and then solubilized with lysis solution. If detection of potential protein–protein interactions in the soluble fraction is not necessary, then it can be adjusted to contain 150 mM NaCl, 0.2% SDS, and 0.5% sodium deoxycholate. If the starting material is from a nonvertebrate, use an isotonic solution containing 0.5% Triton X-100 or another nonionic detergent to lyse the cells or use lysis solution.

B. Preclearing the Sample

After preparation of the sample, any insoluble material and any proteins that bind nonspecifically to protein A/G or *S. aureus* are removed by incubating the sample with the same amount of protein A/G or *S. aureus*, which will be used to precipitate the primary antibody (see Section III,E). First, wash the protein A/G–Sepharose or agarose or *S. aureus*. Add 5–10 vol of the same solution that the sample is in, vortex, centrifuge for 1 min in a microfuge (10,000–12,000 g), and resuspend to the same volume that the protein A/G matrix or bacteria were in originally to maintain the known binding capacity. Then add the protein A/G matrix or *S. aureus* to the sample and incubate the mixture at 4°C for 20 min with shaking to keep the beads or bacteria suspended. Centrifuge the mixture in a microfuge for 3 min (10,000–12,000 g) to pellet the beads/bacteria with nonspecifically bound proteins and any insoluble material. If quantitative immunoprecipitation is desired, this pellet should be assayed for the presence of antigen by immunoblotting (see article 6).

C. Primary Antibody Incubation

The primary antibody is now added to the precleared sample and the mixture incubated for 1 hr at 4°C with shaking. For some antibodies or for samples that are dilute, the incubation time may need to be lengthened to up to 12 hr. As noted in Section II,A, either preimmune or nonimmune immunoglobulin is used as a control.

To determine the amount of antibody required, it is useful to have an estimate of the amount of antigen in the sample and the amount of the antibody of interest in the immunoglobulin preparation. These estimates are usually unknown at first. We assume that about 5% of the IgG is directed against the antigen and that the antigen represents 1% of the protein sample. We then calculate how much IgG to add to make a four- to fivefold excess over the estimated amount of antigen and add that amount. If one wants to do quantitative immunoprecipitations, the postimmunoprecipitation supernatant needs to be checked by immunoblotting to be certain that all antigen is being precipitated.

D. Bridge Antibodies

If the primary antibody is from a species that does not bind protein A or G (see Appendix Table II), a ''bridge'' antibody that binds protein A/G and is directed against the species of the primary antibody needs to be added after the primary antibody incubation. Again, this antibody incubation is done for 1 hr at 4°C with shaking. These antibodies can be purchased from many different suppliers, who also provide information on suggested working concentrations. Alternatively, a secondary antibody that is directed against the species and class of the primary antibody can be covalently coupled to Sepharose or agarose. This can be used in place of a bridge antibody and protein A/G matrix.

E. Precipitation of Primary (or Bridge) Antibody with Protein A/G–Sepharose/Agarose or Formalin-Fixed *Staphylococcus aureus*

As noted in the reagent section, either formalin-fixed *S. aureus* or protein A or G–Sepharose/agarose can be purchased from several commercial firms. The binding capacities of these products are provided; therefore one need only calculate the amount needed for the amount of primary or bridge antibody added. If using a radioactively labeled preparation to immunoprecipitate, *S. aureus* is usually sufficient. However, if the sample is not radioactively labeled and is to be analyzed by SDS-polyacrylamide gel electrophoresis (SDS-PAGE), *S. aureus* generally yield an unacceptably high level of bands visualized by Coomassie blue staining. Thus, for nonlabeled samples, protein A/G–Sepharose/agarose is the better choice.

To prepare the protein A/G matrix or *S. aureus* for precipitation, wash it with the solution that the sample is in (as described in Section III,B). Pellet the beads or bacteria in a microfuge for 1 min, resuspend and wash once or twice with the solution, and finally resuspend the beads or bacteria in the same original volume to maintain a known binding capacity. Add the protein A/G matrix or *S. aureus* to the sample and incubate with agitation for 30 min at 4°C. Pellet the precipitate by centrifuging for 1 min in a microfuge.

F. Washing Immunoprecipitate

To remove any proteins that are interacting nonspecifically with the immunoprecipitate, are trapped between the beads and bacteria, or are simply sticking to the tube, the immunoprecipitate must be washed. This is done by pelleting the beads or bacteria in a microfuge for 1 min (as described at the end of Section III,E) and resuspending them by vortexing in 0.5–1 ml of the same solution the sample was in. This needs to be repeated two or three times. Often, to obtain clean immunoprecipitates, it is necessary to increase the ionic strength of the wash solution. If the antibody has a high affinity for the antigen, the wash solution can usually be adjusted to 0.5 M NaCl.

G. Removal of Antibody–Antigen Complex from Protein A/G

If the peptide composition of the antigen or antigen plus associated proteins is desired, the removal of the antibody–antigen complex from protein A/G is required. This is usually done with a combination of 1% SDS and 15 mM 2-mercaptoethanol in a buffered solution or with 8 M urea. If SDS-PAGE is to be used to analyze the immunoprecipitates, then the immunoprecipitates can be removed with sample buffer that contains SDS and 2-mercaptoethanol. After suspending the immunoprecipitate in hot sample buffer, heat it at 90–100°C for 5–20 min and vortex occasionally during heating. Pellet the protein A/G–Sepharose or agarose or *S. aureus* by centrifuging in a microfuge for 1–2 min. The supernatent is then ready to be analyzed by SDS gel electrophoresis. If two-dimensional gels are to be used for analyzing the immunoprecipitate, use a urea-based sample buffer.

H. Analysis of Immunoprecipitates

If the immunoprecipitates are to be analyzed by gel electrophoresis, the samples (as prepared in the preceding section) can be loaded directly onto a gel. For antigens that are present in low quantities, it may be necessary to immuno-blot the gel in order to detect the antigen (see article 6). If the sample was labeled radioactively, the gel will need to be subjected to autoradiography (for ^{125}I and ^{35}S) or fluorography (^{35}S and most other isotopes used in cell biology).

IV. Troubleshooting

If the immunoprecipitate is assayed by SDS-PAGE, the following problems may be encountered.

1. *No bands other than those from the immunoglobulin are visible on the Coomassie blue-stained gel.* There are several potential causes for this result:

a. The antigen is present in low amounts. Either silver stain the gel or blot a similar gel to determine if a minor band of the appropriate mass is present. If it is, increase the amount of starting material used for immunoprecipitation.

b. The solution conditions are not conducive to the antibody–antigen inter-action or for the conformation of the antigen that is recognized by the antibody. Vary the salt concentration and composition, the pH, and detergents system-atically to try to obtain a precipitate.

c. The antibody recognizes only denatured antigen and the immunoprecip-itation is being done with native antigen. (Antibody recognition of denatured antigen occasionally occurs if denatured antigen was used for immunization.) Test this by separately dotting equal amounts of denatured and native antigen on nitrocellulose and determine whether the antibody recognizes them equally well. Alternatively, an enzyme-linked immunosorbent assay (ELISA) could be used.

2. *No specific bands, including those from immunoglobulin, are present on the Coomassie blue-stained gel.* There are several possible causes for this:

a. The protein A or G or formalin-fixed *S. aureus* has been degraded. Although these reagents have reasonable shelf lives (months to years), they are subject to proteolysis from contaminants (e.g., bacteria, fungus). They should be stored with a preservative present.

b. The immunoglobulins remain tightly bound to the protein A/G. Increase the time for heating the precipitate or try eluting with a different reagent such as urea.

c. The reagent bound to the matrix used to isolate the primary antibody is inappropriate for the species of primary antibody. Refer to Appendix Table II for the species and immunoglobulin specificity of the secondary reagents.

3. *Many bands are present in the immunoprecipitate, with most present in the control precipitate as well:* This result suggests that the washing of the immunoprecipitate was insufficient either in number of washes or stringency. Increase the salt and/or detergent concentrations as a first step. Wash the immunoprecipitates at least twice. Alternatively, the immunoprecipitate can be separated by centrifugation through a sucrose cushion and then washed as described by Firestone and Winguth (1990).

Acknowledgments

We thank Julia Wulfkuhle and Richard Heil-Chapdelaine for their helpful comments on the manuscript. The research in our laboratory that uses these methods is supported by American Cancer Society Grant CD-108 and NSF Grant MCB-9012165.

References

Bennett, V., and Stenbuck, P. J. (1980). Human erythrocyte ankyrin. *J. Biol. Chem.* **255,** 2540–2548.

Ben-Zeev, A., Duerr, A., Solomon, F., and Penman, S. (1979). The outer boundary of the cytoskeleton: A lamina derived from plasma membrane proteins. *Cell* **17,** 859–965.

Chard, T. (1982). An introduction to radioimmunoassay and related techniques. *In* "Laboratory Techniques in Biochemistry and Molecular Biology" (T. S. Work and E. Work, eds.), Elsevier Biomedical Press, Amsterdam.

Firestone, G. L., and Winguth, S. D. (1990). Immunoprecipitation of proteins. *In* "Guide to Protein Purification" (M. Deutscher, ed.), Methods in Enzymology, Vol. 182, pp. 688–700. Academic Press, San Diego.

Kessler, S. W. (1975). Rapid isolation of antigens from cells with a staphylococcal protein A–antibody adsorbent: Parameters of the interaction of antibody–antigen complexes with protein A. *J. Immunol.* **115,** 1617–1624.

Laemmli, U. K. (1970). Cleavage of structural proteins during the assembly of the head of bacteriophage T4. *Nature (London)* **227,** 680–685.

Parker, C. W. (1990). Immunoassays. *In* "Guide to Protein Purification" (M. Deutscher, ed.), Methods in Enzymology, Vol. 182, pp. 700–718. Academic Press, San Diego.

Van Vunakis, H. (1980). Radioimmunoassays: An overview. *In* "Immunochemical Techniques," Part A (H. Van Vunaki and J. Langone, eds.), Methods in Enzymology, Vol. 70, pp. 201–209. Academic Press, New York.

CHAPTER 8

A Quantitative Solid-Phase Binding Assay for Tubulin

Douglas Thrower, Mary Ann Jordan, and Leslie Wilson

Department of Biological Sciences
University of California, Santa Barbara
Santa Barbara, California 93106

METHODS IN CELL BIOLOGY, VOL. 37

I. Introduction

A variety of immunoassays have been developed to quantitate tubulin levels in cells, including radioimmunoassays (RIAs) and enzyme-linked immunosorbent assays (ELISAs) (Hiller and Weber, 1978; Samuel *et al.*, 1983; Das *et al.*, 1989; Seyfert and Sawatzki, 1986; Binder *et al.*, 1986; Thrower *et al.*, 1991). By the selective nature of antigen–antibody interactions such assays are much more specific than sodium dodecyl sulfate-polyacrylamide gel electrophoresis (SDS-PAGE) gels (Spiegelman *et al.*, 1977) and binding to DEAE-cellulose (Bulinski *et al.*, 1980), two methods that have been used to quantitate tubulin levels. Antibody-based assays also offer advantages over the colchicine-binding assay (Wilson, 1970) because the latter requires temporal extrapolations to correct for decay of colchicine-binding activity and does not detect denatured or polymerized tubulin. The first immunological assays for quantitating tubulin were RIAs (see, e.g., Hiller and Weber, 1978). Such assays are based on the competition of radiolabeled and unlabeled antigen for antibody binding (Van de Water *et al.*, 1982). One useful variation of this technique was designed to measure polymerized tubulin in cells by direct binding of a radiolabeled antibody to detergent-extracted, fixed cytoskeletons (Ball *et al.*, 1986). Enzyme-linked immunosorbent assays offer sensitivities and specificities similar to RIAs and have the additional advantage of not requiring the handling, storage, and disposal of radionuclides. In addition, the shelf lives of enzyme-labeled reagents are longer than those of commonly used radiolabeled components and the equipment required to quantitate enzyme-based assays is less expensive.

The basic ELISA methodology has undergone many variations since its initial description (Engvall and Perlmann, 1971). Enzyme-linked immunosorbent methods that have been used to quantitate tubulin include an antigen capture "sandwich" ELISA (Samuel *et al.*, 1983), a direct assay (Das *et al.*, 1989), and two types of competitive assays, the assay of Seyfert and Sawatzki (1986) and those of Binder *et al.* (1986) and Thrower *et al.* (1991).

The antigen capture "sandwich" ELISA of Samuel *et al.* (1983) employs two tubulin antibodies. One tubulin antibody is adsorbed to a microtiter plate to capture antigen in solution. The captured antigen is then quantitated by an "immune sandwich" consisting of a second antibody directed against tubulin, followed by addition of an enzyme–antibody conjugate directed against the second antibody. This type of assay requires that the binding sites for the two tubulin antibodies be sufficiently separated that the first antibody does not hinder binding of the second tubulin antibody. Also, the two tubulin antibodies must be generated in different species so that the capture antibody is not detected by the antibody–enzyme conjugate. This assay was developed to measure tubulin levels in rat cardiac myocytes. In this cell type, tubulin is present at only 0.01% of the soluble protein, necessitating a sensitive assay. This ELISA was shown to detect as little as 10 ng of tubulin.

The direct ELISA of Das *et al.* (1989) was designed to address a problem sometimes encountered with direct antigen-binding assays, that is, the failure of antigens to bind in a consistent manner to microtiter plates. The authors used poly-L-lysine to provide a better binding surface for tubulin present in tissue extracts. As in the antigen capture "sandwich" method, bound tubulin is detected by sequential incubations with a tubulin-specific antibody and an enzyme–antibody conjugate directed against the tubulin antibody. This method requires fewer steps than the others but its accuracy is dependent on the binding of tubulin to the substrate being directly proportional to its concentration in solution, and on the binding not being affected by the presence of other proteins in solution. A sensitivity similar to the ELISA of Samuel *et al.* (1983) was observed, with a lower limit of detection of 5 ng.

The competitive ELISA of Seyfert and Sawatzki (1986) is based on a competition between unlabeled and enzyme-labeled tubulin for a plate-bound anti-tubulin antibody. A sensitivity of 10 ng/ml was reported. This assay requires that tubulin be purified and conjugated to enzyme. It is important that the conjugation step not block binding of tubulin to the antibody and that the binding not affect enzyme activity. Also, failure to remove unconjugated enzyme following the conjugation reaction may result in nonspecific labeling of proteins in the immunoassay. The assay was developed to measure the soluble tubulin pool in *Tetrahymena* cells. In addition, they used this assay to quantitate tubulin in noncytoplasmic microtubule-based structures.

This article will describe the competitive ELISA of Thrower *et al.* (1991) for the quantitation of tubulin. This competitive immunoassay utilizes preincubation of an anti-tubulin monoclonal antibody with an unknown quantity of tubulin in cell extracts to quantitatively reduce the antibody available to bind to a tubulin-coated microtiter plate. Binding of the tubulin antibody to the microtiter plate is detected by an alkaline phosphatase-conjugated secondary antibody. Bound enzyme is reacted with *p*-nitrophenylphosphate to produce a colored product that is measured spectrophotometrically. Quantitation is accomplished by comparison with a known quantity of bovine brain tubulin.

The method described here has several advantages over other published tubulin ELISA methods. It requires only a single antibody directed against tubulin, and utilizes a commercially available enzyme–antibody conjugate directed against the tubulin antibody. Also, the method does not require quantitative binding to the microtiter plate of tubulin in the cell extracts.

We had two primary reasons for developing such an ELISA. First, we required an assay that would be compatible with measuring tubulin in both lysates and cytoskeletal extracts of cultured mammalian cells. Also, we needed a method that would facilitate the processing of large numbers of samples within a reasonable period of time. These criteria were met by this assay and have allowed us to evaluate the effects of several antimitotic drugs on the levels of soluble tubulin and microtubules in cultured mammalian cells (Jordan *et al.*, 1991; M. A. Jordan *et al.*, 1992). By substituting appropriate antibodies this method could be used to quantitate changes in tubulin isotypes or posttranslationally modified tubulins during the cell cycle or during development.

II. Materials

Phosphocellulose P-11 was purchased from Whatman (Maidstone, England), tissue culture flasks from Corning (Palo Alto, CA), and microtiter plates from Costar (Cambridge, MA). All chemicals were reagent grade unless otherwise noted.

The primary antibody used in the ELISA was 1-1.1, a mouse IgM (κ) antitubulin monoclonal antibody. This antibody was one of a group of monoclonal antibodies raised against sea urchin axonemes by Dr. D. Asai, Purdue University, West Lafayette, IN. The production and characterization of these antibodies have been described (Asai *et al.*, 1982; Ball *et al.*, 1986). Antibody 1-1.1 specifically recognizes the β subunit of HeLa cell tubulin (Thrower *et al.*, 1991). An SDS-PAGE gel and Western blot of phosphocellulose-purified bovine brain tubulin, HeLa cytoskeletal extract, and HeLa whole-cell lysate, using antibody 1-1.1, is shown in Fig. 1.

III. Methods

A. Protocol A: Culture and Enumeration of Cells

The assay conditions for HeLa cells will be described. A single flask of HeLa cells provides sufficient material to assay in duplicate both total cell tubulin and tubulin in microtubules. Cells are harvested at 75–85% of confluent growth, as both total tubulin levels and microtubule mass have been reported to vary with cell confluency (Ostlund *et al.*, 1980).

1. Grow cells in 225-cm^2 flasks at 37°C to a density of 1×10^5 cells/cm^2 in 15 ml of medium consisting of 90% Dulbecco's modified Eagle's medium and 10% fetal bovine serum, in an atmosphere containing 5% CO_2.

2. Release cells from the flasks by gently scraping with a rubber policeman.

3. Transfer the contents of each flask into a 15-ml polypropylene centrifuge tube, dilute duplicate 25-μl samples 1:10 into cell counting solution (Table I), and enumerate cells with a hemacytometer counting chamber.

4. Centrifuge each 15-ml tube (5 min, room temperature, speed setting 2) in a clinical centrifuge (IEC; Damon, Needham Heights, MA).

5. Aspirate medium and resuspend each pellet in 10 ml of MEM buffer (Table I) that has been prewarmed to 37°C. Take 8 ml of each suspension and proceed immediately to step 1 of protocol B. Hold the remaining 2 ml on ice until use in step 1 of protocol C.

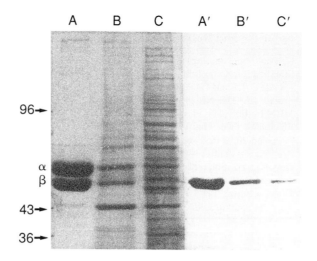

Fig. 1 Specificity of antibody 1-1.1 binding to β-tubulin from bovine brain and HeLa cells. *Left:* Coomassie blue-stained polyacrylamide gel. *Right:* Western blot with antibody 1-1.1. Lanes A and A', bovine brain tubulin, 20 μg of the same preparation used as the ELISA standard; lanes B and B', HeLa cytoskeletal fraction, 10 μg; lanes C and C', HeLa whole-cell lysate, 10 μg. α- and β-tubulin bands and positions of molecular weight standards are labeled. Samples were electrophoresed in duplicate on 10% polyacrylamide gels (Laemmli, 1970). One gel was stained with Coomassie blue, and the other gel was transferred to nitrocellulose according to the procedure of Towbin *et al.* (1979), with the addition of 1% sodium dodecyl sulfate to the transfer buffer to improve the transfer of high molecular weight proteins (Erickson *et al.,* 1982). The nitrocellulose was washed in phosphate Tween buffer (Table 1), incubated in phosphate Tween buffer containing 0.1% nonfat milk (PTB-NFM) (30 min) to block nonspecific binding, incubated with antibody 1-1.1 (1:800 in PTB-NFM, 3 hr), rinsed in PTB-NFM, followed with goat anti-mouse IgM–alkaline phosphatase conjugate (1:500, 3 hr), and developed with nitroblue tetrazolium (220 μg/ml) and 5-bromo-4-chloro-3-indoyl phosphate (160 μg/ml) in alkaline Tris buffer (Table 1).

B. Protocol B: Isolation of Microtubules from Cells and Depolymerization of Microtubules

Eighty percent of the cells is used for determining the tubulin content in microtubules in stabilized cytoskeletons, and 20% of the cells is used for determining total cell tubulin in whole-cell lysates. Such a ratio of cells yields microtubule and total tubulin extracts that contain similar tubulin concentrations, eliminating the need for special dilutions of the fractions before proceeding with the ELISA. The isolation of microtubules is accomplished by detergent lysis of cells in a microtubule stabilization buffer. The composition of stabilization buffer (Table I) is based on buffers previously used to stabilize and extract microtubules from cells (Pipeleers *et al.,* 1977; Solomon *et al.,* 1979). Stabilized cytoskeletons are then incubated in depolymerization buffer (Table I), containing excess calcium, for 1 hr at 0°C to depolymerize the microtubules (Borisy *et al.,* 1974; Ostlund *et al.,* 1980).

Table I
Solutions and Buffers

Cell counting solution: 0.2% trypan blue, 0.15 M NaCl, 0.02% sodium azide

Stabilization buffer (pH 6.9): 0.1 M PIPES,[a] 1 mM EGTA, 1 mM MgSO$_4$, 30% glycerol, 5% DMSO,[b] 5 mM GTP,[c] 1 mM DTT,[c, d] 1 mM TAME,[c, e] aprotinin[c] (0.05mg/ml), 0.02% sodium azide, 0.125% NP-40[f]

MEM buffer (pH 6.9): 0.1 M MES,[g] 1 mM EGTA,[h] 1 mM MgSO$_4$, 0.02% sodium azide

Phosphate Tween buffer (pH 7.2): 1.5 mM KH$_2$PO$_4$, 10.8 mM Na$_2$HPO$_4$, 2.7 mM KCl, 0.05% Tween-20,[i] 0.02% sodium azide

Alkaline Tris buffer (pH 9.5): 0.1 M Tris,[j] 0.1 M NaCl, 1 mM MgCl$_2$

Depolymerization buffer (pH 6.9): 0.1 M MES, 1 mM MgSO$_4$, 10 mM CaCl$_2$, 5 mM GTP,[c] 1 mM DTT,[c, d] 1 mM TAME,[c, e] aprotinin[c] (0.05 mg/ml), 0.02% sodium azide

Dilution buffer (pH 6.9): 0.1 M MES, 1 mM MgSO$_4$, 10 mM CaCl$_2$, 0.1% BSA,[k] 0.05% NP-40,[f] 0.02% sodium azide

---

[a] PIPES, Piperazine-N,N-bis(2-ethanesulfonic acid).

[b] DMSO, Dimethylsulfoxide.

[c] Added immediately before use.

[d] DTT, Dithiothreitol.

[e] TAME, Tosyl arginine methyl ester.

[f] NP-40, Nonidet P-40 (trademark Shell Oil).

[g] MES, 2-(N-Morpholino)ethanesulfonic acid.

[h] EGTA, Ethyleneglycol-bis(β-aminoethyl ether) N,N,N',N'-tetraacetic acid.

[i] Tween-20, Polyoxethylene-sorbitan monolaurate.

[j] Tris, Tris(hydroxymethyl)aminomethane.

[k] BSA, Bovine serum albumin.

1. Take 8 ml of each cell suspension in step 5 of protocol A, and centrifuge as in step 4 of that protocol. Save the remaining 2 ml on ice for preparation of whole-cell lysate.

2. Aspirate and discard supernatants. Add 8 ml of stabilization buffer (Table I) containing 0.125% Nonidet P-40 (NP-40) to each tube. Vigorously pipette to mix and incubate at 37°C for 20 min. Examine this mixture to assure cell lysis has occurred by mixing 10 μl of the suspension with an equal volume of cell counting solution (Table I) and scanning for intact cells with the low-power objective of the light microscope.

3. To each tube add an additional 4 ml of stabilization buffer without NP-40 and centrifuge at 200,000 g, 90 min at 37°C (SW-41 rotor; Beckman, Palo Alto, CA) to collect cytoskeletons. (Begin protocol C during this centrifugation.)

4. Aspirate supernatants and discard. Add 0.3 ml of depolymerization buffer (Table I) to each tube. Place tubes on ice and homogenize pellets with a glass pestle. Allow to incubate on ice for 1 hr.

5. Centrifuge tubes at 48,000 g, 15 min at 4°C.

6. Remove supernatants, measure their volumes, and store supernatants at −70°C until assay.

C. Protocol C: Preparation of Whole-Cell Lysates for Determination of Total Tubulin Levels

1. Take the remaining 2 ml of each suspension from step 5 of protocol A, and centrifuge as in step 4 of that protocol. Discard supernatants.

2. Resuspend pellets by vortexing in 0.3 ml of depolymerization buffer (Table I).

3. Sonicate resuspended pellets at 0°C (Biosonic IV sonicator, low setting, 100-W output; VWR Scientific, San Francisco, CA) for 15 sec, followed by a 2-min pause, and sonicate once again for 10 sec. Keep tubes at 0°C for 1 hr following sonication.

4. Centrifuge at 48,000 g for 15 min at 4°C.

5. Remove supernatants and measure their volumes. Remove two 10-μl aliquots from each lysate for determination of total protein. Store remainder of supernatant at −70°C until determination of tubulin by ELISA. Discard pellets.

6. To assay total protein, take two 10-μl aliquots from each lysate; dilute each with 190 μl of distilled water and determine the protein content by the method of Lowry *et al.* (1951).

We chose to lyse the HeLa cells by sonication on ice because this method has been found to result in less protein loss than homogenization (Bulinski and Borisy, 1979). Sonication was performed on ice at 0°C, and the sonication time was minimized to prevent heating of samples. The duration of sonication was determined by examination of lysates for the presence of intact cells by light microscopy.

We attempted to treat tubulin in cytoskeletons and in whole-cell lysates similarly so that any degradation of protein or change in tubulin antigenicity would occur to similar extents in the two preparations. For this reason, sonication of cells was performed in the buffer designed to depolymerize microtubules in the isolated cytoskeletal preparation (depolymerization buffer). Similarly, both whole-cell lysates and stabilized cytoskeletons were incubated at 0°C for 1 hr.

Total protein was determined in the lysate fractions, in duplicate, by the method of Lowry *et al.* (1951). The depolymerization buffer itself reacted positively with the Lowry reagent. To correct for falsely elevated absorbance values of lysates, a sample containing depolymerization buffer in place of lysate was included in the assay.

D. Protocol D (ELISA): Coating Primary Plate with Tubulin

This section describes the preparation of the microtiter plate on which the binding reactions occur. High protein-binding microtiter plates are recommended because of the greater consistency of protein binding as compared with standard polystyrene plates. Bovine brain microtubule protein is used to coat the plates. This protein is isolated by three cycles of temperature-dependent polymerization and depolymerization, and contains 70% tubulin and 30% microtubule-associated proteins (Farrell and Wilson, 1984). The optimal plate coating concentration is determined by a checkerboard assay, as described in Development of a Competitive ELISA (below).

1. Prepare a solution of bovine brain microtubule protein (6 μg/ml) in MEM buffer (Table I).

2. Add 100 μl of microtubule protein to each well of a Costar 96-well, flat-bottom, high protein-binding microtiter plate (the primary plate), except for wells in the first vertical column (column 1), to which 100 μl volumes of MEM buffer are added. Column 1 is the "no tubulin" primary plate control. Incubate for 2 hr at room temperature, or overnight, covered, at 4°C.

3. Decant plate and tap inverted plate firmly onto a paper towel to remove residual liquid. Rinse three times with phosphate Tween buffer (Table I), each time decanting and tapping plate on a paper towel, and proceed immediately to step 5 of protocol E. Do not allow plate to dry between buffer changes.

E. Protocol E (ELISA): Sample Dilution, Competition Reaction, and Color Development

A standard 96-well plate is used to dilute cell extracts and for incubation of diluted extracts with primary antibody. Antibody dilutions are made just before antibody is added to the microtiter plate. All incubations are done at room temperature unless otherwise noted and all buffers are brought to the appropriate temperature before use.

1. Add 60 μl of dilution buffer (Table I) to wells of column 1 (no tubulin primary plate coat control), columns 3 through 10 (unknowns and standards), and column 11 (no soluble tubulin control). Column 12 wells (no primary antibody control) each receive 120 μl of dilution buffer.

2. Add 60 μl of each sample to be assayed to wells 2 and 3 of a given horizontal row. Assay each sample in duplicate, that is, use two rows for each sample.

3. Serially transfer 60 μl from wells 3 through 10 for each sample. Discard 60 μl from well 10.

4. To columns 1–11 add 60 μl of antibody 1-1.1, diluted 1:3200 in dilution buffer. The final antibody dilution is now 1:6400. Incubate for 1 hr.

5. Transfer 100 μl from each secondary plate well to its corresponding position in the primary plate. Discard secondary plate and incubate primary plate for 1 hr.

6. Decant primary plate and rinse three times with phosphate Tween buffer.

7. Fill all wells with 100 μl of goat anti-mouse IgM–alkaline phosphatase conjugate (Fisher Scientific, Los Angeles, CA), diluted 1:500 in dilution buffer. Incubate for 1 hr.

8. Decant and rinse with phosphate Tween buffer. Dissolve p-nitrophenyl-phosphate into diethanolamine buffer (Bio-Rad, Richmond, CA) at 37°C. Add 100 μl to each well and incubate for 30 min at 37°C or until the absorbance at 405 nm (A_{405}) of column 11 wells reaches a value of approximately 1.5.

9. Add 100 μl of 0.5 M NaOH to each well to stop color development (this step also intensifies color slightly).

10. Measure absorbance values at a wavelength of 405 nm, using a microtiter plate reader or a conventional spectrophotometer.

IV. Typical Results and Discussion

A. Calculation of ELISA Results

The mean absorbance value of the column 1 wells is first subtracted from all other wells. Corrected absorbance values obtained for each sample are then plotted as a function of sample dilution. We have used Sigmaplot software (Jandall Scientific, Sausalito, CA) to graph unknown and standard absorbance values. The resulting curves contain a steep portion of relatively constant slope, the midpoint of which is described by the formula $A_{50\%} = (A_{min} + A_{max})/2$. A_{min} and A_{max} are the absorbance values of the flattened regions of each curve, which correspond to the sample dilutions at which soluble tubulin is in excess or depleted, respectively, relative to the concentration of primary antibody. The dilution at $A_{50\%}$ was determined for each curve by drawing a verticle line to the x axis at $A_{50\%}$. An example of data plotted in this manner is shown in Fig. 2. The tubulin concentration for each unknown sample can be calculated by the formula

$$[\text{Tubulin}]_{unknown} = ([\text{tubulin}]_{standard})(\text{dilution at } A_{50\%, \, unknown})/ \\ \text{dilution at } A_{50\%, \, standard}$$

An alternative method for analyzing the results of a competitive ELISA has been described by Harrington *et al.* (1990). This method requires a V_{max} micro-plate reader (Molecular Devices Corp., Palo Alto, CA) interfaced with a computer, and Softmax (Molecular Devices) software. The advantages of this method are time reduction (kinetic readings are taken over a 2-min period instead of 30 min) and $A_{50\%}$ values are automatically calculated by the Softmax software.

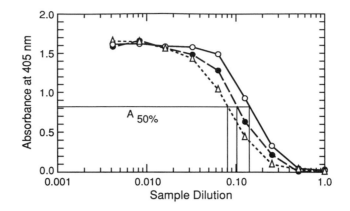

Fig. 2 ELISA of tubulin in a HeLa cytoskeletal extract ($\triangle$), a HeLa whole-cell lysate ($\bullet$), and a bovine brain tubulin standard ($\bigcirc$). Shown are absorbance measurements at 405 nm as a function of sample dilution. Standard errors ranged from 0.01 to 0.08 absorbance units. The $A_{50\%}$ values were determined as described in text. Vertical lines indicate the dilutions corresponding to the $A_{50\%}$ value for each curve. Tubulin concentration of standard was 50 μg/ml prior to dilution.

B. Standards and Controls

Phosphocellulose-purified bovine brain tubulin (Mitchison and Kirschner, 1984) is used as the ELISA standard. Protein concentration is determined by the method of Lowry *et al.* (1951) with bovine serum albumin as a standard. Microtiter plate columns 1, 11, and 12 are used as controls. Column 1 wells lack a tubulin precoat. This provides a control for nonspecific binding of antibodies to the microtiter plate. A low A_{405} value (less than 0.2) is indicative of adequate blocking of nonspecific binding by the bovine serum albumin (BSA) and detergent components of the dilution buffer and of a satisfactory washing procedure. The mean absorbance value of these wells is used as the zero absorbance value for the remaining wells. Column 11 is incubated in the absence of soluble antigen and should give a value equal to the A_{max} plateau of the unknowns and standards. This provides a positive control for the primary antibody. Column 12, lacking primary antibody, has absorbance values slightly above those of column 1 (less than 0.1 units above column 1), corresponding to the A_{min} region of antigen dilution curves. Periodically, pools of lysate and microtubule fractions should be assayed with no secondary antibody incubation step to test for the presence of any endogenous alkaline phosphatase activity in these fractions.

C. Sensitivity and Linearity

The sensitivity of the ELISA may be determined by serially diluting a bovine brain phosphocellulose-purified tubulin standard and assaying duplicate samples. Under the conditions of this assay, the minimal concentration for which an $A_{50\%}$ could be determined was 6.5 μg/ml. This allows detection of as little as 800 ng of tubulin per assay. The assay is linear over a range of 6.5–200 μg/ml (correlation coefficient, 0.993). Linearity is determined by plotting the dilution corresponding to $A_{50\%}$ vs concentration of standard.

We found that sensitivity of the assay could be increased approximately twofold for each corresponding increase in the dilution of antibody 1-1.1. An assay using the primary antibody at a final dilution of 1:25,600 (rather than 1:6400, as described in protocol E) detected a tubulin concentration of 0.78 μg/ml. This gives a lower limit of detection of 94 ng. Using the 1:25,600 dilution of antibody 1-1.1, the assay is linear for antigen concentrations between 100 and 0.78 μg/ml. At this level of sensitivity, it is necessary to increase the primary plate coating concentration to 24 μg/ml, and to increase the secondary antibody incubation time to 2 hr. Also, an increase in nonspecific background color (detected as an elevated A_{min}) may require the use of purified tubulin in place of microtubule protein in the primary plate antigen coating solution. Although the more sensitive procedure would theoretically allow the use of smaller aliquots of cells for each experimental data point, we found in practice that the reproducibility of the extraction procedure was diminished when fewer cells were used.

D. Reproducibility

The reproducibility of the standard ELISA (1:6400 primary antibody dilution) and reproducibility of the entire procedure is shown in Table II. Pooled cytoskeletal extracts and lysates were assayed in triplicate on three different days

Table II
Reproducibility of Method

	Coefficient of variation[a]	
	Cytoskeletal fraction (%)	Lysate fraction (%)
Intra-ELISA variation ($n = 3$)	4.0	7.0
Inter-ELISA variation ($n = 3$)[b]	8.7	3.7
Overall variation for entire procedure ($n = 32$)[c]	24.3	27.3

[a] Coefficient of variation = (standard deviation/mean) $\times$ 100.

[b] Calculated from three independent ELISAs of pooled cytoskeletal and lysate extracts stored frozen at $-70°$ C.

[c] Overall variation for entire procedure is based on 32 samples from 16 independent extracts and ELISAs completed over a 6-month period.

and the coefficients of variation were calculated for assays performed on one day (intra-ELISA variation) and on different days (inter-ELISA variation). Overall variation in the methodology was examined by determining the coefficients of variation for 32 samples assayed during a period of 6 months. The coefficients of variation for quantitation of both microtubules (cytoskeletal fractions) and total cell tubulin (lysate fractions) were 24.3 and 27.3%, respectively (Table II). Most of the variation for the entire procedure was probably due to variable loss of tubulin during cell extraction. In addition, some variation was contributed by the ELISA (Table II). For the greatest accuracy it is necessary to perform multiple, completely independent extractions and assays for each experimental condition.

E. Comparison with Results Obtained by Other Methods

Total HeLa tubulin was $3.6 \pm 0.2\%$ of total cell protein (32 samples from 16 independent extractions and ELISAs). This value is similar to the value of 4.3% reported by Bulinski *et al.* (1980) for HeLa cell tubulin quantitated by colchicine binding and RIA, using a HeLa tubulin standard. We found that tubulin in microtubules was $1.1 \pm 0.1\%$ of total cell protein; thus, approximately one-third of the tubulin was in polymer. This is similar to the range of 34–41% of tubulin in microtubules in cultured cells, as determined by colchicine-binding assay (Ostlund *et al.*, 1979).

F. General Considerations

The buffers used for the ELISA are designed to be compatible with the cell extracts and to support specific immune binding reactions. To avoid changing solution conditions during dilution, the ELISA dilution buffer and microtubule depolymerization buffer (Table I) both contain the same concentrations of 2-(*N*-morpholino)ethanesulfonic acid (MES), $MgSO_4$, and $CaCl_2$. Bovine serum albumin and NP-40 were added to prevent nonspecific protein interactions. A Tris-based buffer was originally tried as a microtiter plate washing buffer but it yielded inconsistent results. It was replaced with phosphate-buffered saline containing Tween-20 (phosphate Tween buffer, Table I).

Rinsing of microtiter plates is accomplished with a Vaccu-pette/96 (Fisher), a device designed to change medium in 96-well plates. It is suspended about 1 cm above the plates, preventing its 96 tips from being contaminated by the solution contained in the wells. Buffer is dispensed simultaneously and equally into all wells. Buffer is then decanted manually (protocol D, step 3). This method for rinsing yields more reproducible results than the method using a single port wash bottle and saves considerable time. It has an advantage over automated systems that employ a vacuum aspirator, because plates do not become dry between rinses, which may lead to significant well-to-well variation.

In the original published procedure (Thrower *et al.*, 1991), both the primary plate and secondary plates were incubated with a blocking buffer containing BSA subsequent to primary plate antigen coating (step 3 of protocol D) and prior to the addition of dilution buffer to the secondary plate (step 1 of protocol E). These blocking steps have subsequently been found to be unnecessary. This was demonstrated by running parallel ELISAs, in triplicate, with or without the blocking incubations. Mean values and standard errors of the fractional degree of dilution at $A_{50\%}$ were 0.070 ± 0.004 for assays performed under original blocking conditions, 0.078 ± 0.001 when only the secondary plate was blocked, and 0.082 ± 0.002 when only the primary plate was blocked. The mean values obtained under the different blocking conditions were not significantly different ($p < 0.05$).

V. Development of a Competitive ELISA

This section contains suggestions that should be helpful in adapting the ELISA to use with other antibodies or antigens. First, as described in Materials (above), the primary antibody should be characterized by immunoblotting to screen for any cross-reactivity with other proteins in the solutions to be assayed.

A. Protocol F: Determination of Optimal Antibody Dilution

The sensitivity of the competitive ELISA described here is determined by the concentration of primary antibody. For this reason, it is necessary to determine the appropriate dilution of antibody that will provide adequate sensitivity for the range of antigen concentrations in the cell or tissue extracts being assayed. This is most easily accomplished with a checkerboard assay that varies extract dilution in one direction (horizontally across the plate) and antibody concentration in the other direction (vertically down the plate).

1. Prepare the primary plate as described in protocol D, except that only the first four rows are coated with microtubule protein.

2. Follow step 1 of protocol E, but fill only the first four rows with buffer.

3. Add 60-μl aliquots of a cell extract pool to wells 2 and 3 of the first four rows of the secondary plate and dilute as described in protocol E, step 3.

4. To the first row (row A) add 60-μl aliquots of primary antibody at a 1:100 dilution; to row B, 1:800; to row C, 1:6400; to row D, 1:51,200.

5. Follow the standard protocol given in protocol E for the remainder of the ELISA procedure, again using only the first four rows of the microtiter plates. Control wells are as described in protocols D and E.

The optimal antibody dilution is that giving absorbance vs dilution curves that have $A_{50\%}$ points in the central part of the extract dilution range.

B. Protocol G: Optimal Concentration of Antigen for Coating Microtiter Plates

The accuracy with which $A_{50\%}$ values are determined greatly influences the accuracy of the ELISA. Errors in this determination can be minimized by maximizing the slopes of the linear portion of absorbance vs dilution curves. The steepness of absorbance vs dilution curves for an assay incorporating a given primary antibody concentration is dependent on the concentration of antigen used to coat the primary plate. The optimal concentration is one that results in the greatest ratio of A_{max} to A_{min}. The value can be determined by an ELISA assay in which the concentration of the primary plate coating is varied. Only the primary plate is used in this assay.

1. Distribute 100 μl of MEM buffer to each of wells 1 and 3–10 of the first two rows (rows A and B) of a primary plate.

2. Add a solution of bovine brain microtubule protein (or other antigen) (100 μg/ml) in MEM buffer to wells 2 and 3. One hundred microliters of solution is transferred from well 3 to 4 and serially diluted through well 10. One hundred microliters is removed from well 10 and discarded. Incubate the plate, rinse, and decant the wash buffer as described in protocol D, steps 2 and 3.

3. Add 100 μl of primary antibody to the first row (row A) of the primary plate at a dilution twofold greater than that normally used. This antibody dilution is the same as the final dilution that results from the addition of primary antibody to an equal volume of cell extract in the standard protocol (protocol E, step 4). Add 100 μl of dilution buffer to the wells of row B. Incubate for 1 hr.

4. Subsequent steps are carried out as previously described in protocol E, steps 6–10, with the exception that no reagents are added to wells 11 and 12.

Mean absorbance values of column 1 wells are subtracted from all other wells. The wells in row A give A_{max} values and the corresponding wells in row B give A_{min} values. The difference in absorbance between each vertical pair of wells is determined, and the optimal antigen concentration for coating the primary plate is chosen based on the pair of wells that gives the greatest change in absorbance at 405 nm.

C. Choice of Standard

Bovine brain tubulin works well as a standard. It is easy to prepare in large quantity and it reacts specifically with the primary antibody (see Fig. 1). Immediately following purification, the standard is drop frozen in liquid nitrogen and stored at $-70°C$. Aliquots are thawed just once before use. A tubulin standard obtained from a heterologous cell or tissue source can be used to determine the relative changes in cellular tubulin from one source of cells. However, the absolute amount of tubulin cannot be determined if the tubulin in the cells or tissues being examined does not compete for primary antibody to the same degree as the tubulin being used as a standard.

To ascertain whether bovine brain tubulin reacts with antibody 1-1.1 in a fashion quantitatively similar to HeLa cell tubulin, we prepared a HeLa cell tubulin standard. HeLa microtubule protein was isolated by three cycles of polymerization by the method of Bulinski and Borisy (1979), followed by phosphocellulose chromatography to remove microtubule-associated proteins (Mitchison and Kirschner, 1984). The purity of this material was confirmed by SDS-PAGE gels, with bands being visualized by the silver stain method of Merril *et al.* (1981), as shown in Fig. 3. HeLa tubulin was found to compete equally with brain tubulin for binding to antibody 1-1.1 (see Fig. 4). Not all tubulins from cells can be readily purified by the temperature dependent cycling method, but other methods may be used. These include cycling in the presence of exogenous polymerizing agents such as glycerol (Nagle *et al.,* 1977), or taxol (Collins, 1991), or purification by DEAE-cellulose chromatography (Weisenberg *et al.,* 1968).

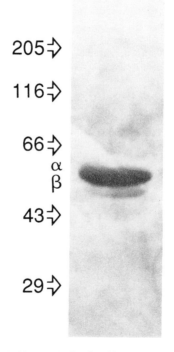

Fig. 3 Purity of HeLa tubulin. A 20-μg sample of purified HeLa cell tubulin was electrophoresed on a 10% SDS-PAGE gel (Laemmli, 1970) and silver stained (Merril *et al.,* 1981). Positions of α- and β-tubulin bands and molecular weight standards are labeled. HeLa tubulin bands stain with silver unevenly. Both bands stained equally with Coomassie blue (not shown).

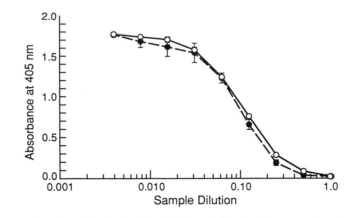

Fig. 4 ELISA of bovine brain tubulin (●) and HeLa tubulin (○). Shown are absorbance measurements at 405 nm as a function of sample dilution. Initial tubulin concentrations were 50 μg/ml. Data points represent the mean of triplicate determinations. Standard errors of the mean are indicated with brackets when standard error ranges exceed the mean value symbols.

Acknowledgment

This work was supported by American Cancer Society Grant CH381.

References

Asai, D. J., Brokaw, C. J., Thompson, W. C., and Wilson, L. (1982). Two different monoclonal antibodies to alpha-tubulin inhibit the bending of reactivated sea urchin spermatozoa. *Cell Motil.* **2,** 599–614.

Ball, R. L., Carney, D. H., Albrecht, T., Asai, D. J., and Thompson, W. C. (1986). A radiolabeled monoclonal antibody binding assay for cytoskeletal tubulin in cultured cells. *J. Cell Biol.* **103,** 1033–1041.

Binder, L. J., Frankfurter, A., and Rebhun, L. I. (1986). Differential localization of MAP-2 and tau in mammalian neurons *in situ. Ann. N.Y. Acad. Sci.* **466,** 145–166.

Borisy, G. G., Olmsted, J. B., Marcum, J. M., and Allen, C. (1974). Microtubule assembly *in vitro. Fed. Proc.* **33,** 167–173.

Bulinski, J. C., and Borisy, G. G. (1979). Self-assembly of microtubules in extracts of cultured HeLa cells and the identification of HeLa microtubule-associated proteins. *Proc. Natl. Acad. Sci. U.S.A.* **76,** 293–302.

Bulinski, J. C., Morgan, J. L., Borisy, G. G., and Spooner, B. S. (1980). Comparison of methods for tubulin quantitation in HeLa cell and brain tissue extracts. *Anal. Biochem.* **104,** 432–439.

Collins, C. (1991). Reversible assembly purification of taxol-treated microtubules. *In* "Molecular Motors and the Cytoskeleton" (R. Vallee, ed.), Methods in Enzymology, Vol. 196, pp. 246–253. Academic Press, San Diego.

Das, S., Banerjee, S. K., Sil, M., and Sarkar, P. (1989). An ELISA method for quantitation of tubulin using poly-L-lysine coated microtiter plates. *Indian J. Exp. Biol.* **27,** 972–976.

Engvall, E., and Perlmann, P. (1971). Enzyme-linked immunosorbent assay (ELISA). Quantitative assay of IgG. *Immunochemistry* **8,** 871–874.

Erickson, P. F., Minier, L. N., and Lasher, R. S. (1982). Quantitative electrophoretic transfer of polypeptides from SDS polyacrylamide gels to nitrocellulose sheets: A method for their re-use in immunoautoradiographic detection of antigens. *J. Immunol. Methods* **51**, 241–249.

Farrell, K. W., and Wilson, L. (1984). Tubulin–colchicine complexes differentially poison opposite microtubule ends. *Biochemistry* **23**, 3741–3748.

Harrington, C. R., Edwards, P. C., and Wischik, C. M. (1990). Competitive ELISA for the measurement of tau protein in Alzheimer's disease. *J. Immunol. Methods* **134**, 261–271.

Hiller, G., and Weber, K. (1978). Radioimmunoassay for tubulin: A quantitative comparison of the tubulin content of different established tissue culture cells and tissues. *Cell* **14**, 795–804.

Jordan, M. A., Thrower, D., and Wilson, L. (1991). Mechanism of inhibition of cell proliferation by *Vinca* alkaloids. *Cancer Res* **51**, 2212–2222.

Jordan, M. A., Thrower, D., and Wilson, L. (1992). Effects of vinblastine, podophyllotoxin, and nocodazole on mitotic spindles. Implications for the role of microtubule dynamics in mitosis. *J. Cell Sci.* **102**, 401–416.

Laemmli, U. K. (1970). Cleavage of structural proteins during the assembly of the head of bacteriophage T-4. *Nature (London)* **227**, 680–685.

Lowry, O. H., Rosebrough, N. J., Farr, A. L., and Randall, R. J. (1951). Protein measurement with the Folin phenol reagent. *J. Biol. Chem.* **193**, 265–275.

Merril, C. R., Goldman, D., Sedman, S. A., and Ebert, M. H. (1981). Ultrasensitive stain for proteins in polyacrylamide gels shows regional variation in cerebrospinal fluid proteins. *Science* **211**, 1437–1438.

Mitchison, T., and Kirschner, M. (1984). Microtubule assembly nucleated by isolated centrosomes. *Nature (London)* **312**, 232–237.

Nagle, B. W., Doenges, K. H., and Bryan, J. (1977). Assembly of tubulin from cultured cells and comparison with the neurotubulin model. *Cell* **12**, 573–586.

Ostlund, R. E., Leung, J. T., and Vaerewyck Hajek, S. (1979). Biochemical determination of tubulin–microtubule equilibrium in cultured cells. *Anal. Biochem.* **96**, 155–164.

Ostlund, R. E., Leung, J. T., and Vaerewyck Hajek, S. (1980). Regulation of microtubule assembly in cultured fibroblasts. *J. Cell Biol.* **85**, 386–391.

Pipeleers, D. G., Pipeleers-Marichal, M. A., Sherline, P., and Kipnis, D. M. (1977). A sensitive method for measuring polymerized and depolymerized forms of tubulin in tissues. *J. Cell Biol.* **74**, 341–350.

Samuel, J. L., Schwartz, K., Lompre, A. M., Delcayre, C., Marotte, F., Swynghedauw, B., and Rappaport, L. (1983). Immunological quantitation and localization of tubulin in adult rat heart isolated myocytes. *Eur. J. Cell Biol.* **31**, 99–106.

Seyfert, H. M., and Sawatzki, G. (1986). An estimation of the soluble tubulin content in *Tetrahymena* cells of normal and of size-altered phenotype. *Exp. Cell Res.* **162**, 86–96.

Solomon, F., Magendantz, M., and Salzman, A. (1979). Identification with cellular microtubules of one of the coassembling microtubule-associated proteins. *Cell* **18**, 431–438.

Spiegelman, B. M., Penningroth, S. M., and Kirschner, M. W. (1977). Turnover of tubulin and the N site of GTP in Chinese hamster ovary cells. *Cell* **12**, 587–600.

Thrower, D., Jordan, M. A., and Wilson, L. (1991). Quantitation of cellular tubulin in microtubules and tubulin pools by a competitive ELISA. *J. Immunol. Methods* **136**, 45–51.

Towbin, H., Staehelin, T., and Gordon, J. (1979). Electrophoretic transfer of proteins from polyacryamide gels to nitocellulose sheets: Procedures and some applications. *Proc. Natl. Acad. Sci. U.S.A.* **76**, 4350–4354.

Van de Water, L., III, Guttman, S. D., Gorovsky, M. A., and Olmsted, J. B. (1982). Production of antisera and radio-immunoassays for tubulin. *Methods Cell Biol.* **24**, 79–96.

Weisenberg, R. C., Borisy, G. G., and Taylor, E. W. (1968). The colchicine-binding protein of mammalian brain and its relation to microtubules. *Biochemistry* **7**, 4466–4479.

Wilson, L. (1970). Properties of colchicine binding protein from chick embryo brain. Interactions with *Vinca* alkaloids and podophyllotoxin. *Biochemistry* **9**, 4999–5007.

CHAPTER 9

Confocal Immunofluorescence Microscopy of Microtubules in Oocytes, Eggs, and Embryos of Algae and Amphibians

David L. Gard and Darryl L. Kropf

Department of Biology
University of Utah
Salt Lake City, Utah 84112

METHODS IN CELL BIOLOGY, VOL. 37

I. Introduction

In the 50 years since Coons pioneered the use of fluorescently labeled antibodies as stains for light microscopy (Coons *et al.*, 1941, 1942; Coons, 1961), immunofluorescence microscopy has revolutionized the study of cellular organization. Until recently, however, the full potential of immunofluorescence microscopy could be realized only with small or flattened cells, such as fibroblasts or epithelial cells grown in culture. In larger cells or tissues, the superposition of specific and nonspecific fluorescence from outside the focal plane substantially decreased image contrast and often masked the fine cellular details of antigen localization. For this reason, immunofluorescence microscopy of larger cells or complex tissues was practical only after involved procedures of embedding and sectioning.

Two technological advances have significantly extended our ability to view large cells by immunofluorescence microscopy. First, Agard and co-workers (1989) developed techniques for computer-aided "deconvolution" of standard epifluorescence images, subtracting fluorescence contributed from outside of the focal plane. Second, the development of practical confocal microscopes has provided the ability to section large cells "optically," eliminating directly the contribution of fluorescence from outside the focal plane (White *et al.*, 1987). In recent years, both technologies have become increasingly available to cell biologists, through moderately priced software for image deconvolution (capable of running on personal computers) and commercially available confocal (spinning disk) or confocal laser-scanning microscopes.

Numerous protocols outlining the preparation of cultured cells or tissue sections for immunofluorescence microscopy have been published [discussions of the theory and practice of immunofluorescence can be found in Larsson (1988) and Beltz and Burd (1989)]. It is not our intent to add to this list, or to discuss the general background of immunofluorescence microscopy. Rather, we shall summarize our experiences using immunofluorescence combined with confocal laser-scanning microscopy (CLSM) to examine cytoskeletal organization in algal and amphibian oocytes, eggs, and early embryos (Kropf *et al.*, 1990; Gard, 1991, 1992; Schroeder and Gard, 1992; Allen and Kropf, 1992). Their large size (0.1 to 2 mm in diameter) and yolk- or pigment-filled cytoplasm have previously rendered these cells difficult, or nearly impossible, to examine by high-resolution immunofluorescence microscopy. In the following discussion we shall point out some of the difficulties we have encountered, and how these problems have been overcome. We would stress that the procedures presented have been arrived at empirically, in our own laboratories or by others who have faced similar difficulties in examining large or otherwise intractable cells. With this in mind, the procedures presented are in a continual state of revision and optimization and should be viewed as starting points, rather than as finalized protocols. We hope that the techniques presented will prove useful to others using immunofluorescence microscopy to examine cells that have previously proved intractable.

Discussion of antibody production and assay will be found in other chapters in this volume. Immunocytochemical techniques for whole-mount or sectioned amphibian embryos have been published (Klymkowsky and Hanken, 1991; Kelly *et al.*, 1991). More complete discussions of confocal microscopy and its application to biological samples can be found in Pawley (1989) and Matsumoto (1993).

II. Immunofluorescence Microscopy of *Pelvetia* and *Xenopus*

Preparation of *Pelvetia* or *Xenopus* eggs for immunofluorescence microscopy follows much the same procedure as is used for more typical samples, such as cultured cells grown on coverslips. However, some modifications are required to accommodate the large size and adhesive (or nonadhesive) nature of the cells.

Pelvetia embryos, for example, construct a cell wall soon after fertilization, which allows them to adhere to many natural and unnatural substrata (such as a glass coverslip). This characteristic makes it convenient to handle *Pelvetia* embryos fixed to glass coverslips, in much the same manner that one would handle cultured cells. However, *Xenopus* oocytes, eggs, and embryos, and *Pelvetia* zygotes prior to cell wall assembly, are nonadherent and are more easily processed in "suspension." These cells are fixed, processed for immunofluorescence, dehydrated, and cleared in 1.5-ml microcentrifuge tubes, and only then are mounted on slides for observation. Detailed protocols for the preparation of *Pelvetia* and *Xenopus* cells for immunofluorescence are provided in Tables I–III. Key features of these protocols are discussed in the following sections.

A. Fixation

Many factors enter into the design or choice of a fixation protocol for immunofluorescence. Two, however, are of paramount importance. First, the fixation protocol must be compatible with the antibodies or staining technique to be used. Second, the fixation procedure must adequately immobilize and preserve the antigen under study and the cellular structure(s) with which it is associated.

Fixation regimes that are commonly used for immunofluorescence and other techniques of light microscopy can be loosely divided into two categories: those that immobilize proteins by precipitation, using organic solvents or acids; and those using chemical cross-linking agents, the most common of which are aldehydes such as formaldehyde (or paraformaldehyde) and glutaraldehyde. An excellent discussion of the applications and advantages of these differing fixation strategies can be found in Larsson (1988).

Table I
Indirect Immunofluorescence of Microtubules in *Xenopus* Eggs[a]

1. Samples should be as free of extraneous tissue or extracellular material as possible. Small ovaries or tissue samples can be fixed whole. Larger oocytes are isolated by collagenase digestion followed by manual removal of residual follicle material (Gard, 1991). Jelly coats must be removed from eggs and early embryos
2. Fix oocytes for 2–8 hr in approximately 1 ml of FGT or FG fix (see Appendix I) at room temperature
3. Postfix in 100% methanol overnight. Samples can be stored in methanol indefinitely at room temperature or at $-20°C$
4. Bleach with peroxide–methanol (Dent and Klymkowsky, 1989) for 24–72 hr at room temperature (optional)
5. Rehydrate in PBS (several changes at room temperature)
6. Use a sharp scalpel blade to bisect the oocytes or eggs (optional)
7. Incubate in 100 mM NaBH$_4$ (in PBS; no detergents) for 4–6 hr at room temperature, or overnight at 4°C
8. Carefully remove NaBH$_4$, and wash samples with TBSN for 60–90 min at room temperature (three or four washes, approximately 1 ml each)
9. Incubate samples for 16–36 hr at 4°C in 75–150 μl of primary antibody diluted in TBSN plus 2% BSA, with gentle rotation
10. Wash in TBSN for 30–48 hr at 4°C, changing buffer at 8- to 12-hr intervals
11. Incubate 16–36 hr at 4°C in 75–150 μl of fluorochrome-conjugated secondary antibody diluted in TBSN plus 2% BSA (with gentle rotation)
12. Wash with TBSN, as in step 9.
13. Dehydrate in 100% methanol (three or four changes over 60–90 min at room temperature) (graded series of methanol can be used if shrinkage or distortion is excessive). Samples can be stored indefinitely in methanol
14. Remove methanol, and add approximately 1 ml of BA:BB clearing solution. *Do not mix.* Methanol-dehydrated oocytes will float. Tap the tubes on a bench to break surface tension. Oocytes will clear as they sink through the BA:BB (takes 5–10 min). Remove the clearing solution and replace with fresh BA:BB
15. Mount in BA:BB
16. If samples retain a milky appearance, they may not have been adequately dehydrated. Poorly dehydrated samples can be salvaged by repeating the dehydration and clearing. Borohydride reduction can be eliminated and incubation times can be shortened for FT- or methanol-fixed samples (see Appendix I). Ethanol or acetone can be substituted for the dehydration preceding clearing

[a] All incubations and washes are performed in 1.5-ml microcentrifuge tubes.

Immunofluorescence microscopy of microtubules provides an informative test of the efficacy of these different fixation protocols. Obtaining adequate fixation of microtubules has proved problematic, due in large part to their dynamic nature and their sensitivity to cold, micromolar Ca^{2+}, and ionic strength. The larger sizes of whole-mounted samples, such as amphibian or algal eggs and embryos, make adequate fixation of microtubules even more difficult, due to the increased time required for penetration of fixative. We have evaluated a number of different fixation protocols for their ability to preserve both cy-

Table II
Indirect Immunofluorescence for Microtubules in *Pelvetia* Zygotes

1. Grow fertilized eggs, attached to No. 1 coverslips, to desired stage. Zygotes attach firmly to coverslips by 4 hr development, and solutions can be added or removed without disturbing the cells. Prior to this time, cells are nonadherent, and all manipulations must be performed in microcentrifuge tubes
2. Incubate the cells 10 min in MTSB at 14°C
3. Fix the cells in freshly prepared FGT/MeOH fix for 20 min at room temperature
4. Postfix the cells in 100% methanol for at least 3 hr at −20°C. Change the methanol three times during this period. Cells can be left for several days in methanol at −20°C, if necessary
5. Rehydrate the cells into PBSG (room temperature)
6. Incubate in PBSG plus 100 mM NaBH$_4$ for 4 hr at room temperature (O/N at 4°C)
7. Rinse the cells three times (10 min each) in solution C at room temperature. The cells can be left in solution C for 24 hr, or longer
8. Digest cell walls using enzyme solution C. Add just enough enzyme solution to cover the cells, and shake gently on a rotary platform. Check the morphology of the cell wall every hour, and stop the digestion when the cell walls become visibly distended. The time required for digestion varies from 4 to 8 hr at room temperature
9. Rinse the cells three times (5 min each) in PBSG (room temperature)
10. Incubate in PBSG containing 2.5% (w/v) Carnation nonfat dry milk for 12–24 hr at 4°C
11. Rinse the cells three times (5 min each) in PBSG (room temperature)
12. Incubate with primary antibody diluted in PBSG for 12–24 hr at 4°C. Pipette 25 μl (for 18-mm^2 coverslip) onto Parafilm in the bottom of a petri dish, and place the coverslip cell side down onto the drop. Seal the dish with parafilm to prevent evaporation
13. Rinse the cells three times (5 min each) in modified PBS (room temperature)
14. Incubate in fluorochrome-conjugated secondary antibody diluted in PBSG for 12–24 hr at 4°C (see step 12)
15. Rinse three times (5 min each) in PBSG (room temperature)
16. Dehydrate in 100% methanol [three times (10 min each) at room temperature]
17. Mount the coverslips with 25 μl BA:BB clearing solution, and seal with fingernail polish

Modifications for rapid fixation in methanol–glycerol

1. After preextraction in MTSB, fix the cells in methanol–glycerol fix for approximately 30 min, or until the cells turn from brown to green
2. Rehydrate the cells in solution C. To the cells in fix solution, add a total of 1 vol (original fix volume) of solution C as four equal additions (1/4 vol each) spaced at 5-min intervals. Then add three additional volumes of solution C as three equal additions (1 vol each) spaced at 5-min intervals. Incubate for 5 min. Pour off solution, and add solution C
3. Proceed to step 8 above. Incubations in blocking solution and antibodies can be reduced to 4 hr at room temperature, if desired

toplasmic and spindle microtubules (Gard, 1991, 1992; Kropf *et al.*, 1990; Allen and Kropf, 1992), and have found that compromises are often necessary between the quality of microtubule preservation and reactivity with particular antisera. As with most procedures, the most useful fixation protocol is best determined empirically.

Table III
Dual-Fluorescence Microscopy of Microtubules and Chromosomes in *Xenopus* Eggs

1. Fix samples overnight in 100% methanol (see discussion of methanol fixation). No consistent difference was noted between cells fixed at room temperature or at −20°C
2. Rehydrate in TBSN (no BSA)
3. Use a sharp scalpel blade to bisect eggs
4. Incubate 16–24 hr in primary antibody diluted in TBSN plus 2% BSA (4°C with gentle rotation)
5. Wash 16–36 hr in TBSN at 4°C (with gentle rotation), changing buffer every 8–12 hr
6. Incubate 16–24 hr in secondary antibody diluted in TBSN plus 2% BSA (4°C with gentle rotation)
7. Wash in TBSN as in step 5[a]
8. Stain oocytes with propidium iodide (PI; 5–10 μg/ml; Molecular Probes) dissolved in 100% methanol (two changes, 15–20 min each)
9. Wash twice with 100% methanol (two changes, 30 min each)
10. Clear and mount in BA:BB (see above)

[a] Alternatively, add PI to the last TBSN wash, and dehydrate in methanol as described in Table I. Occasionally, high levels of apparent cross-channel fluorescence are apparent. This might be caused by excessive PI concentration or staining time. Try decreasing the concentration (to 1 μg/ml), staining time, or increasing the wash times.

The primary advantage of using precipitation with methanol (or another organic solvent) as the primary fixation for immunofluorescence is simplicity and speed, requiring but a single step and utilizing one (or a few) relatively stable reagents that rapidly penetrate cells and tissues. In our experience, however, organic solvents did not provide optimum preservation of microtubules in large cells such as *Pelvetia* or *Xenopus* eggs. Cytoplasmic microtubules in methanol-fixed cells were often collapsed into twisted bundles or appear fragmented (see Fig. 1A and B). This effect was most noticeable in the interior of large cells, and preservation of surface or cortical detail was often acceptable. Addition of dimethylsulfoxide (DMSO) (Dent's fixative; Dent and Klymkowsky, 1989), taxol (to 1 μM), formaldehyde (to 3.7%), or acetic acid (10%) to methanol did not significantly improve microtubule preservation (D. L. Gard, unpublished observations). In fact, acetic acid–methanol and Carnoy's fixative (ethanol:ether:acetic acid) proved the least satisfactory fixative for preserving microtubule structure in *Xenopus* oocytes (D. L. Gard, unpublished observations). In addition, cells fixed in methanol or other organic solvents often appeared shrunken and distorted, due to dehydration.

Despite its drawbacks, fixation in methanol (or other solvents) remains useful for immunofluorescence microscopy of samples or antibodies that are not compatible with aldehyde fixation. For example, although cytoplasmic microtubules were poorly preserved in methanol-fixed cells, spindles were often preserved to a greater extent (although not as well as in aldehyde-fixed cells; see following).

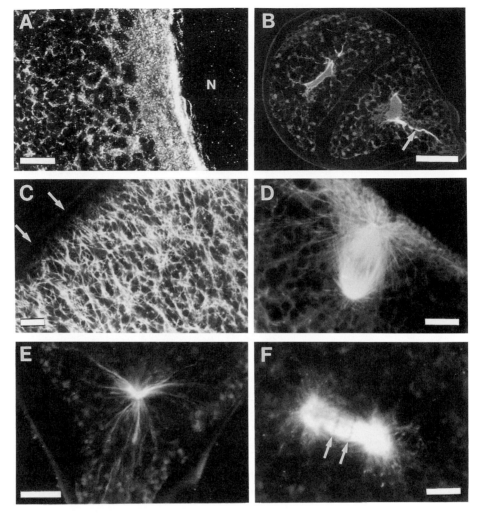

Fig. 1 A comparison of microtubule preservation by methanol-based or aldehyde fixation. (A) No microtubules remain in the interior of this stage VI *Xenopus* oocyte fixed with methanol at −20°C. The punctate staining observed with anti-tubulin represents fragmented microtubule remnants (N denotes the oocyte nucleus). (B) Spindle and cytoplasmic microtubules are poorly preserved in this *Pelvetia* embryo fixed in methanol–glycerol. Microtubules have collapsed into twisted bundles (the arrow denotes a collapsed bundle of microtubules extending to the growing tip of the rhizoid cell). (C and D) Cytoplasmic and spindle microtubules are well preserved in a stage VI *Xenopus* oocyte (C) and unfertilized *Xenopus* egg (D) fixed in formaldehyde–glutaraldehyde–taxol (FGT-fix). These deep cross-sections were obtained by bisecting the large cells with a scalpel, and optically sectioning below the knife-damaged region [see text; arrows in (C) denote the oocyte surface]. (E and F) Individual cytoplasmic and spindle microtubules are apparent in these images of interphase (E) and anaphase (F) *Pelvetia* zygotes [compare to (B) above; arrows in (F) denote the location of the anaphase chromosomes]. All images were obtained using a Bio-Rad MRC-600 laser-scanning assembly fitted to a Nikon optiphot, using ×40 (NA 1.0) or ×60 (NA 1.4) plan apochromatic objectives. Bars: 25 μm (A–C) and 10 μm (D–F). (Figure 1D from Gard, 1992.)

This allowed us to examine the distribution of chromosomes in meiotic spindles of maturing *Xenopus* oocytes, staining microtubules with anti-tubulin and chromosomes with propidium iodide (which does not stain well after aldehyde fixation; see Color Plate 1; Gard, 1992). We have included protocols for methanol fixation and immunofluorescence microscopy of microtubules in both *Pelvetia* embryos (Table II) and *Xenopus* eggs (Table III).

Fixation with formaldehyde (or paraformaldehyde) provided significantly better preservation of cellular shape and structure, and is a commonly used procedure for light microscopy (Larsson, 1988). However, formaldehyde alone did not optimally preserve cytoplasmic microtubules in *Xenopus* oocytes, requiring the use of a microtubule-stabilizing agent such as glycerol or taxol (Gard, 1991). Moderate concentrations of taxol ($0.1-1.0~\mu M$), insufficient to promote microtubule assembly (Schiff *et al.*, 1979, and references therein), stabilized microtubules during fixation without the osmotic distortion attending the use of glycerol as a microtubule-stabilizing agent.

The most consistent preservation of microtubule structure and organization in *Xenopus* oocytes and eggs was obtained by using combinations of formaldehyde and glutaraldehyde (see Fig. 1C and D). Our current protocol for fixing microtubules in *Xenopus* uses a modification of Karnovsky's fixative (Karnovsky, 1965) combining 3.7% formaldehyde, 0.25% glutaraldehyde, and $0.1-0.5~\mu M$ taxol in a microtubule assembly buffer containing the nonionic detergent Triton X-100 (FGT fix). Although taxol is included routinely in our protocols, identical results have been obtained in the absence of taxol (Gard, 1991; Schroeder and Gard, 1992).

A modified version of FGT fix (FGT/methanol fixation; see Table II and Section V) has been developed for fixing microtubules in *Pelvetia* zygotes and embryos (Allen and Kropf, 1992). This protocol uses a lower final concentration of formaldehyde (1.4%), which provides good microtubule preservation (see Fig. 1E and F) while facilitating later enzymatic digestion of the cell wall (see Section II,C). Methanol (20%) is included during fixation of *Pelvetia*, serving both as a fixative as well as extracting some of the pigments found in algal cells (see Section II,B).

Unfortunately, a number of secondary problems are introduced by the use of glutaraldehyde as a fixative. First, glutaraldehyde can destroy some antigenic epitopes (Larsson, 1988), making it incompatible with some antibodies or staining protocols. Although none of the commercial tubulin antibodies we have used have suffered from this problem, each antibody must be screened individually for its reactivity with glutaraldehyde-fixed cells. Fixation with methanol alone, or gentle fixation with formaldehyde or paraformaldehyde, is less likely to block antibody binding (Larsson, 1988). Glutaraldehyde penetrates tissues more slowly than methanol or formaldehyde, slowing the fixation reaction. This characteristic of glutaraldehyde can be offset, in part, by including formaldehyde during fixation. Antibody penetration into aldehyde-fixed cells is also noticeably slower than into methanol-fixed cells, requiring increased incubation times. In some instances, we have found it useful to preextract cells or tissues with microtubule-stabilizing buffers containing nonionic detergents (Triton X-100 or Nonidet P-40), to speed penetration of the fixative, and subsequent penetration of antibodies. Such preextraction has been particularly useful in *Pelvetia*. However, it is important to determine that the antigen of interest is not solubilized during the preextraction. Finally, glutaraldehyde-fixed samples must be treated with sodium borohydride, to reduce unreacted aldehydes and diminish glutaraldehyde-induced autofluorescence (Weber *et al.*, 1978; see also Section II,B). Despite these considerations, combinations of formaldehyde and glutaraldehyde remain the fixative of choice for immunofluorescence microscopy of microtubules in *Xenopus* and *Pelvetia* eggs and embryos.

Fixation in formaldehyde or formaldehyde–glutaraldehyde (FT, FGT, or FGT/methanol) is routinely followed by postfixation in 100% methanol. Postfixation was required for good preservation of microtubule structure in formaldehyde-fixed (FT fix) *Xenopus* oocytes (A. Roeder and D. L. Gard, unpublished observations). Although not absolutely required for microtubule preservation in glutaraldehyde-fixed oocytes (FGT fix), we include postfixation in our protocol as a matter of routine. In *Pelvetia*, postfixation in 100% methanol extracts additional pigment, helping to minimize cytoplasmic autofluorescence.

B. Pigmentation and Autofluorescence

Heavy deposits of pigment in the cortex and cytoplasm of amphibian oocytes and eggs hamper immunofluorescence microscopy of these cells. Dent and Klymkowsky (1989) demonstrated that a solution of hydrogen peroxide in methanol could be used to bleach *Xenopus* eggs, effectively eliminating the interfering pigment. Extended bleaching of *Xenopus* oocytes (up to 72 hr) does not appear to affect either the structural preservation or reactivity of microtubules with tubulin antibodies. Similar treatment of *Rana* eggs with peroxide for only 18–24 hr releases pigments, which can then be removed in subsequent washes with phosphate- or Tris-buffered saline (D. L. Gard, unpublished observations).

Despite reports of severe autofluorescence from yolk platelets in whole mounts of *Xenopus* eggs (Klymkowsky and Hanken, 1991), we observed little yolk autofluorescence during confocal microscopy of *Xenopus* or *Rana* oocytes and eggs (other than that induced by glutaraldehyde fixation). This discrepancy may result from the fact that confocal microscopy samples a much smaller volume of the cell, limiting the contribution of autofluorescence from the yolk to unnoticeable levels.

Autofluorescence resulting from the use of glutaraldehyde-containing fixatives can be minimized by treating cells with sodium borohydride prior to incubating with antibodies (Weber *et al.*, 1978). We have tested borohydride concentrations ranging from 50 to 200 m*M*, without observing evidence that borohydride treatment decreased binding of the tubulin antibodies used in our studies. However, the reactivity of other antibodies with borohydride-reduced cells should be evaluated. Detergents should not be included during the borohydride reduction, to avoid foaming of the effervescent borohydride solution. Borohydride is a strong reducing agent, and due caution should be used during its preparation and use.

In plant or algal cells, pigmentation and autofluorescence pose a more serious problem. Many of the photosynthetic pigments, as well as secondary polyphenolic compounds, fluoresce at the same wavelengths as commonly used fluorochromes. Some of the offending compounds can be extracted with methanol both during and after fixation (*Pelvetia* cells turn from natural brown to green after methanol extraction). Even so, considerable autofluorescence remains, much of it due to unextracted chlorophylls that fluoresce in the far red part of the spectrum ($\lambda > 650$ nm).Chlorophyll autofluorescence (as well as fluorescence from other compounds) often can be reduced substantially through the use of a narrow bandpass filter, centered on the emission wavelength of the fluorochrome being used for antibody detection. In *Pelvetia*, we have successfully used a dichroic filter centered at 580 nm (10-nm band width; Pomfret Research Optics, Orange, VA) in conjunction with rhodamine-conjugated secondary antibodies ($\lambda_{em} = 570$–600). This secondary filter, when placed in the pathway of the emitted light in a standard epifluorescence microscope or just ahead of the photomultiplier in a Bio-Rad (Richmond, CA) MRC-600 confocal laser-scanning microscope (in place of the second filter block), preferentially passes rhodamine fluorescence while reducing autofluorescence from shorter (polyphenolics) and longer (chlorophylls) wavelengths. This additional filtration significantly improves contrast between antigen-specific fluorescence and autofluorescence. Last, we find that the optical sectioning capabilities of the confocal microscope (see Section III,C) also dramatically improve detection of microtubule detail in *Pelvetia* zygotes, by eliminating autofluorescence from outside the focal plane.

C. Antibody Incubations

The penetration of antibodies becomes a limiting factor in the preparation of large cells, such as eggs and embryos, for whole-mount immunofluorescence microscopy. We find that antibodies (IgG) penetrate only about 100–150 μm into FGT-fixed *Xenopus* oocytes in an overnight incubation (D. L. Gard, unpublished observations). However, the speed of antibody penetration into cells varies considerably with the fixation protocol. For example, antibodies penetrate more rapidly in cells fixed with methanol alone than in FGT-fixed cells.

The reduced rate of antibody penetration (even in methanol-fixed cells) and large cell size require extending the antibody incubations and intermediate washes, considerably lengthening the time required for sample preparation. Antibody incubations for *Xenopus* oocytes (up to 1.2-mm diameter) are typically 16–48 hr (depending on the cell size), whereas *Pelvetia* embryos (typically 100-μm diameter) require 4–24 hr. In some cases, antibody penetration can be enhanced by extracting membranes and cytoplasmic proteins prior to fixation. However, care should be taken that preextraction does not alter the antigen distribution.

Plant and algal cells pose a unique problem with regard to antibody penetration, in that their cell wall acts as a molecular sieve that effectively excludes immunoglobulins. For this reason, the cell wall must be enzymatically removed or loosened prior to incubating with antibodies. In higher plants, treatment with cellulase alone is often sufficient to loosen the cell wall (Wick *et al.*, 1981). However, in *Pelvetia*, a combination of cellulase, hemicellulase, and alginate lyase is required for optimum wall digestion (Table II; Kropf *et al.*, 1990; Allen and Kropf, 1992; modified from Kloareg and Quatrano, 1987). Because the structure of the cell wall varies with species and cell type, the digestion protocol for each will have to be determined empirically.

Enzymatic wall digestion is most effective prior to aldehyde fixation. However, plant cells exhibit rapid physiological responses to treatments that compromise the cell wall, including rearrangements of their cytoskeletal architecture (Shibaoka, 1991; V. Allen and D. L. Kropf, unpublished observations). For this reason, wall digestion is best performed after fixation. Loosening of the wall is usually sufficient to allow antibody penetration, and complete removal of the cell wall is not necessary. Addition of calcium chelating agents, such as ethylene glycol-bisβ-aminoethyl ether)-N,N,N',N'-tetraacetic acid (EGTA), can aid in loosening the cell wall by depleting Ca^{2+} involved in ionic bonding between polysaccharide wall components.

D. Clearing and Mounting

The yolk-filled cytoplasm of postvitellogenic amphibian oocytes and eggs renders these cells completely opaque. The development of a technique for clearing *Xenopus* eggs has greatly facilitated immunofluorescence microscopy of these cells. Murray and Kirschner (as cited in Dent and Klymkowsky, 1989) found that a mixture of benzyl alcohol and benzyl benzoate (1:2, by volume) could be used to match the refractive index of yolk platelets, rendering oocytes and eggs nearly transparent. This BA:BB clearing solution has subsequently been used with a wide variety of animal (Klymkowsky and Hanken, 1991) and plant cells (E. King and D. L. Gard, unpublished observations), and we routinely use BA:BB for clearing and mounting *Pelvetia* embryos.

A major disadvantage of BA:BB clearing solution (and other clearing agents as well) stems from its immiscibility with aqueous solutions. Samples to be cleared with BA:BB must be completely dehydrated through several changes of methanol, ethanol, or acetone prior to clearing. This requirement prevents the use of BA:BB on samples that are sensitive to methanol (or any other dehydrating agent) [see Sect. V and Klymkowsky and Hanken (1991) for discussions of other hazards associated with BA:BB].

After dehydration and clearing in BA:BB, smaller cells (such as *Pelvetia* embryos and small amphibian oocytes) are mounted in BA:BB between a standard glass coverslip (#1 or #1-1/2 coverslip) and microscope slide. Coverslips are then sealed with clear fingernail polish. Fingernail polishes or other sealants should be tested for their compatibility with BA:BB before use [clear Sally Hanson's "Hard as Nails" works well as a sealant].

Small transparent cells, which do not require clearing, can be mounted in aqueous mounting solutions (avoiding dehydration and clearing steps). Aqueous mountants are particularly useful when the antigen or staining reaction is not compatible with dehydration in organic solvents.

Larger *Xenopus* oocytes are either bisected (normally before processing) and mounted in BA:BB in glass well slides (0.5-mm wells), or are mounted whole in custom-made chamber slides machined from aluminum. Coverslips are cemented to each side of the aluminum slide, allowing observation of both sides of the cell. Both glass well slides and aluminum chamber slides can be cleaned with acetone and reused.

Unlike smaller cells, large *Xenopus* oocytes and eggs cannot be directly mounted in aqueous mountants containing glycerol. Despite disruption of the plasma membrane with detergents, these large cells remain sensitive to the osmotic effects of glycerol. *Xenopus* eggs appear particularly sensitive to these effects, with shrinkage and collapse of the animal pole a common effect of mounting in 90% glycerol (A. Roeder and D. L. Gard, unpublished observations). Osmotic distortion can be minimized by first placing eggs in 10–20% glycerol and then slowly evaporating water from this solution, to a final concentration of approximately 90% glycerol, over a period of 5–10 days (A. Roeder *et al.*, unpublished observations).

E. Troubleshooting

1. Eliminating Nonspecific Background

Often, what appears to be unacceptably high background fluorescence when whole-mount specimens are observed by standard epifluorescence microscopy is reduced or eliminated by the optical sectioning techniques of confocal microscopy (see Section III,C). In many case, specific fluorescence from outside of the focal plane is misinterpreted as background or nonspecific staining. For this reason, judgments regarding nonspecific background are best made after examining samples with a confocal microscope.

One common cause of unacceptable background fluorescence is insufficient wash time between antibodies or after the secondary antibody. Given that antibody diffusion appears to be a limiting factor in preparing large cells for immunofluorescence, rinses should be as long or longer than the antibody incubations themselves. Note, however, that shorter rinses have proved sufficient for *Pelvetia* embryos.

Background fluorescence can also result from nonspecific binding of the fluorochrome-conjugated second antibody to cellular structures, which can appear either as diffuse fluorescence or as staining of specific cellular structures. The contributions of nonspecific binding of secondary antibody to the overall fluorescence pattern is best ascertained by eliminating the specific primary antibody, treating cells with the secondary alone. Under these conditions, all fluorescence will be nonspecific, and several strategies can be followed to minimize or eliminate it, including (1) increasing the salt concentration of the antibody buffer to minimize electrostatic interactions between antibody and cellular proteins, (2) the inclusion of inert carrier proteins such as bovine serum albumin (BSA) or nonfat milk, and nonionic detergents during antibody incubations, (3) preadsorbing the secondary antibody with nonspecific proteins (such as rat liver acetone powder), (4) preadsorbing the secondary antibody with a preparation of fixed cells identical to the sample, (5) preincubating the sample with unconjugated serum identical to the conjugated secondary (prior to incubating with the conjugated secondary antibody), and (6) trying a different secondary antibody entirely (nonspecific binding of conjugated secondary antibodies can vary with vendor, as well as with different batches from the same vendor).

Elimination of nonspecific staining due to primary antibody is more problematic, because it is often difficult to distinguish between nonspecific background and antibody-specific staining in an unexpected pattern. Commonly used approaches involve preblocking samples with "inert" proteins such as BSA, gelatin, or nonfat milk, and including carrier proteins (BSA or nonfat dry milk) and nonionic detergents in the primary antibody solutions (see Larsson, 1988; Beltz and Burd, 1989, for more complete discussions).

Determining whether the observed staining pattern is specific for the antibody and reflects the true distribution of antigen can also be problematic. Antibody "specificity" has been demonstrated most commonly by Western immunoblotting (see article 6, this volume). Although blots can demonstrate the monospecificity of an antibody, the binding conditions used for Western blotting are quite often different from those used in immunofluorescence. Thus, it still must be shown that the staining pattern observed by immunofluorescence in fixed cells reflects the distribution of antigen, rather than some nonspecific binding of antibody. In general, controls for immunofluorescence include lack of staining when specific antibody is omitted, lack of staining by preimmune or nonimmune antisera (when polyclonal antisera are used), or abrogation of the staining pattern after preadsorption or blocking of the primary antibody with the corresponding antigen (Larsson, 1988).

2. Lack of Staining

In some regards, it is probably more disconcerting when an antibody does not stain cells at all. It is not unprecedented to find that some antibodies that give good reactions by immunoprecipitation or Western blots do not stain fixed cells. Two possible explanations are that (1) the antigen is extracted during sample preparation and fixation, or (2) the epitope against which the antibody is directed is destroyed during fixation. In both cases, a number of different fixation protocols can be tested until one is found to be compatible with the antibody in use. Proteases or protein denaturants can, in some cases, be used to unmask antigens after fixation (discussed at length in Larsson, 1988). However, these treatments can have dramatic effects on cell structure, and should be used with care.

3. Fading and Photobleaching

Commonly used fluorochromes, such as fluorescein and rhodamine, are bleached rapidly when illuminated with the high-intensity light sources used for standard epifluorescence or confocal microscopy. To minimize photobleaching, samples should be protected from unnecessary exposure to light. Many protocols include strict precautions about keeping samples in the dark during incubation with the fluorescent second antibody and subsequent steps. However, we have never observed any effect on sample brightness caused by exposure to ordinary room lights (either incandescent or fluorescent). Rather, most photobleaching occurs under the microscope during observation and image collection. With this in mind, care should be taken during examination of the samples to avoid unnecessary exposure to the mercury arc or laser light source.

Several compounds have been reported to prevent or inhibit photobleaching of samples during immunofluorescence microscopy, by acting as antioxidants or radical scavengers (Giloh and Sedat, 1982). Commonly used antifade agents include p-phenylenediamine (Johnson et al., 1982), DABCO [1,4-diazabicyclo(2.2.2)octane; Johnson et al., 1982], and N-propyl gallate (Giloh and Sedat, 1982). Sodium azide has also been reported to slow photobleaching, as well as inhibiting microbial growth in aqueous mounting solutions. We have found that BA:BB clearing solution itself provides some protection against photobleaching, and generally have not found it necessary to include other antifade agents when mounting samples in BA:BB.

III. Confocal Microscopy: Advantages and Considerations

Figure 2 demonstrates the advantages afforded by confocal microscopy for observing large cells such as *Pelvetia* or *Xenopus* oocytes or embryos. Little or no cellular detail is discernable when these cells, prepared for immunofluorescence with a monoclonal antibody to α-tubulin, are observed by standard tech-

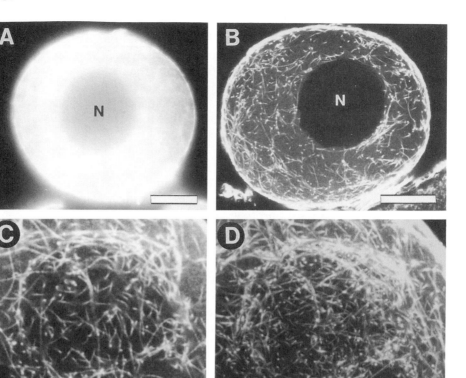

Fig. 2 A comparison of epifluorescence versus confocal laser-scanning microscopy of microtubules in state I *Xenopus* oocytes. (A) Fluorescence from outside the focal plane degrades this image of microtubules in a state I *Xenopus* oocyte photographed by standard epifluorescence microscopy (×63 NA 1.4 planapochromatic objective; N denotes the oocyte nucleus). (B) Significantly more microtubule detail is apparent in this 1-μm thick "optical section" of a stage I oocyte viewed by CLSM (with a ×60 NA 1.4 planapochromatic objective). (C) A 1-μm optical section through the mitochondrial mass (Balbiani body) of a stage I *Xenopus* oocyte stained with anti-tubulin. (D) A projection (summation) of three 1-μm serial optical sections provides an image with extended depth of focus, revealing many more microtubules (sections were projected with a maximum-brightness algorithm). Bars: 25 μm (A and B) and 10 μm (C and D).

niques of epifluorescence microscopy (Fig. 2A). Fluorescence from outside of the focal plane is superimposed on the focused image, degrading contrast and completely masking the details of microtubule distribution.

In contrast, the view obtained by confocal microscopy in Fig. 2B (in this instance, confocal laser-scanning microscopy, or CLSM) reveals in striking detail the distribution of microtubules in a similar sample. The difference between Fig. 2A and B stems from the ability of the confocal microscope to suppress fluorescence from outside of the objective focal plane, thereby taking an "optical section" of the specimen. A complete discussion of the theory and practice of confocal microscopy is beyond the scope of this article. Rather, we shall address some of the factors that must be considered when preparing samples for confocal immunofluorescence microscopy (confocal laser-scanning microscopy in particular). More detailed discussions of the theory and application of confocal microscopy can be found in Inoué (1989) and Matsumoto (1993).

A. Choice of Fluorochromes

One disadvantage of laser-scanning microscopes, compared to standard epifluorescence or confocal microscopes using mercury arc lamps, or other light sources with a broad emission spectrum, is the limited number of excitation wavelengths available. The argon ion lasers commonly found among the first generation of commercially available confocal laser-scanning microscopes have strong emission lines of 488 and 514 nm, which limits the choice of useful fluorochromes. Fortunately, this limitation has little impact on most applications for immunofluorescence. Both fluorescein and rhodamine-conjugated secondary antibodies, which are widely available from a number of commercial vendors, have proved satisfactory for CLSM. Fluoresceins (λ_{ex} = 490 nm; Haugland, 1991; Tsien and Waggoner, 1989) are efficiently excited by the 488-nm line of argon ion lasers. In practice, the poor excitation efficiency of rhodamines (λ_{ex} = 540–575 nm; Haugland, 1991; Tsien and Waggoner, 1989) by the 514-nm line of argon ion lasers is, in part, offset by the greater resistance of rhodamine to photobleaching. Confocal laser-scanning microscopes equipped with argon–krypton lasers have become available that allow more efficient excitation of rhodamine, Texas Red (Molecular Probes, Inc., Eugene, OR), and other fluorochromes requiring long excitation wavelengths. In addition, new fluorochromes are under development (by Molecular Probes and other vendors), with usefulness in laser-scanning microscopy being a key element in their design.

B. Photobleaching

In our experience, photobleaching has proved no more severe with laser-scanning microscopy than is typically encountered with standard epifluorescence microscopy. Although the intensity of laser illumination can be greater than that provided by more conventional light sources, the scanning nature of CLSM dramatically limits the duration of illumination. As with any fluorescence technique, however, care should be taken to minimize unnecessary exposure of the sample to the laser beam. In practice, we use neutral-density filters to reduce the intensity of the laser illumination to the minimum practical for image collection (typically 1–3% of maximum). In an extreme case, we were able to collect more than 130 optical sections (each an average of five scans) from a rhodamine-labeled *Xenopus* oocyte mounted in BA:BB, without undue photobleaching. Antifade agents should also be used in aqueous mounting solutions, when possible.

C. Optical Sectioning

In confocal laser-scanning microscopy, as with any technique of microscopy, the specifications of the objective lens used for image collection affect many parameters of the collected image, including image scale and field (magnification), brightness (magnification and numerical aperture), and resolution (numerical aperture) (see Inoué, 1986, for an excellent discussion of the theory and practice of light microscopy). The numerical aperture (NA) of the objective, in conjunction with the confocal aperture, also affects the "confocal" characteristics of the image, by establishing the thickness of the optical sections obtained. Typical section thicknesses range from approximately 40 μm (with NA 0.45) to less than 0.7 μm (with NA 1.4). Thus, both magnification (image scale) and numerical aperture should be considered when choosing objectives for confocal microscopy.

The ease with which "serial optical sections" can be collected from large cells is one of the main attractions of confocal microscopy. Series of adjacent optical sections can then be recombined to give information on the three-dimensional distribution of antigen. Perhaps the simplest method of reconstruction involves adding, either linearly or using a weighted algorithm, adjacent sections to produce an image with extended depth (Fig. 2C). Alternatively, projection of serial sections can be used to create stereo pairs, which provide a sense of depth when viewed with a stereoscope, or volume rendering can be used to generate a true three-dimensional representation of antigen distribution from serially collected confocal sections.

In practice, serial optical sectioning with confocal microscopes is limited by a number of independent factors, including penetration of antibodies into the sample, sample transparency, resistance of the sample to photobleaching, the working distance of the objective in use, and cell size. Optimum visualization of individual microtubules requires the use of objectives with large numerical apertures (preferably 1.3–1.4) that have working distances of only 100–170 μm. This, and poor penetration of antibodies, has limited optical sectioning in our studies to approximately 125 μm.

D. General Considerations

The digital nature of images obtained by CLSM makes processing of confocal images, both during and after image collection, a relatively straightforward matter (a more complete discussion of digital image processing can be found in

Inoué, 1986). However, in the interest of image accuracy, we normally limit processing of confocal images to the optimization of contrast and elimination of photon noise through Kalman averaging of successive scans (or other filtering algorithms) during image collection. These raw images are then saved, either to magnetic floppy disks or large-capacity optical disks, for archival storage. Once archived, images can be manipulated without compromising the original data.

Some care must be taken when interpreting images obtained by confocal microscopy. Often, inhomogeneous penetration of antibody (due to the geometry of the cell or tissue, or for any number of reasons) can be mistaken for differential localization of the antigen being studied. Because confocal microscopy is typically used to examine large cells, or thick sections, poor penetration of the fixative or antibodies can result in misinterpretation of the resulting images.

E. Photographing Confocal Images

In contrast to photography of more conventional immunofluorescence images in which film speed must be maximized, photography of the digital images obtained by CLSM is much less demanding. Best results are obtained by using films with moderate speed and fine grain. We have had excellent results photographing confocal images from a high-resolution flat-field monochrome monitor with a 35-mm camera and Kodak (Rochester, NY) TMAX 100 film. Moderate shutter speeds, from 0.5 to 2 sec, should be used, to avoid scanning artifacts from the computer monitor.

Confocal laser-scanning microscopy is, by its nature, a monochrome imaging system. However, pseudocolor imaging is often useful for highlighting antigen distribution or cellular structures. We photograph color images directly from the image monitor by using a 35-mm camera equipped with a 100-mm macrolens. Best results have been obtained using Kodak Ektachrome 100 HC (for color transparencies) and Kodak Ektar 125 (for color negatives).

IV. Conclusion

The advent of confocal microscopy has significantly extended the useful range of immunofluorescence microscopy to large, previously intractable, cells or tissues. In some respects, the technology of microscope development has outstripped currently used techniques for sample preparation. In the discussion above, we have shared some of the techniques we have developed in our own research, hoping that they will aid others developing techniques for immunofluorescence microscopy of difficult cells.

Appendix I: Recipes and Reagents

A. Fixatives

Fix buffer (for *Xenopus*):
   80 mM potassium piperazine-N,N'-bis(2-ethanesulfonic acid((K PIPES) (pH
      6.8), 1 mM MgCl$_2$, 5 mM EGTA, 0.2% Triton X-100
Modified fix buffer (for *Pelvetia*):
   80 mM K PIPES (pH 6.8), 1 mM MgCl$_2$,
      5 mM EGTA, 0.2% Triton X-100, 20% glycerol
Formaldehyde–glutaraldehyde fixative with taxol (FGT fix):
   3.7% formaldehyde (from 37 or 18% stock), 0.25% glutaraldehyde (from a
      50% stock), 0.5 μM taxol (from a 100 μM stock) in fix buffer
FGT fix with methanol for *Pelvetia* (FGT/methanol):
   1.4% formaldehyde, 0.25% glutaraldehyde, 1.0 μM taxol, 20% methanol in
      modified fix buffer

   Methanol/glycerol fix for *Pelvetia:*
   74% methanol, 20% glycerol, 5% DMSO, 1% Triton X-100

All fixatives are used within 1 hr of preparation. FG fix is the identical
formulation *without* taxol. FT fix is the identical formulation *without* glutaraldehyde. Formaldehyde can be AR grade (Mallinkrodt, Paris, KY) or electron
microscopy (EM) grade (methanol free; Ted Pella, Inc., Redding, CA). Glutaraldehyde is EM grade (stored as a 50% stock at 4°C; Ted Pella, Inc.). We have not
tested glutaraldehyde from other sources. Taxol is from a 100 μM stock in
DMSO, stored at −20°C (from National Cancer Institute. Commercially available from CalBiochem, La Jolla, CA).

B. Other Buffers

Phosphate-buffered saline (PBS):
   128 mM NaCl, 2 mM KCl, 8 mM Na$_2$HPO$_4$, 2 mM KH$_2$PO$_4$ (pH 7.2)

Modified PBS with glycerol (PBSG):
   137 mM NaCl, 2.7 mM KCl, 1.7 mM KH$_2$PO$_4$, 8 mM Na$_2$HPO$_4$, 0.1% NaN$_3$,
      0.1% BSA, 20% glycerol

Tris-buffered saline with NP-40 (TBSN):
   155 mM NaCl, 10 mM Tris-HCl (pH 7.4), 0.1% NP-40

Triton X-100 can be substituted for NP-40. Tris-buffered saline (*without
detergent*) can be substituted for PBS during NaBH$_4$ reduction.

C. Solutions for Removing Cell Walls

Solution C (modified from Kloareg and Quatrano, 1987):
10 mM 2-(N-morpholino)ethanesulfonic acid (MES) (pH 5.8), 100 mM NaCl, 20 mM MgCl$_2$, 1 mM EGTA, 2 mM KCl, 0.2% BSA (w/v), 1 M sorbitol, store at -20°C.

Enzyme solution C (modified from Kloareg and Quatrano, 1987):
7 mg/ml cellulase (CELF; Worthington Biochemical Corp, Freehold, NJ), 40 mg/ml hemicellulase (Sigma, St. Louis, MO), 6.25 units/ml alginate lyase [a gift from B. Kloareg, Centre National de la Recherche Scientifique (CNRS), France], 0.1 mM phenylmethylsulfonyl fluoride (PMSF) in Solution C.

Abalone gut acetone powder (Sigma) can be substituted for purified alginate lyase, but activity is reduced.

D. Clearing and Mounting Solutions

Benzyl alcohol–benzyl benzoate clearing solution (Dent and Klymkowsky, 1989):
1 part benzyl alcohol to 2 parts benzyl benzoate

Benzyl alcohol (BA) and benzyl benzoate (BB) are natural products found in flower oils (BA) and balsams (BA and BB). Both are mildly aromatic, and are used in the perfume and confection industries. The LD$_{50}$ values for ingestion of BA and BB in mice are 3.1 and 1.4 g/kg, respectively (Merck index). However, both are skin irritants, and due caution should be used to avoid contact (especially with the eyes).

Most laboratory plasticware (microcentrifuge tubes, etc.) is compatible with BA:BB, notable exceptions being cellulose acetate and the plastic used in computer keyboards. We have heard rumors of microscope objectives being destroyed (''unglued'') by careless use of BA:BB. Although we have not had any severe mishaps, caution is advised when mounting and examining samples prepared with BA:BB to avoid costly damage to microscopes and other equipment.

Glycerol mounting solution:
10 mM Tris-HCl (pH 8.0), 10 mM NaN$_3$, in 90% glycerol

NaN$_3$ provides some protection against photobleaching. Other antifade agents include phenylenediamine, DABCO, or n-propyl gallate (see text).

E. Miscellaneous Solutions

Peroxide–methanol bleach (Dent and Klymkowsky, 1989):
1 part hydrogen peroxide (30%) to 2 parts 100% methanol (100%)

Peroxide is an excellent bleaching agent for oocyte pigments, clothing, and skin. Use due caution to avoid damage to clothing or painful chemical "burns."

Appendix II: Antibodies

We have used the following anti-tubulin antibodies with good results.

Antibody	Antigen	Species	Source	Ref.
DM1A	α-Tb	Mouse	ICN (Lisle, IL)	Blose *et al.* (1984)
DM1B	β-Tb	Mouse	ICN	Blose *et al.* (1984)
N-357	β-Tb	Mouse	Amersham (Arlington Heights, IL)	Blose *et al.* (1984)
6-11B-1	α-Tb[a]	Mouse	NCA[b]	Piperno *et al.* (1987)
Tub-1A2	α-Tb[c]	Mouse	Sigma	Kreis (1987)

[a] Specific for acetylated α-tubulin. Preferentially stains nondynamic microtubules.
[b] Not commercially available.
[c] Specific for tyrosinated α-tubulin. Stains dynamic microtubules.

Fluorescein- and rhodamine-conjugated secondary antibodies were obtained from Organon Teknika-Cappel (Malvern, PA).

Appendix III: Supplier Information

Supplier	Reagent
Amersham Corp. (Arlington Heights, IL)	Antibodies
Calbiochem (San Diego, CA)	Taxol
ICN Immunologicals (Lisle, IL)	Antibodies
Molecular Probes, Inc. (Eugene, OR)	Conjugated secondary antibodies, propidium iodide, labeled phalloidin, protein-labeling reagents
Organon Teknika-Cappel (Malvern, PA)	Fluorescent secondary antibodies
Pomfret Research Optics (Orange, VA)	Custom dichroic filters
Sigma Chemical Corp. (St. Louis, MO)	All other reagents

Acknowledgments

The authors would like to thank D. Affleck, V. Allen, B. Error, A. Friend, J. Jordan, A. Roeder, M. Schroeder, and R. Swope, all of whom have contributed to the work described. Special thanks are due Dr. Ed King for enthusiastic discussions and invaluable assistance with the confocal microscope facility. The work described has been supported by a grant from the National Institute of General Medical Studies (D.L.G.) and the National Science Foundation (D.L.K.; DCB 8904770). Acquisition of the confocal microscope was made possible by an award from the University of Utah.

References

Agard, D. A., Hiraoka, Y., Shaw, P., and Sedat, J. W. (1989). Fluorescence microscopy in three dimensions. *Methods Cell Biol.* **30**, 353–377.

Allen, V., and Kropf, D. L. (1992). Nuclear rotation and lineage specification in *Pelvetia* embryos. *Development* **115**, 873–883.

Beltz, B. S., and Burd, G. D. (1989). "Immunocytochemical Techniques." Blackwell, Oxford.

Blose, S. H., Meltzer, D. I., and Feramisco, J. R. (1984). 10-nm filaments are induced to collapse in living cells microinjected with monoclonal and polyclonal antibodies against tubulin. *J. Cell Biol.* **98**, 847–858.

Coons, A. H. (1961). The beginnings of immunofluorescence. *J. Immunol.* **87**, 499–503.

Coons, A. H., Creech, H. J., and Jones, R. N. (1941). Immunological properties of an antibody containing a fluorescent group. *Proc. Soc. Exp. Biol. Med.* **47**, 200–202.

Coons, A. H., Creech, H. J., Jones, R. N., and Berliner, E. (1942). The demonstration of pneumococcal antigen by the use of a fluorescent antibody. *J. Immunol.* **45**, 159–170.

Dent, J., and Klymkowsky, M. W. (1989). Wholemount analysis of cytoskeletal reorganization and function during oogenesis and early embryogenesis in *Xenopus*. *In* "The Cell Biology of Development" (H. Schatten and G. Schatten, eds.), pp. 63–103. Academic Press, San Diego.

Gard, D. L. (1991). Organization, nucleation, and acetylation of microtubules in *Xenopus laevis* oocytes: A study by confocal immunofluorescence microscopy. *Dev. Biol.* **143**, 346–362.

Gard, D. L. (1992). Microtubule organization during maturation of *Xenopus* oocytes: Assembly and rotation of the meiotic spindles. *Dev. Biol.* **151**, 516–530.

Giloh, H., and Sedat, J. W. (1982). Fluorescence microscopy: Reduced photobleaching of rhodamine and fluorescein protein conjugates by *n*-propyl gallate. *Science* **217**, 1252–1255.

Haugland, R. P. (1991). "Handbook of Fluorescent Probes and Research Chemicals." Molecular Probes, Inc., Eugene, Oregon.

Inoué, S. (1986). "Video Microscopy." Plenum, New York.

Inoué, S. (1989). Foundations of confocal scanned imaging in light microscopy. *In* "The Handbook of Biological Confocal Microscopy" (J. Pawley, ed.), pp. 1–12. IMR Press, Madison, Wisconsin.

Johnson, G. D., Davidson, R. S., McNamee, K. C., Russell, G., Goodwin, D., and Holborow, E. J. (1982). Fading of immunofluorescence during microscopy: A study of the phenomenon and its remedy. *J. Immunol. Methods* **26**, 231–242.

Karnovsky, M. J. (1965). A formaldehyde–glutaraldehyde fixative of high osmolarity for use in electron microscopy. *J. Cell Biol.* **27**, 131a.

Kelly, G. M., Eib, D. W., and Moon, R. T. (1991). Histological preparation of *Xenopus laevis* oocytes and embryos. *Methods Cell Biol.* **36**, 389–417.

Kloareg, B., and Quatrano, R. S. (1987). Isolation of protoplasts from zygotes of *Fucus distichus* (L.) Powell (Phaeophyta). *Plant Sci.* **50**, 189–194.

Klymkowsky, M., and Hanken, J. (1991). Whole-mount staining of *Xenopus* and other vertebrates. *Methods Cell Biol.* **36**, 419–441.

Kreis, T. E. (1987). Microtubules containing detyrosinated tubulin are less dynamic. *EMBO J.* **6**, 2597–2606.

Kropf, D. L., Maddock, A., and Gard, D. L. (1990). Microtubule distribution and function in early *Pelvetia* development. *J. Cell Sci.* **97**, 545–552.

Larsson, L.-I. (1988). "Immunocytochemistry: Theory and Practice." CRC Press, Boca Raton, Florida.

Matsumoto, B., ed. (1993). *Methods Cell Biol.* **38.**

Pawley, J. (1989). "The Handbook of Biological Confocal Microscopy." IMR Press, Madison, Wisconsin.

Piperno, G., LeDizet, M., and Chang, X.-J. (1987). Microtubules containing acetylated α-tubulin in mammalian cells in culture. *J. Cell Biol.* **104,** 289–302.

Schiff, P. B., Fant, J., and Horwitz, S. B. (1979). Promotion of microtubule assembly *in vitro* by taxol. *Nature (London)* **277,** 665–667.

Schroeder, M. M., and Gard, D. L. (1992). Organization and regulation of cortical microtubules during the first cell cycle of *Xenopus* eggs. *Development* **114,** 699–709.

Shibaoka, H. (1991). Microtubules and the regulation of cell morphogenesis by plant hormones. *In* "The Cytoskeletal Basis of Plant Growth and Form" (C. W. Lloyd, ed.), pp. 159–168. Academic Press, San Diego.

Tsien, R. W., and Waggoner, A. (1989). Fluorophores for confocal microscopy: photophysics and photochemistry. *In* "The Handbook of Biological Confocal Microscopy" (J. Pawley, ed.), pp. 153–161. IMR Press, Madison, Wisconsin.

Weber, K., Rathke, P. C., and Osborne, M. (1978). Cytoplasmic microtubular images in glutaraldehyde-fixed tissue cells by electron microscopy and immunofluorescence microscopy. *Proc. Natl. Acad. Sci. U.S.A.* **75,** 1820–1824.

White, J. G., Amos, W. B., and Fordham, M. (1987). An evaluation of confocal versus conventional imaging of biological structures by fluorescence microscopy. *J. Cell Biol.* **105,** 41–48.

Wick, S. M., Seagull, R. W., Osborn, M., Weber, K., and Gunning, B. E. S. (1981). Immunofluorescence microscopy of organized microtubule arrays in structurally stabilized meristematic plant cells. *J. Cell Biol.* **89,** 685–690.

CHAPTER 10

Immunolabeling of Antigens in Plant Cells

Susan M. Wick

Department of Plant Biology
Biological Sciences Center
University of Minnesota
St. Paul, Minnesota 55108

I. Introduction

A. Aspects of Plant Cells and Tissues That Influence Handling Procedures for Immunolabeling

The manner of incubating specimens with antibodies for the purpose of localizing intracellular constituents is essentially identical for all types of cells.
However, the processing steps required before reaching this stage vary considerably between cells of plants, fungi, and algae and between those of animals
and the animal-like protists. This article focuses on the use of antibodies to label
plant cells that are derived from tissues, although some of the procedures
described here will be applicable to studies of plant cell suspension cultures,
fungi, and algae, as well. Discussion will be limited to labeling for light microscopy examination; postembedding immunoelectron microscopy labeling of
plant cells is not drastically different from techniques used on animal cells.

With the exception of endosperm, generative cells, and sperm cells, all plant
cells have a cellulosic cell wall, which, when intact, presents a barrier to
antibody entry. Epidermal cells from the aerial parts of a plant usually also have
a cuticle that resists penetration of aqueous solutions. Consequently, in order
to make plant cell cytoplasm accessible to antibodies, the cell wall must be
breached in some fashion. This often involves weakening or removing components of the cell wall enzymatically (Sections III,B, IV,B, and VI,G) or
sectioning through it (Section V). Alternative methods that have been shown to
work in specialized cases include cutting into large cells with a razor blade
(Flanders *et al.*, 1990), damaging pollen wall integrity via repeated freeze-thaw
cycles (Tiwari and Polito, 1988), or freeze-fracturing pollen grains (Tiwari and
Polito, 1990). As a general rule, samples need to be fixed with cross-linking
agents such as aldehydes in order to withstand the considerable handling that
these methods entail and still maintain their *in vivo* intracellular structure.

Another factor with which one needs to contend when immunolabeling plant
cells is autofluorescence. The most obvious source in green samples is chloro-

plasts, which display red autofluorescence. In this case, a simple solution is to choose a nonred fluorochrome for immunolabeling and to prevent red autofluorescence from reaching the microscope oculars or camera by use of the appropriate barrier filters (Section III,E). Autofluorescence that can be more difficult to deal with is that due to secondary metabolites of many cells. Sometimes this can be at least partially alleviated by extracting part of the cytoplasm before immunolabeling (Section VI,B) or circumvented by using immunogold procedures (Section VI,C). Lignified cell wall and components of pollen grain walls can also cause considerable problems; cutting sections through a sample in order to eliminate observation through an autofluorescent wall may be of some help (Section V), as is immunogold labeling.

The thickness of plant cells usually precludes visualizing the entire cell depth in focus at a single focal plane, especially when using oil immersion objective lenses or other high numerical aperture lenses. If one is trying to assess the distribution of an antigen throughout a cell, this may be considered a drawback, whereas if one is attempting to visualize spatial interrelationships within a complex network of slender elements of the cytoskeleton, the ability to focus on a thin layer of cytoplasm without superimposition of information from the bulk of the cell is clearly a benefit. Plant cell thickness normally does not result in problems with antibody penetration as long as high concentrations of highly cross-linking fixatives are avoided and IgG antibodies are used (rather than the much larger IgM). Even antibodies conjugated to 1- or 5-nm gold are able to penetrate appropriately fixed plant cells (Section VI,C).

B. Preparations Suited to Different Questions

The type of information that is being sought has a bearing on which route to take with specimen preparation for immunolabeling. If the question to be addressed does not require being able to determine labeling patterns in neighboring cells in a tissue or between tissues, then isolation of cells (Section III) may be a suitable choice. However, if identification of the tissue of origin of a cell is important (and cell shape alone does not allow this discrimination), or if comparisons of neighboring cells or of cells along a developmental gradient is essential to the study, then sectioning of material (Section V) needs to be considered. Cutting sections may also be appropriate for the study of organs that are resistant to enzymatic separation of cells, such as the mature leaves of many species or the leaves of any age of some species.

If the epidermis or the cells in one or two cortical layers immediately adjacent to the epidermis are specifically of interest, and information obtained from only part of the depth of a cell (such as would be obtained in a section) is not suitable or optimal, then preparation of epidermal sheets plus or minus cortical layers (Section IV) is useful. In at least some species, the epidermis can be stripped off mature leaves without apparent damage to cells, but this does not work for all species, or usually for young developing leaves. Free-hand paradermal slices are also an option for examining the epidermis, but unless one has access to a confocal microscope one must be willing to limit observation to the edges of the slice, where there may be regions of single-cell deep, undamaged cells. If material is obtained via either of these last two techniques, modification of the procedures outlined in Section IV may be of some use in handling it for immuno-labeling.

II. Reagents

A. Fixatives

The following, all of which contain formaldehyde as the major fixative, have been useful for fixation of plant specimens for light microscopy immunolabeling. All need to be made from paraformaldehyde (PFA), not from formalin solution. Ethylene glycol-bis(β-aminoethylether-N,N,N',N'-tetraacetic acid (EGTA) is added to fixatives 1–3 and 5 as an inhibitor of Ca^{2+}-activated proteases; however, for localization of calmodulin or other antigens that require Ca^{2+} for maintaining *in vivo* conformation or for interaction with other cellular components, it is better to omit the EGTA. For isolated cells, the first four fixatives are expected to give the best results because they avoid potential problems associated with glutaraldehyde, such as induced cytoplasmic autofluorescence, reduced penetration of antibodies, and reduced antigenicity of the molecule to be localized. Fixative 5 is the standard fixative used by us for processing of leaf epidermal sheets.

We store fixatives at room temperature to prevent polymerization of aldehydes and, in our experience, unless noted otherwise, each is stable for up to 1 month at room temperature. Occasionally a batch of PFA powder will be bad; if repeated attempts to obtain good cell preservation fail, or if background fluorescence of samples not exposed to antibodies is high, it may be best to try another batch of PFA.

1. Paraformaldehyde (PFA) (3.7%) in 5 mM EGTA–50 mM potassium phosphate buffer, pH 6.8: Place 1.85 g of PFA plus about 40 ml of EGTA–PO_4 buffer (see below) in a 50-ml volumetric flask, and gently heat to 60–70°C, with stirring, to dissolve. Do not allow to boil. If the solution is still cloudy after about 30 min, a few drops of 1 N KOH may help, but often there will be a small amount of PFA powder that will not dissolve. When cool, adjust the volume to 50 ml with buffer and check the pH, using HCl to adjust if necessary. If there is reason to suspect that the antigen of interest needs Ca^{2+} for its normal conformation, stability, or location within the cell, use 50 mM potassium phosphate buffer without EGTA to make the fixative solution.

a. 50 mM Potassium phosphate buffer (50 mM):

Stock solution A: 0.2 M potassium phosphate (monobasic) $= 27.218$ g
KH_2PO_4 per liter
Stock solution B: 0.2 M potassium phosphate (dibasic) $= 45.646$ g
$K_2HPO_4 \cdot 3H_2O$ per liter

Combine 25 ml of each stock and dilute to 200 ml with H_2O to give 50 mM phosphate buffer. Check that pH is 6.8.

b. EGTA stock (500 mM): Add about 15 ml of H_2O to 3.8 g of EGTA in an acid-cleaned plastic bottle. Slowly add KOH pellets while stirring to dissolve EGTA. When the solution becomes clear, adjust the final volume to 20 ml, and check that the pH is between 6.8 and 8.5. Store in plastic to prevent the EGTA from leaching Ca^{2+} from glass surfaces.

c. EGTA (5 mM)–50 mM potassium phosphate buffer (EGTA–PO_4): Add 1 ml of 500 mM EGTA stock to 99 ml of 50 mM potassium phosphate buffer.

2. Picric acid (0.2%) plus 3.7% PFA in 5 mM EGTA–50 mM potassium phosphate buffer, pH 6.8: For many cells, picric acid improves cellular preservation and can also enhance the antibody reaction. It imparts a yellow color to specimens, which can make them easier to see during processing. To make fixative, proceed as for fixative 1 (above) until the fixative is cooled, then add 7.5 ml of saturated picric acid that has been filtered through paper, and add EGTA–PO_4 to 50 ml. Picric acid is explosive when dry; picric acid salts must always be kept covered with water in the stock bottle, and care must be taken to clean up any spills and flush solutions down the drain with copious amounts of water. We have found this fixative to be stable for at least 1 year at room temperature.

3. PFA (3.7%) in 5 mM EGTA–50 mM borate buffer, pH 9–10: The rate and degree of fixation with PFA can be increased by raising the pH of the fixative solution. Especially when EGTA is omitted from the formulation, this fixative has proved to be useful for immunolabeling of calmodulin in some plant cells (Wick and Duniec, 1986). It may be helpful for localization of other soluble antigens, as well. Proceed as for fixative 1 (above), substituting borate buffer for phosphate.

Borate buffer (50 mM):

Stock solution A: 50 mM sodium tetraborate (contains 0.2 M borate) $= 1.905$ g of borax per 100 ml
Stock solution B: 0.2 N KOH

Combine 25 ml of stock A and 8 ml of stock B and dilute to 100 ml with H_2O. The pH will be about 9.4.

4. PFA–ZnCl$_2$–citrate: 0.5% ZnCl$_2$ plus 3.7% PFA in 50 mM citrate buffer, pH 4.7, for 20 min followed by 3.7% PFA in 5 mM EGTA–50 mM potassium phosphate buffer, pH 6.8.

This fixation combination has given particularly good results for calmodulin and microtubule localization in plant cells (Wick and Duniec, 1986), possibly because zinc salts at low pH act as a mordant of the cytoplasmic matrix and prevent loss of antigens during fixation.

Citrate buffer (50 mM):

Stock solution A: 0.1 M citric acid = 21.01 g of H$_3$C$_6$H$_5$O$_7$ · H$_2$O per liter
Stock solution B: 0.1 M sodium citrate = 29.41 g of C$_6$H$_5$O$_7$Na$_3$·2H$_2$O per liter

5. PFA (4%) ± 0.1% glutaraldehyde plus 0.2% dimethyl sulfoxide (DMSO) in 5 mM EGTA–50 mM piperazine-N,N'-bis(2-ethanesulfonic acid) (PIPES) buffer containing 2 mM MgSO$_4$ and 4% glycerol (PMEG), pH 6.8.

We use this typically for fixing leaf segments prior to preparing sheets of epidermal cells for labeling. The PMEG buffer is believed to be good for stabilization of microtubules. We normally make the fixative by combining 8% PFA with 2 × PMEG buffer in a 1 : 1 ratio and adding DMSO. Prepare a solution of 8% PFA according to the general instructions for fixative 1 (above), using 2 g of PFA in a final volume of 25 ml and substituting distilled water for phosphate buffer. Mix with an equal volume of 2 × PMEG and add 0.1 ml of DMSO/50 ml. If desired, glutaraldehyde (from a stock solution) can be added to a few milliliters of fixative immediately before use to give a final concentration of 0.1%.

PMEG buffer (2×): For a final volume of 100 ml, combine 3.024 g of PIPES, 0.3804 g of EGTA or 2 ml of stock EGTA (500 mM), 0.099 g of MgSO$_4$·7H$_2$O, 8 ml of glycerol

Add KOH pellets while stirring, just until PIPES dissolves. Check that the pH is 6.8 and adjust if necessary with KOH or HCl.

B. Cell Wall-Digesting Enzymes

Commercially available wall-digesting enzymes are available from a number of biochemical companies such as Serva (Paramus, NJ) Calbiochem (La Jolla, CA), Sigma (St. Louis, MO), and ICN (Costa Mesa, CA). They are listed in catalogs as cellulases (or under the proprietary name Cellulysin), hemicellulases, and pectinases (or as the proprietary names Macerase or Macerozyme). They are actually crude mixtures obtained from fungal or bacterial cultures and they contain, in addition to the named enzyme, considerable levels of activity of the other classes of cell wall-digesting enzymes, as well as protease and amylase activity. They may also contain materials that are not readily water soluble. Their pH optimum is typically between 4.0 and 5.5. Use of these enzymes is not an exact science, due to considerable variation from one organismal source of enzyme to another as well as among different batches from the same source, wall composition differences between samples of different ages and from different organs on the same plant, and differences in wall composition based on plant species.

As a general rule, we use these enzyme mixtures, either alone or in combinations, at concentrations of 1–5%, made up in unbuffered solutions containing 5 mM EGTA and a concentration of mannitol that roughly matches the osmotic potential of the cells to be examined (typically 0.2–0.4 M mannitol). Use of an osmoticum in the digestion solution helps to prevent distortion due to cellular swelling in solutions of low osmotic strength, which can occur even after fixation. To determine the appropriate mannitol concentration for a given sample, thin slices of living tissue are bathed in a variety of mannitol solutions and observed with a microscope; a mannitol concentration just below that causing incipient plasmolysis is chosen as the one to be included during cell wall digestion. If the presence of Ca^{2+} is beneficial for the antigen of interest, EGTA can be omitted. Adjust the pH to within the correct range with HCl if necessary. We usually freeze the mannitol $\pm$ EGTA solution at $-20°C$ in 1- and 2-ml aliquots and add lyophilized enzyme just before use. Enzyme solution can be reused several times over the course of a week if stored under refrigeration with NaN_3 added to a concentration of 0.02% (add as a 1:500 dilution of a 10% NaN_3 stock).

  EGTA (5 mM)–0.4 M mannitol: Combine 3.644 g of mannitol plus 0.5 ml of 500mM EGTA stock per 50 ml.
  EDTA (10 mM)–0.3 M mannitol: Combine 2.733 g of mannitol and 0.074 g of ethylenediaminetetraacetic acid (EDTA) disodium dihydrate and add distilled water to about 45 ml. Slowly add KOH while stirring until the EDTA dissolves. Adjust the pH back down to pH 5–6 with HCl, and add distilled water to a final volume of 50 ml.

C. Protease Inhibitors

Sensitive antigens can be damaged by proteolytic activity during fixation or subsequent processing steps. In addition to proteases that contaminate wall-digesting enzyme mixtures, plant cells contain varying amounts of endogenous proteases. EGTA has already been mentioned above as an inhibitor of Ca^{2+}-dependent proteases; the following inhibitors [available from Boehringer Mannheim (Indianapolis, IN), Chemicon (Temecula, CA), Sigma, and others] can provide further protection. Phenylmethylsulfonyl fluoride (PMSF) can be added to any step, but leupeptin and pepstatin are peptides and are probably of little use in the fixative step, during which they could become cross-linked and inactivated by the fixative.

For storage of solutions of peptides, antibodies, and other proteins, it is important to avoid freezers that are part of a combined refrigerator–freezer unit, as these rarely maintain a temperature of −20°C or colder, and protein solutions can thaw at temperatures only slightly higher than this. Likewise, self-defrosting freezers must be avoided, as protein solutions will be subjected to thawing and refreezing at every defrost cycle. As much as 50% of activity can be lost per cycle.

PMSF: A 0.3 *M* stock solution contains 0.052 g of PMSF in 1 ml of 2-propanol. Store at −20°C and just before use dilute at 1:1000 (0.3 m*M*) or 1:300 (1 m*M*). *Caution:* PMSF is extremely toxic. Work in a fume hood and wear protective gear while preparing the stock solution

Leupeptin: A 1-mg/ml stock in H_2O can be frozen in aliquots. Use at 5–10 μl/ml (5–10 μg/ml or 10–20 μM). Avoid multiple freeze-thaw cycles

Pepstatin: Freeze aliquots of a stock solution of 0.5 mg/ml in H_2O and use at 14 μl/ml (7 μ/ml or 10 μM). If a more highly concentrated stock solution is desired, it can be made up in methanol or ethanol. Avoid multiple freeze-thaw cycles

D. Embedding Medium and Subbing Solution for Cryosectioning

Good infiltration of plant specimens and increased ease of cutting thin cryosections for light microscopy can be obtained with sucrose and OCT embedding medium (Miles, Inc., Elkhart, IN) mixed according to the formulation of Barthel and Raymond (1990): 20% sucrose:Tissue Tek OCT compound (2:1). Store refrigerated. For adhering cryosections to slides, gelatin-coated (subbed) slides are traditionally used. Before subbing, clean slides overnight in chromic acid or concentrated HCl, rinse with distilled water, and store in 70–95% ethanol if not using immediately. Rinse with distilled water before subbing. If slides are not completely clean, subbing solution will not coat them evenly.

Subbing Solution

Mix the following until the gelatin dissolves, filter through paper into a small beaker that is wide enough to hold a slide, and use immediately for best results:

Gelatin	0.1 g
Chrome alum (chromic potassium sulfate)	0.01 g
Warm distilled water	10 ml

Dip wet slides into the solution, air dry, and store covered in the refrigerator until ready to use.

E. Primary Antibodies

Commercial antibodies that react with plant antigens are available from a variety of sources. If there is any question about reactivity of a particular primary antibody with the antigen of interest in plant specimens, an immunoblot of plant proteins can help to answer it. Examples of cytoskeletal protein antibodies that we have found useful for plant studies include monoclonal antibodies to α- and β-tubulin from Amersham (Arlington Heights, IL) (N.356 and N.357) and monoclonal anti-actin C4 from Boehringer Mannheim or ICN. All of these are IgG antibodies. Freeze antibodies in small aliquots at $-20°C$ or colder (see comment in Section II,C on freezers). Once it has been thawed, undiluted antibody is usually stable for at least a few months and sometimes for a year or more. Diluted antibodies often can be used for a month or more without noticeable loss of activity.

The appropriate dilution to use for cellular immunolabeling is specific to each antibody, and may range from about 1/50 to as high as 1/20,000. Testing a series of 10-fold dilutions is a useful way to determine the concentration that gives the highest ratio of specific staining to background fluorescence for a given specimen. We dilute antibodies into phosphate-buffered saline (PBS) that contains 0.02% NaN_3 plus 1% bovine serum albumin (BSA; added as a 10% stock solution stored frozen). By raising the protein concentration, added BSA helps to stabilize diluted antibody solutions against aggregation and also serves as a blocking agent to help prevent nonspecific binding of antibodies to cellular components during the labeling step. The working solution of antibody usually also contains 0.05–0.1% Triton X-100 or a similar detergent, which helps to eliminate nonspecific antibody binding.

PBS–NaN_3 (per liter): 8 g NaCl, 0.2 g KCl, 2.16 g $Na_2HPO_4 \cdot 7H_2O$ (or the molar equivalent of other hydrate forms), 0.2 g KH_2PO_4, 0.2 g NaN_3

The pH should be between 7.2 and 7.6, and can be adjusted with KOH if necessary. NaN_3 is an effective bacteriostat, and this solution can therefore be stored at room temperature. For any applications that involve using antibodies in living cells, the NaN_3 must be omitted.

F. Fluorescent, Biotin-Labeled, or Gold-Labeled Secondary or Tertiary Antibodies

Commercially available secondary/tertiary antibodies normally are supplied with guidelines indicating the typical working dilution, which is often within the range of 1/20 to 1/100 for fluorescently labeled antibodies, 1/50 to 1/500 for biotin labels, or 1/100 to 1/400 for antibodies conjugated to 1-nm gold particles. To optimize results for a particular sample, testing a dilution series is again necessary. These antibodies are diluted into the same solution of PBS/NaN_3 plus BSA ± detergent as are primary antibodies. Storage requirements are the same as those for primary antibodies, except that some gold-labeled antibodies cannot be frozen (see comment in Section II,C on freezers). Stability properties are often similar to those for primary antibodies, but a notable exception is the instability of some antibodies to chicken IgG: Dilutions of these need to be used the day they are made. Consult the product data sheet for specific details.

Addition of a third step further increases the amplification effect obtained with a two-step (primary and secondary antibody) procedure and can result in enhanced sensitivity and signal-to-noise ratios. Thus a three-step method may be desirable for antigens present in low concentration. Typical examples include use of biotin-labeled secondary antibodies followed by fluorescently tagged avidin or streptavidin, which react essentially irreversibly with biotin, or followed by gold-conjugated tertiary antibody against biotin. Gold labels are preferred when autofluorescence levels interfere with antigen detection via fluorescent antibodies.

G. Fluorescent Streptavidin

Fluorescent streptavidin is diluted into the same buffer as the antibodies, and typical working dilutions are within the range of 1/100 to 1/400, but again, this needs to be determined for the specific sample being examined.

H. Mounting Media

For fluorescence work, a mounting medium that retards photobleaching is highly desirable. Both of the following have been useful in our work. The first one, which contains a polyvinyl alcohol (Moviol; Hoechst-Roussel, Somerville, NJ), dries to a semipermanent mount that does not require the coverslip to be sealed, and yet can be removed by soaking in an aqueous solution. The second one remains fluid and thus requires that the coverslip be sealed with varnish before viewing with oil immersion objective lenses.

1. Moviol containing n-propyl gallate (modification of mounting medium described in Osborn and Weber, 1982): Place 6 g of analytical-grade glycerol in a 50-ml centrifuge tube. Add about 2 g of Moviol 4-88 (Hoechst-Roussel) or Elvanol 51-05 and stir thoroughly. Add 6 ml of distilled water and leave at room temperature for a few hours with occasional stirring. Add 12 ml of 0.2 M Tris (2.42 g/100 ml), pH 9–10, and heat to 50°C for 10 min with stirring. Cool to room temperature, add n-propyl gallate to a final concentration of about 2% (0.5 g/ 25 ml), and stir until no more n-propyl gallate dissolves. Remove undissolved Moviol and n-propyl gallate by centrifugation at 5000 g for 15 min at 4° and freeze the mounting medium in 0.5- to 1-ml aliquots. Thaw just before use and refreeze the remainder immediately. A vial may be repeatedly thawed and refrozen; discard when the medium becomes yellow. The ability of this medium to retard fading is graphed in Wick and Duniec (1986). Occasionally we find that a batch of this medium must be discarded because fluorescent specimens

mounted in it are unusually dim or bleach rapidly. There is some reason to suspect that this is due to contact with traces of detergent during preparation of the mounting medium.

2. Paraphenylene diamine (PPD) (0.1%) in buffered glycerol (Johnson and Araujo, 1981): To 0.025 g of PPD, add 2.5 ml of PBS–NaN_3 (Section II,E) or Tris-buffered saline (TBS)–NaN_3 containing 0.5 M $NaHCO_3$ (0.42 g/10 ml of buffer). Check with pH paper that the pH is 8.0 and adjust if necessary. Add glycerol to 25 ml, freeze in 0.5- to 1-ml aliquots, and thaw just before use. Coverslips must be ringed with varnish or nail polish before viewing on a microscope. Use care whenever handling PPD.

TBS–NaN_3 (per liter): 20 mM Tris (2.422 g), 130 mM NaCl (7.597 g), 0.05% NaN_3 (0.5 g); adjust pH to 7.4 if necessary

III. Immunofluorescence Microscopy of Cytoskeletal Elements in Isolated Root Tip Cells

The procedure outlined here summarizes refinements and modifications that have been made since our initial work on microtubules in onion root tip cells (Wick *et al.*, 1981). We have used it for localization of tubulin, selected tubulin isoforms, and actin in meristematic root tip cells of a variety of species. Examples of labeled, isolated cells are seen in Fig. 1a and b. Variations that optimize localization of Ca^{2+}-sensitive proteins such as calmodulin can be found in Section VI,D. We normally use roots of seedlings grown on moist filter paper in petri dishes in order to avoid particles of soil or potting medium, which can interfere with the squashing step. Care must be taken to avoid fungal contamination of germinating seeds, as this can result in depolymerization of cytoskeletal elements, and presumably could also disrupt the *in vivo* distribution of other cellular components. For seeds that are not treated with a fungistat, this usually requires surface sterilization of seeds with a 1/10 dilution of commercial bleach and growth under sterile conditions.

A. Fixation

If actively dividing cells are of interest, select roots that are between 1 and 2 cm long for small-seeded species or slightly longer for large-seeded species such as maize or peas. Avoid very short roots, because their cells often contain many starch grains that can severely damage cells during the squashing step. Using a scalpel or razor blade, cut the terminal 2–3 mm of the root into a small pool of fixative (Section II,A) in a flat dish. In nearly all instances, the best results are obtained with fixatives that do not include glutaraldehyde (i.e., with fixatives 1–4 of Section II,A). For ease of handling of small specimens through multiple processing steps, we then transfer the root tips into a vial of fixative that holds a cylindrical container with a wire mesh at the bottom (electron microscopy specimen preparation basket, BU 011 133-T; Baltec, Middlebury, CT) to allow exchange of fluids without loss of specimens. If possible, handle samples via pipetting rather than with tweezers throughout processing, to avoid damage. The ratio of fixative volume to sample volume should be at least 10:1. Fix small roots for 1 hr at room temperature; roots with a diameter greater than 1 mm may require 1.5–2 hr. If proteases are a concern, PMSF can be added (Section II,C).

Use a pipette to remove fixative from the space between the wall of the basket and the wall of the vial, and flush the fixative down a drain. Wash the sample for 15–30 min with the same buffer that was used to make the fixative (EGTA–PO$_4$, EGTA–borate, etc.; Section II,A), changing the wash solution two to three times during this period. If PFA–ZnCl$_2$–citrate (fixative 4) was used as the primary fixative, use EGTA–PO$_4$ for washing.

B. Cell Wall Weakening

The wall digestion step involves the greatest degree of variability from one sample to another, and may require considerable experimentation to find a combination of enzymes and digestion time that allows release of cells at the squashing step. Also, levels of proteases appear to vary from one species to another. For onion root tips, we routinely use 1–2% Onozuka R-10 cellulase (Serva) in 5 mM EGTA–0.4 M mannitol with no added protease inhibitors for 15–20 min at room temperature. Our standard regime for maize roots is 5% Onozuka R-10 cellulase plus 2% Macerozyme R-10 (Serva) in 5 mM EGTA–0.4 M mannitol for 1 hr at room temperature, also with no protease inhibitors. To speed up this step, digestion can be done at 37–40°C, cutting the time to one-third or one-fourth of that required at room temperature. After digestion, wash the samples free of enzymes, using the same buffer as was used to make fixative. The length of this wash step depends on the total fixation and wash time that has elapsed so far; the ease of cell separation during the squashing step appears to depend to some extent on how long specimens have been exposed to the chelating effects of EGTA, which presumably helps to break down Ca^{2+}–pectates of the middle lamella between adjacent cells. For onion roots that are subjected to 1 hr of fixation, a 30-min wash after fixation, and a 15- to 20-min digestion period, the wash after digestion normally needs to be at least 1 hr in order to obtain good squashes. For large roots that have had a 2-hr fixation period, washing for a few minutes to remove the enzymes and mannitol may be sufficient. A long wall digestion step (Section VI,G) is not likely to have much effect on the time needed for this wash, because the chelating ability of EGTA will be negligible at the low pH of the wall-digesting enzyme solutions.

C. Separation and Immobilization of Cells

The most convenient surfaces to use for squashing of root tips and subsequent antibody incubations with isolated cells are slides coated with a Teflon pattern that forms a series of shallow wells (Cel-Line, Newfield, NJ or Polysciences, Warrington, PA). We find that slides with a frosted end are the most useful. Slides should be rinsed in acetone and distilled water before use. When working with meristematic cells, which are nonvacuolate, we find that cells can be dried down to adhere directly to the glass with no distortion of cellular architecture that is discernable at the light microscope level when compared to cells that have been kept hydrated throughout processing. Immobilization onto uncoated glass is possible only if the humidity is relatively low.

When the humidity is high or the cells of interest are vacuolate and would collapse if allowed to dry, we use poly-L-lysine (Sigma) as an adhesive. Polylysine hydrobromide that has a molecular weight greater than 300,000 works well. A few microliters of a 1-mg/ml solution is spread in each slide well that will be used and the slide is incubated for 1 hr at room temperature in a moist chamber (such as a petri dish with saturated filter paper at the bottom) to prevent the solution from drying. Squashing of root tips to separate the cells is done between two slides; therefore, to facilitate this step it is best to coat with polylysine the wells in pairs of slides, one slide being coated in the mirror image of its partner. The polylysine solution can then be removed with a micropipette, returned to its original container, refrigerated, and reused repeatedly over the course of several months before it begins to deteriorate. Once coated, slides should be used within about 1 hr, before the charge the polylysine imparts to the slide surface is neutralized. A film of polylysine solution will remain in the wells after the bulk has been removed, and if this is allowed to dry down it can leave a thick coating that will obscure cell visualization with the microscope. To prevent this, polylysine solution remaining on the glass surface can be rinsed off briefly with a stream of distilled water or, if samples are added immediately to the wells for squashing, the buffer in which samples are carried can serve to dilute it.

The aim of the squashing step is to release individual cells, free of their cell walls. It is a procedure that some people find difficult at first, but one that can be mastered with some practice. If the various wells on a slide are to contain samples that have received different treatment or will be labeled with different concentrations of antibodies, label the frosted end of the slide with this information before beginning squashing. Use pencil for this if using fluorescent antibodies; all waterproof ink we have seen is fluorescent. If some samples are to be labeled with different primary or secondary antibodies, it is a good idea to put these on separate slides to avoid any cross-contamination during the washing steps after antibody incubations. Place one or two root tips in each of the wells of one of the slides to be used, and remove most but not all of the buffer with a syringe and hypodermic needle or with a fine-tip pipette. Lay the other slide of the pair face down on top of this one, matching up the wells. Press straight down on the upper slide until the sample flattens, then pull the slides apart without twisting or sliding them against each other, so that shear forces, which can damage cells, are not created. If when pushing down on the upper slide the root tips are rubbery and resist being squashed, this is an indication that wall digestion has not progressed far enough. In particular, the middle lamella has not been weakened enough to allow separation of the cells to either side of it. If this occurs, try including a pectinase (macerase) or higher concentrations of pectinase in the digestion step.

If the samples are successfully flattened, and if the correct amount of buffer was left in the wells before squashing, then after pulling the slides apart there will be a droplet of liquid (in which are suspended isolated cells), along with the squashed debris which contains many more cells that can be released by dragging this material around on the slide. Working quickly, before the wells dry out, pick up the squashed pieces of root tip with fine-pointed tweezers and use this debris to spread the isolated cells over the entire surface of the pairs of wells. If different wells represent different samples, it is important to remember that one slide of a pair is the mirror image of the other slide. If too much liquid was left in the wells, it will be difficult to remove cells from the debris by this swabbing technique; with less liquid present, there seems to be a greater tendency for cells to come in contact with the glass and adhere to it. If liquid was forced out of the wells during the squash, that is an indication that far too much buffer was left around the samples prior to squashing.

After spreading cells, discard all large pieces of tissue; they will act as sinks for added antibodies and can prevent proper placement of a coverslip. If polylysine-coated slides are used, and cells are to be kept hydrated, add a droplet of buffer immediately to each well containing cells and let the slides sit undisturbed for 30–60 min to allow cells to settle onto the polylysine. This may need to be done in a humidified chamber to prevent drying. However, if uncoated slides are used, the wells must be allowed to dry out completely to allow cells to adhere to the glass; when dry, there will be a dull, opaque film of cells and dried buffer in the wells. This may take as little as 10 min in a dry atmosphere. It is also possible to dry cells overnight in a refrigerator or for 20–30 min in a dry, 30–40°C incubator. We have been able to store dry slides in the refrigerator for a week or more without significant loss of tubulin antigenicity, but other antigens should be tested for their stability on storage before attempting this. We have found that optimal labeling for calmodulin is obtained only if the entire processing, from fixation to antibody incubation, is done in a single day.

Extremely thin roots, such as those of *Arabidopsis,* are not effectively squashed by using slides with wells formed by a Teflon mask. For these samples, we squash roots between two cleaned plain slides (coated with polylysine, if desired). To have multiple samples on each slide and to prevent antibody solutions from creeping across large areas of the slide (or even off the slide), we use a PAP pen (Research Products International, Mount Prospect, IL) to draw a waterproof, nonfluorescent ring around each group of cells after roots have been squashed, thus defining shallow wells on the slide. The tip of this pen is fairly broad, so it is advisable to leave at least 1 cm of space between adjacent groups of cells on a slide. If the ease of handling Teflon-welled slides is preferred, it may also be possible to use a combination of one of these slides and one plain one for the squashing step.

Before application of antibodies, it is helpful to examine the slides to determine if intact cells were released during the squashing step. We normally do this at high magnification on a dissecting microscope, both with hydrated cells and with dried cells, without adding a coverslip to the slides. Intact root cells look like boxes or rectangular cushions, whereas free nuclei from ruptured cells are spherical. By using one or two onion root tips per pair of wells, we typically end up with a few hundred intact cells per well and sometimes a few free nuclei. If the proportion of ruptured to intact cells is high, this may indicate that there were twisting forces during the squash, or that the subsequent spreading of cells was done too vigorously or in the presence of too little liquid.

It may also be desirable to mount a sample in water and check it on a compound microscope, using phase-contrast optics, to determine if cell walls remain. (Optimally, cells pop out of weakened cell walls during the squashing step.) In experiments to determine wall digestion conditions necessary to obtain isolated cells of wheat, rye, and maize root tips, we found cells that displayed poor immunofluorescence contrast and others that showed uniform bright fluorescence after antibody application, even though examination of unstained cells showed low autofluorescence levels. On examination in phase-contrast mode, a bright refractile ring was seen around those cells that showed high background fluorescence. We interpret this to mean that, although the middle lamella had been weakened enough to allow cell separation during squashing, individual cells retained at least part of the cellulose and hemicellulose components of the cell wall, and this was interfering with antibody penetration into the cell and subsequent removal from the cell wall during wash steps. A higher concentration of cellulase or longer digestion time is called for if this is seen. If cells appear to be free of wall material, the coverslip can be floated off with an excess of water and the sample can be immunolabeled.

D. Indirect Immunofluorescence Labeling

Indirect immunofluorescence labeling can be done either with a fluorochrome-labeled secondary antibody that reacts with the primary antibody, or with a biotin-labeled secondary antibody followed by fluorochrome-labeled streptavidin. We have found the latter method to be more sensitive than the former. When purchasing secondary antibodies, not only must the species of animal from which the primary antibody is derived be specified, but whether the primary antibody is an IgG or an IgM must be specified as well. A secondary antibody supplied as an IgG (about 150–160 kDa) will be much easier to work with than an IgM about 600 kDa. If, in the interest of increasing the speed of penetration into a specimen a reagent somewhat smaller than the intact IgG molecule is preferable, it is also possible to purchase secondary antibodies that consist of a label attached to F(ab)$_2$ fragments of IgG.

The most commonly used fluorochromes are fluorescein, which fluoresces in the green to yellow range, and rhodamine and Texas Red, both of which fluoresce red. Choice of fluorochrome depends on the color of any autofluorescence in the sample of interest. Some people have a harder time seeing red fluorescence than green at the microscope, especially if not fully dark adapted; therefore if there is no strong green autofluorescence, fluorescein might be preferable to use as the fluorochrome. Antibodies conjugated to blue fluorochromes, which may circumvent some autofluorescence problems, are also available: 7-amino-4-methylcoumarin-3-acetic acid (AMCA) (available from Jackson ImmunoResearch, West Grove, PA and Molecular Probes, Eugene, OR) and Cascade Blue (Molecular Probes). For special applications, high-sensitivity, bright fluorophores such as Cy3 (Jackson ImmunoResearch), BODIPY (Molecular Probes), and phycoerythrins (Jackson ImmunoResearch and Molecular Probes) are also a possibility.

We normally do antibody and streptavidin incubations in a 37°C incubator, with slides propped above saturated paper in a closed container. It is critical that antibodies and streptavidin do not dry down at any stage of their application or incubation, because this will result in strong background fluorescence. Place the slides on short glass rods or pieces of wooden sticks above the wet paper before adding solutions, work quickly while pipetting, and cover the container immediately. For each incubation, add 3–8 μl of the appropriate solution, depending on the size of the well to be filled. For a typical experiment we use a 1/200 to 1/500 dilution of mouse monoclonal antitubulin TU-01 (Czechoslovak Academy of Sciences, Prague) or a 1/1000 dilution of Amersham monoclonal anti-α- or anti-β-tubulin. We use the pipette tip to spread out the added droplet (without scraping the bottom of the well) so that it covers the entire surface of the well. We normally incubate slides with the primary antibody for 30–45 min, although if time is at a premium 20 min is probably sufficient for most samples. Some people recommend incubating at 4°C overnight, but with the antibody and specimen combinations that we have used this usually results in unacceptably high background fluorescence.

To wash away excess antibody, place slides in a small beaker of PBS–NaN$_3$ (Section II,E) for about 10 min. Two slides may be placed back to back in a single beaker, but slides incubated with different antibodies should be washed in separate beakers. In spite of the enormous dilution of antibodies that occurs during the wash steps, we have documented cross-contamination by antibodies from one slide to another when they are washed together. Likewise, if different wells on a slide have been exposed to different concentrations of the same antibody, exercise some precautions during the wash step. In this case, we usually find it sufficient to go through one or two cycles of adding a large drop of

PBS–NaN$_3$ to each well and then drawing this off with a pipette or filter paper before placing the slides into the wash beaker.

At the end of the wash period, place the slides back into the incubation container and remove any liquid on the front surface of the slide with small wedges of filter paper. A thin film of buffer may be left in the wells that contain cells. Add 3–8 μl of the secondary antibody and incubate as for the primary antibody. For experiments in which the primary antibody was made in mouse, the secondary antibody might be a 1/50 dilution of fluorescein-labeled goat anti-mouse IgG (F-GAM) or a 1/100 dilution of biotin-labeled goat anti-mouse IgG (B-GAM). After incubation, wash off the excess antibody as described above. If the secondary antibody was fluorescently tagged, slides are ready for mounting at this stage: wash off the back of each slide with distilled water for a few seconds to remove buffer salts that interfere with transmitted light visualization of the cells, and remove excess liquid from the face of each slide. If cells are to remain hydrated, remove excess liquid from the face of the slide with filter paper, as above, and proceed with mounting. If dehydration of cells is not a problem, slides can be propped upright against a beaker to drain.

If the secondary antibody was biotin tagged, a third incubation step with fluorescently tagged avidin or streptavidin is required, for example, a 1/200 dilution of rhodamine-labeled streptavidin. This incubation needs to be only about 15 min long, after which slides are again washed in PBS–NaN$_3$, the back of each is washed with distilled water, and excess liquid is removed from the face.

E. Mounting and Observation of Cells

Whichever mounting medium is used, it is important to avoid bubbles. This is done relatively easily at the time that the coverslip is added if the medium is applied with a wooden stick rather than with a pipette, and if the coverslip is lowered into place slowly, from one edge to the opposite edge. The volume of Moviol mounting medium decreases as the medium hardens, so if too little is used bubbles can also form at this time. However, if too much medium is used, it will be difficult to make the coverslip lie flat and the optics will be suboptimal. If there is excess mounting medium around the edge of the coverslip, remove with filter paper before allowing the Moviol mounting medium to harden or before sealing the edges of a coverslip mounted with a glycerol-based medium. Moviol-based mounting medium will take from a few hours to overnight to harden, but samples can be viewed soon after mounting if care is taken to prevent the mixing of immersion oil with any nonsolidified medium at the edges of the coverslip. With either mounting medium, we normally wait at least 15–20 min to allow the antifade agent (n-propyl gallate or PPD) to diffuse into the cell and make contact with the fluorescent molecules it is protecting.

A typical fluorescence filter set for observing fluorescein-labeled material consists of an exciter filter that transmits blue light in the 450- to 490-nm range and a barrier filter that transmits green and other longer wavelength light above about 520 nm. Older versions of fluorescein filter sets sometimes have an exciter filter that transmits light below 450 nm. Depending on the sample to be observed, this may not be optimal due to autofluorescence induced by the 435-nm emission peak of mercury vapor lamps. If eliminating red fluorescence due to chlorophyll or other autofluorescing molecules is important, a fluorescein filter set that contains a bandpass barrier filter will be needed, such as one that transmits only in the 515- to 565-nm range. Depending on the microscope setup, it is sometimes possible to create the same effect by adding a red cutoff filter (i.e., one that eliminates light longer than, e.g., 580 or 590 nm) into the light path after the standard fluorescein barrier filter. Filter sets used for rhodamine and Texas Red routinely have an exciter filter that transmits green light in a narrow bandpass centered around 546 nm, and a barrier filter that transmits red light of wavelengths longer than 590 nm.

With epifluorescence optics, the brightest fluorescence images will be seen with oil immersion lenses of high numerical aperture and fairly low magnification (i.e., ×40 rather than ×100). Highly corrected lenses contain more lens elements than simpler, less corrected lenses and, because light is lost at each lens-to-air interface inside the lens, a highly corrected lens will tend to give a dimmer fluorescence image than its simpler (and usually less expensive) counterpart of the same magnification. Because fluorescence work often involves observation of a single fluorochrome at a time, there is little or no sacrifice of optical quality involved in using a lens that is not color corrected. Even in the presence of antifade agents, some photobleaching can be expected after exposure to the intense beam of light that is transmitted by a high numerical aperture objective lens, so it is good practice to close off the light impinging on the microscope stage whenever the specimen is not being observed directly.

The use of fairly low magnification objective lenses that give the brightest image usually means that the cells of interest occupy only a fraction of the field of view, and considerable enlargement of the image is often required for routine working prints and publication prints. Because of the numbers of prints we produce, we routinely use black-and-white film for recording. For most applications, Kodak (Rochester, NY) T-Max 400 film exposed at 1600 ASA for fluorescein and developed in T-Max developer gives good results. For fine detail, Hypertech film (Microfluor, Stony Brook, NY) exposed at 800 ASA for fluorescein and developed in Kodak D-19 is recommended. Many microscope camera light meters are relatively insensitive to red light, and their use as a guide for correct film exposure to red light results in drastic overexposure of the negative. Therefore, when working with rhodamine or Texas Red fluorochromes, we typically use ASA settings of 3200 for both T-Max and Hypertech film and develop them as usual.

After viewing, always clean the immersion oil from the coverslip with a tissue and then with a small amount of acetone. (Besides making the slides harder to handle, oil that is left can creep under the edge of a coverslip mounted with Moviol, soften the mounting medium, and ultimately dissociate antibodies from their antigens.) Cleaned slides stored in the dark in a refrigerator can show stable fluorescence for many years.

If, after observation of a specimen mounted in Moviol-based medium, it is decided that further staining or other manipulation is to be done, or if air bubbles have formed during hardening in critical places over the specimens, the coverslip can be removed by soaking the slide in PBS until the coverslip floats free. All traces of immersion oil must be removed before doing this, or a film of oil coating the specimen could result. Never try to drag the coverslip off, because the shear forces will damage the cells.

IV. Preparation of Leaf Epidermal Sheets for Immunolabeling

Preparation of leaf epidermal sheets is somewhat more complicated than the procedure for isolating cells from roots or other plant organs, and successful application involves mastery of some delicate sample manipulation skills, particularly in the immobilization step and in the removal of extra tissue layers. We have applied the procedure described here to primary leaves of rye, wheat, barley, and maize (Wick *et al.*, 1989; Cho and Wick, 1989, 1990), which are furled within a coleoptile, but variations of the procedure have also been used to examine epidermal and subepidermal cortical cells in leaves of other plants (Wick *et al.*, 1989) and for other organs such as the shoot apex (Marc and Hackett, 1989).

A. Fixation and Immobilization of Tissue

The fixative we use contains PFA with or without glutaraldehyde in PMEG buffer (Section II,A, fixative 5), although for microtubule localization similar results can be obtained with other fixatives as well. To find cells that are actively dividing, use seedlings in which the shoot is 1 cm long or slightly longer. In a flat dish that contains a small amount of fixative, use a scalpel or razor blade to excise the shoot from the seed close to the base and remove the coleoptile. Fix (for 1 hr at room temperature) the 5-mm segment of primary leaves from the region at the base of the shoot, which contains the meristematic region. Because leaves that remain tightly rolled up are difficult to handle at the immobilization step, it is useful to cut segments into narrow strips at this stage. As with root tips, samples can be handled easily if they are collected into a cylindrical

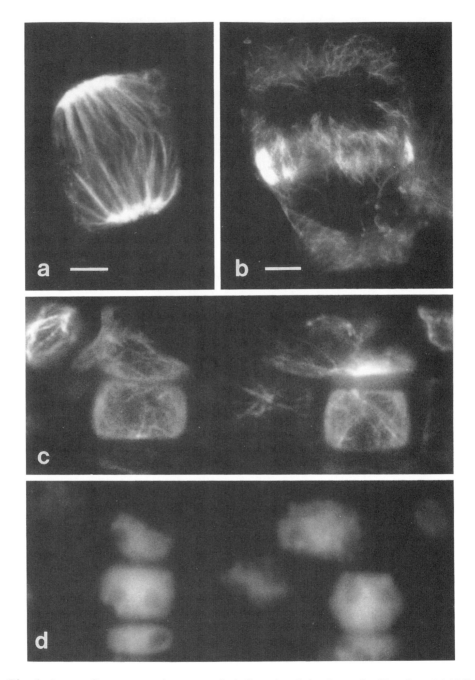

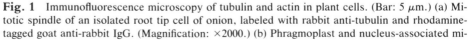

Fig. 1 Immunofluorescence microscopy of tubulin and actin in plant cells. (Bar: 5 μm.) (a) Mitotic spindle of an isolated root tip cell of onion, labeled with rabbit anti-tubulin and rhodamine-tagged goat anti-rabbit IgG. (Magnification: ×2000.) (b) Phragmoplast and nucleus-associated mi-

"basket" with a mesh bottom to allow exchange of liquids (Section III,A). If the sample processing is to begin on a certain day and be finished the next day (e.g., if samples require several hours of drug treatment or other manipulation before they are ready for fixation), this is a possible stopping point. We have found that if, after the first hour of fixation, we store samples in one-quarter strength fixative overnight at room temperature before continuing their processing, visualization of the cytoskeleton is still possible the following day. After fixation, wash in $1\times$ PMEG for 1 hr.

Extensive amounts of cuticular waxes at the epidermal surface may interfere with immobilization of samples onto a slide; these samples may be immersed in acetone at $-20°$C for 10–20 min at this point. Conformational changes that occur during denaturation with organic solvents may render some antigens unrecognizable by certain antibodies; if there is any reason to suspect this, skip this step. After dewaxing specimens, wash them briefly in distilled water.

Place leaf segments into 4% glycerol for 15–20 min. Glycerol that penetrates the cell wall at this stage appears to help prevent glue from penetrating through the wall and into the cytoplasm at the immobilization step. While working at a dissecting microscope, blot excess glycerol from a leaf segment with a piece of moist laboratory tissue and place the sample, epidermis side down, into the well of a slide (See Section III,C) that has been cleaned with acetone and distilled water. Take care that the sample lies flat, without folds, twisting, or buckling. Add a small droplet of cyanoacrylate glue (such as Krazy Glue) and use a sharply tapered wooden stick or the head of a pin fastened to the end of a stick to draw the glue quickly along the length of the leaf segment before the glue hardens and the sample dehydrates. If the samples are likely to lie flat as soon as they are placed in the well, try adding the glue first and then the specimen, as a way to lessen the amount of time the sample must dry out. It is also possible to tack down just the ends of the sample with glue, but great care must then be used in subsequent steps to avoid dislodging or damaging it.

If too much glue is used, the segment may not lie flat on the surface of the slide, or excess glue may infiltrate into the cells and harden, preventing subsequent entry of antibodies. Immediately add water or PBS to the samples once the glue has dried and wait about 20 min to allow the glue to harden completely. Use a tube of glue that has been opened within the last few months, store in the

crotubules of a cytokinetic onion cell, labeled with mouse monoclonal anti-tubulin and fluorescein-tagged rabbit anti-mouse IgG. The root tip from which this cell was isolated was subjected to a 2.5-hr cell wall digestion period in the presence of protease inhibitors. (Magnification: ×1800.) (c and d) Sheet of epidermal cells from a rye leaf, in the region where stomatal complexes are forming. (c) is labeled with mouse monoclonal anti-actin and fluorescein-tagged rabbit anti-mouse IgG. (d) is the corresponding view of the same cells stained with the DNA indicator dye Hoechst 33258. (Magnification: ×1800.) (Courtesy of Soon-Ok Cho, University of Minnesota, St. Paul.)

refrigerator between uses, and allow the glue to warm to room temperature before dispensing it. If samples do not adhere well, consider trying a fresh tube of glue.

B. Removal of Extraneous Tissue Layers prior to Labeling

The first step in removing extra tissue involves carefully shaving off cell layers above the glued-down epidermis. This is done at the dissecting microscope with water covering the specimen. We often use a surgical razor blade holder and sharp microblades (Fine Science Tools, Bellingham, WA), but a less expensive substitute is a small chip of a double-edged razor blade glued to a stick or clamped into a scalpel blade holder with a narrow-diameter tip, which makes it possible to slice with the blade at a relatively flat angle.

At this point, specimens need to be subjected to cell wall digestion so that the remaining extra cell layers can be removed. We digest rye leaf samples for 1 hr at room temperature or 30 min at 37°C in 1% Onozuka R-10 cellulase plus 1% Macerozyme R-10 (both from Serva) in 0.3–0.4 M mannitol containing 10 mM EDTA, 10 μg/ml (20 μM) leupeptin, 7 μg/ml (10 μM) pepstatin, 1 mM PMSF, 1% BSA, and 0.05% Triton X-100. We digest maize leaves for about 20 min at 37°C in a solution containing 5% cellulase plus 2% Macerozyme in 0.3 mM mannitol, with all other components as for rye. With slides in a humidified chamber, the digestion solution can be added as a large drop covering each well that contains a sample. After washing the slides briefly with distilled water or PBS, return to the dissecting microscope and use a fine-tipped camel hair paint brush to carefully remove unwanted cell layers interior to the epidermis. If cells are difficult to remove, the degree of wall digestion needs to be increased.

At this stage, we plunge the slides into methanol at −20°C for 10 min in an attempt to rapidly stop the activity of any remaining wall-digesting enzymes, and then place the slides in PBS at room temperature for 10 min. If the antibody to be used recognizes only denatured antigen (Section VI,F), the methanol step is necessary unless a cold acetone step for dewaxing specimens has been used before glueing. If glutaraldehyde was included in the fixative, samples are now exposed to 0.1 M glycine (0.7507 g/100 ml) for 10 min to help block any remaining reactive groups on the glutaraldehyde, which might lead to nonspecific sticking of antibodies, and are again washed in PBS for 10 min. When applying this procedure to other systems or with other antibodies, it is recommended that one check to determine if these steps are necessary because, in general, the simplest and quickest handling procedures are likely to result in the best preservation of antigens.

Antibody incubations, slide mounting, observation, and photography procedures are the same as for isolated cells (Section III,E).

V. Cryosectioning for Light Microscopy Immunolabeling

The first step in cryosectioning is to obtain well-fixed tissue. For aerial organs, a low concentration (0.01–0.1%) of a detergent such as Triton X-100 may be needed to obtain penetration of the fixative; in addition, gentle vacuum at early stages of fixation may be helpful. Fixatives without glutaraldehyde (Section II,A) are the best starting point, and use of one containing picric acid, which turns samples yellow, will make it easier to find and section the specimen once it is frozen into an opaque white block of embedding compound. For samples that are no more than 1 mm in diameter, 1 hr of fixation is probably sufficient. After a brief rinse in buffer, the samples may be infiltrated into Tissue Tek OCT compound, or 1.5–2.0 M sucrose, or a 2:1 ratio of 20% sucrose and Tissue Tek (Barthel and Raymond, 1990). Sucrose seems to help support the tissue and can make sectioning easier, especially if the sections are to be thinner than 8–10 μm. We use the sucrose–Tissue Tek combination and usually do a gradual infiltration over the course of 2–3 hr, starting with a roughly 20–30% concentration of embedding medium in buffer and ending with 100% fresh embedding medium for 1 hr.

At this stage material should be frozen in the cryostat without letting it sit for many extra hours, because the relatively mild fixation attained with PFA will eventually allow deterioration of the cytoplasm. We have stored some frozen specimens in liquid nitrogen for several weeks before sectioning, and have found that antigenicity was retained. In fact, for some samples, a few weeks in liquid nitrogen actually improves the ease of sectioning and the appearance of immunolabeled microtubules (J. Hush, personal communication).

We generally cut sections at about 5–10 μm and collect them on slides or coverslips that either have been subbed (Section II,D) or treated with poly-L-lysine (Section III,C). Encircling the sections with a PAP pen (Section III,C) will reduce problems with antibodies spreading beyond the region occupied by the sections. After a brief wash with PBS to remove embedding medium from the sections, antibody application is as for isolated cells (Section III,D), except that antibody incubations as short as 20 min may be possible. Also, greater gentleness may be required to avoid losing sections during the washing steps. Mounting of slides, observation with epifluorescence microscopy, and photography are performed according to the procedures outlined in Section III,E.

If the use of a cryostat is not possible, but the advantages afforded by sectioned material is still desired, polyethylene glycol (PEG) (Van Lammeren et al., 1985; Holubowicz and Goffinet, 1988; McKerracher and Heath, 1986), diethylene glycol distearate (McKerracher and Heath, 1986), and butyl-methyl methacrylate (Gubler, 1989) can be used for embedment prior to sectioning for light microscopy. Our experience with PEG is that it is difficult to section, with sections often curling severely as they come off the knife; they are also prone to fly away due to static electricity. McKerracher and Heath (1986) concluded that neither PEG nor diethylene glycol distearate gave adequate preservation of microtubules at the electron microscope level, and even the light microscopy images obtained with PEG may not be optimal according to Gubler (1989). Whereas PEG is removed from cut sections with water, the other two reversible embedding media are removed from sections with organic solvents, so denaturation of antigens and lack of recognition by an antibody that was raised against the native antigen may be a problem.

VI. Troubleshooting/Variations on Protocols

A. Glutaraldehyde Fixation

Whenever possible, we avoid the use of glutaraldehyde with plant specimens because of its tendency to increase cytoplasmic autofluorescence (which becomes even more pronounced on storage of cells fixed in the presence of glutaraldehyde), because it tends to make cytoplasm "sticky," thus increasing background fluorescence (i.e., besides autofluorescence, the fluorescence observed when only fluorescently labeled secondary antibodies or biotin-labeled secondary antibodies followed by fluorescent streptavidin have been used, without any primary antibody) (Wick and Duniec, 1986), because it can render some antigens unrecognizable by antibodies that are able to interact with the same antigens when they have not been exposed to glutaraldehyde, and because it can cross-link (fix) the cytoplasm too well, so that antibody penetration is impeded. However, there are circumstances in which the superior fixation achieved with glutaraldehyde is desirable. In this case, use of the lowest effective concentration, which may be as little as 0.1–0.25% glutaraldehyde, is recommended.

Depending on the type of antigen to be localized, it may be useful to extract some of the soluble cytoplasm with detergent (Section VI,B) to try to lower cytoplasmic autofluorescence, as is done during processing of sheets of epidermal cells (Section IV,B). Also, the use of $NaBH_4$ (sodium borohydride) after cells are immobilized may reduce some of the autofluorescence. Treatment consists of three successive additions (3 min each) of a solution of 0.05% $NaBH_4$ in PBS. Add the $NaBH_4$ just at the point of adding the solution to the wells of the slides. Because of the violent bubbling that will occur on the slides, it is best if isolated cells are attached with polylysine. Some cytoplasmic damage may occur during this treatment.

If glutaraldehyde has been used as a fixative, it is particularly important to include BSA in antibody solutions to help prevent nonspecific sticking of antibodies to cellular components. If the background fluorescence that results from nonspecific sticking of antibodies appears to be a major problem (instead of or in addition to any problems with autofluorescence), include a blocking step before antibody application. We have found that incubation at 37°C for 30 min with

1–10% BSA or normal serum from the species in which the second antibody was raised (or from any species other than the one from which the first antibody was obtained) works well for blocking. A brief incubation in 0.1 M glycine can also help to block reactive aldehyde groups (Section IV,B).

B. Extraction of Cytoplasm to Reduce Autofluorescence

Some autofluorescence is species specific, whereas in some instances there is reason to believe it is also growth condition specific: our experience with the water fern *Azolla* indicated that root tips of plants growing relatively slowly at cool temperatures had much higher autofluorescence levels than those growing more quickly at warmer temperatures. There are procedures that can help to reduce autofluorescence. If not trying to localize small, soluble antigens, addition of detergent during fixation, wall digestion, or all processing steps including antibody incubation steps can often be useful. We have used up to 1% Triton X-100 or Nonidet P-40 and up to 0.2% saponin for this purpose (Wick and Duniec, 1986). An investigation of maize root cells revealed significant autofluorescence of chromatin when viewed with fluorescein filters, but we were able to eliminate this by incubation of cells in methanol at −20°C for a few minutes. Extraction of some chlorophyll with organic solvents after fixation may also be a possibility. As mentioned above (Section VI,A), treatment with $NaBH_4$ can diminish some types of autofluorescence. We have also tried fixation in the presence of caffeine as a way to precipitate highly fluorescent phenolic products, but have found this to be ineffective in reducing autofluorescence.

Autofluorescence should not be confused with background fluorescence that results from antibodies binding nonspecifically to samples (Section VI,A).

C. Immunogold Labeling for Light Microscopy

In situations in which autofluorescence is so severe as to make immunofluorescent labeling undesirable or impossible, immunogold labeling of walled plant cells for light microscopy is an attractive option. Although antibodies conjugated to 5-nm gold particles (Satiat-Jeunemaitre, 1990) and even 20-nm gold (S. M. Wick, unpublished observations) will penetrate plant cells for light microscopy observation, the recently available 1-nm gold–antibody conjugates, followed by silver enhancement of the gold, hold the most promise for widespread usefulness. According to data supplied by one manufacturer (Goldmark Biologicals, Phillipsburg, NJ), 5-nm gold conjugates have a molecular weight of about 1 million, whereas 1-nm gold conjugates have only slightly greater mass than an unconjugated antibody molecule, and can give greater labeling intensity than larger conjugates.

We have done preliminary work to visualize microtubule arrays in onion root tip cells by using the following immunogold labeling/silver enhancement scheme and our normal handling procedures for obtaining isolated cells (Section III):

Blocking in undiluted, heat-inactivated normal goat serum, 15 min at room temperature.

1/500 dilution of mouse monoclonal anti-tubulin TU-01

1/250 dilution of biotinylated sheep IgG anti-mouse IgG (B-SAM) (RPN.1001; Amersham)

1/200 dilution of goat anti-biotin conjugated to 1-nm gold (Au$_1$-GAB) (BioCell gold conjugates; available from Goldmark Biologicals)

Slides are then washed five times (1 min each) in deionized, distilled water. It is critical to remove all ions, especially chloride ions, before silver enhancement, or there will be spurious development of silver grains. Equal amounts of initiator and enhancer (both components of the BioCell silver enhancing kit; Goldmark) are mixed on a clean surface and applied as large drops to microscope wells containing cells. Results obtained to date suggest a 15-min enhancement period is adequate, but microtubule images are faint and further testing is needed to optimize the antibody concentrations and enhancement conditions. Polylysine-coated or subbed slides are considered incompatible with silver enhancement procedures because of their ionic charge, although it may be possible to neutralize this before proceeding with silver enhancement.

D. Antigens Sensitive to Calcium Ions

Use of fixatives and other solutions that contain Ca^{2+} chelators, such as EGTA, can make it difficult to localize calmodulin and other proteins whose conformation and cellular location may require the presence of Ca^{2+}. We have found it useful to add micromolar Ca^{2+} to the fixative to help stabilize these proteins. EGTA apparently helps to dissolve the Ca^{2+} pectates of the middle lamella between adjacent cells; therefore, when EGTA is omitted from fixative and buffer solutions, we find that a longer wall digestion step is necessary, typically about three times longer than if EGTA is present (Wick and Duniec, 1986). See Section VI,G for precautions to take against protease activity during this longer digestion period.

E. Antigens Sensitive to Aldehyde Fixation

Actin microfilaments can be depolymerized by aldehydes and they are particularly sensitive to the action of glutaraldehyde. We find that immunolabeling of actin is greatly facilitated by pretreatment of samples with the protein cross-linking agent m-maleimidobenzoyl-N-hydroxysuccinimide ester (MBS; Pierce, Rockford, IL). A 0.1 M stock solution of MBS in DMSO (3.142 mg/0.1 ml) is diluted 1/1000 in PIPES buffer containing 0.05% Triton X-100 to give 0.1 mM

MBS, in which samples are pretreated for 1 hr before fixation with PFA (Cho and Wick, 1990). Apparently the MBS prefixes the actin well enough so that it can withstand the effects of mild aldehyde fixation. If glutaraldehyde fixation is desired, it probably is best to wait until the PFA fixation has proceeded for at least 30 min before adding a low concentration of glutaraldehyde.

If the specimen can be manipulated in such a way as to permit examination without aldehyde fixation (such as examination of a single cell layer at the edge of a paradermal slice), a more detailed image of actin distribution can sometimes be obtained with rhodamine-labeled phalloidin after MBS pretreatment (but no aldehyde fixation) rather than with antibodies to actin. Without a fixation step, however, the cytoplasm in general is not stable, and specimens probably will need to be examined and photographed immediately after application of the labeled phalloidin.

F. Antibodies That Require Denaturation of Antigen with Organic Solvents

Some antibodies, such as C4 monoclonal antibody for actin, are unable to react with their corresponding antigen when it is in its native conformation. Presumably this is a result of using denatured antigen as the immunogen when raising the antibodies. Exposure of samples to acetone or methanol at $-20°C$ for 10–20 min is an effective way to denature proteins to allow immunolabeling. If samples are glued down with Krazy Glue, avoid the use of acetone, which will redissolve the glue.

G. Specimens That Require More Extensive Cell Wall Digestion

If a prolonged wall digestion period is necessary to attain cell separation at the squashing step, addition of protease inhibitors and/or 1% BSA, which serves as a substrate for proteases (and thereby protects cellular proteins from protease action), is recommended. We have digested onion root tips for as long as 16 hr at room temperature in the presence of PMSF and leupeptin and found that, although cells are very fragile after this treatment, the microtubules are remarkably stable. Nearly all samples we have handled yield to squashing after a much shorter period of digestion than this, usually within 1 hr.

H. Double Labeling

Often there is interest in determining the relative locations of two antigens within the same cell; if so, double-labeling procedures are in order. This is achieved most readily if the two primary antibodies to be used are from different animal species, such as mouse and rabbit. In this case, unless there is reason to think that the binding of one antibody to its antigen will interfere with binding of the other antibody, both primary antibodies can be applied simultaneously, as can both secondary antibodies. However, if it is suspected that one antigen may be internal to the other, so that antibody binding to the more external antigen may impede antibody penetration to the site of the internal antigen, optimal results may depend on successive application of the two primary antibodies, with the one for the innermost antigen being used first.

Choosing which fluorochrome to use with each secondary antibody may depend on the pattern of autofluorescence seen with the two relevant sets of fluorescence filters. For instance, if it is necessary to contend with autofluorescence that appears uniform throughout the cell when observed with fluorescein filters, and one of the antigens of interest is a fibrous protein whereas the other has a more amorphous form or is diffusely distributed through the cell, you might consider using fluorescein-labeled antibodies to localize the fibrous protein and a red fluorochrome for the other antigen, because fibers will be more readily discerned against the autofluorescence than will a diffuse pattern. Of course, if other fluorochromes and filter sets are available that allow one to avoid autofluorescence wavelengths altogether, that is the optimum situation. For instance, to label material that has strong red autofluorescence due to chloroplasts but does not have other significant autofluorescence, double labeling with a blue-emitting fluorochrome such as AMCA and a fluorescein conjugate might be possible.

Recording with black-and-white photography the information on double-labeled slides typically involves taking two separate pictures of the same field of view, each with one of the two pertinent filter sets, and then mounting prints side by side for comparison, sometimes along with a fluorescence image of the nuclei or a transmitted light image showing cell outlines or other features. Superposition of the images obtained with two fluorochromes, such as fluorescein and rhodamine, onto a single color slide or print is easily achieved by taking a double exposure, one with each filter set (Color Plate 2). However, it is also possible to excite both fluorescein and red light-emitting phycoerythrin conjugates with a standard fluorescein exciter filter; therefore if a bandpass barrier filter for fluorescein is not in place, (i.e., provided that a barrier filter that eliminates transmission of red light is not in place), both of these fluorochromes can be visualized simultaneously in double-label experiments (Jackson ImmunoResearch, technical data).

VII. Perspectives

Immunolabeling techniques have opened a new door to the study of plant cells, allowing visualization at the light microscope level of some components that otherwise can be seen only with electron microscopy, and rapid assessment of large numbers of cells. Almost all immunolabeling work of the past decade on plants has involved immunofluorescence microscopy with standard epifluorescence microscopes. Plant immunocytochemistry and immunohistochemistry of the future hold the promise of an increasing range of options. The availability of 1-nm gold particles conjugated to antibodies may provide a way to immunolabel plant specimens that have heretofore proved to be recalcitrant due to autofluorescence. Also, the upsurge of confocal microscopy and computer programs for reconstruction and rotation will allow a more detailed examination of the three-dimensional aspects of the distribution of an antigen and interaction with other cellular components. It is likely that we will see more attention paid in future to the production and application of antibodies tailor made to detect plant antigens, and the beginnings of the use of antibodies to investigate the function of certain components in plant cells.

Acknowledgments

I thank past and present colleagues who taught me and developed with me techniques for applying the power of antibodies to the study of plant cell development: Mary Osborn and Klaus Weber of the Max Planck Institute (Göttingen, Germany), Brian Gunning and Jadwiga Duniec of the Australian National University, Canberra, and Soon-Ok Cho and Russell Goddard of the University of Minnesota, St. Paul.

References

Barthel, L. K., and Raymond, P. A. (1990). Improved method for obtaining 3-μm sections for immunocytochemistry. *J. Histochem. Cytochem.* **38**, 1383–1388.

Cho, S.-O., and Wick, S. M. (1989). Microtubule orientation during stomatal differentiation in grasses. *J. Cell Sci.* **92**, 581–594.

Cho, S.-O., and Wick, S. M. (1990). Distribution and function of actin in the developing stomatal complex of winter rye (*Secale cereale* cv. Puma). *Protoplasma* **157**, 154–164.

Flanders, D. J., Rawlins, D. J., Shaw, P. J., and Lloyd, C. W. (1990). Nucleus-associated microtubules help determine the division plane of plant epidermal cells: Avoidance of four-way junctions and the role of cell geometry. *J. Cell Biol.* **110**, 1111–1122.

Goddard, R. H., and La Claire, J. W., II (1991). Calmodulin and wound healing in the coenocytic green alga *Ernodesmis verticillata* (Kützing) Børgesen: Immunofluorescence and effects of antagonists. *Planta* **183**, 281–293.

Gubler, F. (1989). Immunofluorescence localisation of microtubules in plant root tips embedded in butyl-methyl methacrylate. *Cell Biol. Int. Rep.* **13**, 137–145.

Holubowicz, R., and Goffinet, M. C. (1988). Polyethylene glycol embedment for histological studies of bean seed testa of low moisture content. *Stain Technol.* **63**, 33–37.

Johnson, G. D., and Araujo, G. M. N. (1981). A simple method of reducing the fading of immunofluorescence during microscopy. *J. Immunol. Methods* **43**, 349–350.

Marc, J., and Hackett, W. P. (1989). A new method for immunofluorescent localization of microtubules in surface layers: Application to the shoot apical meristem of *Hedera*. *Protoplasma* **148**, 70–79.

McKerracher, L. J., and Heath, I. B. (1986). Comparison of polyethylene glycol and diethylene glycol distearate embedding methods for the preservation of fungal cytoskeletons. *J. Electron Microsc. Tech.* **4**, 347–360.

Osborn, M., and Weber, K. (1982). Immunofluorescence and immunocytochemical procedures with affinity purified antibodies: Tubulin-containing structures. *Methods Cell Biol.* **24**, 97–131.

Satiat-Jeunemaitre, B. (1990). Visualization of microtubules in walled plant cells by immunogold silver-enhancement. *Cell Biol. Int. Rep.* **14**, 1109–1118.

Tiwari, S. C., and Polito, V. S. (1988). Organization of the cytoskeleton in pollen tubes of *Pyrus communis:* A study employing conventional and freeze-substitution electron microscopy, immunofluorescence, and rhodamine–phalloidin. *Protoplasma* **147**, 100–112.

Tiwari, S. C., and Polito, V. S. (1990). The initiation and organization of microtubules in germinating pear (*Pyrus communis* L.) pollen. *Eur. J. Cell Biol.* **53**, 384–389.

Van Lammeren, A. A. M., Keijzer, C. J., Willemse, M. T. M., and Kieft, H. (1985). Structure and function of the microtubular cytoskeleton during pollen development in *Gasteria verrucosa* (Mill.) H. Duval. *Planta* **165**, 1–11.

Wick, S. M., and Duniec, J. (1986). Effects of various fixatives on the reactivity of plant cell tubulin and calmodulin in immunofluorescence microscopy. *Protoplasma* **133**, 1–18.

Wick, S. M., Seagull, R. W., Osborn, M., Weber, K., and Gunning, B. E. S. (1981). Immunofluorescence microscopy of organized microtubule arrays in structurally stabilized meristematic plant cells. *J. Cell Biol.* **89**, 685–690.

Wick, S. M., Cho, S.-O., and Mundelius, A. R. (1989). Microtubule deployment within plant tissues: Fluorescence studies of sheets of intact mesophyll and epidermal cells. *Cell Biol. Int. Rep.* **13**, 95–106.

CHAPTER 11

Antigen Localization in Fission Yeast

Caroline E. Alfa, Imelda M. Gallagher, and Jeremy S. Hyams

Department of Biology
University College London
London WC1E 6BT, England

I. Introduction

Our aim in this article is to review applications of immunolocalization methodology to the cytology and ultrastructure of fission yeast *Schizosaccharo-*

myces pombe and its distant relative *Schizosaccharomyces japonicus* var. *versatilis*. *Schizosaccharomyces pombe* has achieved prominence primarily as a model for investigations of eukaryotic cell division cycle control mechanisms (Nurse, 1990a). This work has stimulated investigations into a number of other areas of basic cell biology of the organism. What is particularly appealing about *S. pombe*, and indeed the budding yeast *Saccharomyces cerevisiae*, is its amenability to manipulation by both classic genetics and molecular biology. In the context of this volume this means that pure antigens for antibody production can be obtained by expressing the product of a cloned gene in either an *Escherichia coli* or a baculovirus expression system. The cellular expression of a particular protein can be manipulated by mutation, deletion, or overexpression of the respective gene. Proteins for which no antibody is available can be localized *in situ* by "epitope tagging" a cloned gene with a DNA sequence encoding a peptide for which specific antibodies are already available.

The Organism

Schizosaccharomyces pombe is a bacilliform, unicellular ascomycete with a length at division of ~14 μm and a width of ~3.5 μm. Diploid cells are bigger (both longer and wider) but not as large as *S. japonicus*, which offers much greater cytological clarity (Robinow and Hyams, 1989). Fission yeast grow by length extension at one or both tips; cells therefore lengthen as a function of their position in the cell cycle but they do not get fatter. The cell is bounded by a galactomannan-rich cell wall that provides useful topological markers for some types of cell cycle analysis but that complicates the application of antibodies for immunofluorescence. Mitosis occurs without breakdown of the nuclear envelope and it is difficult to distinguish the three chromosomes (this is not true in *S. japonicus*). Cytokinesis occurs by septation at the volume midpoint of the cell. A wide variety of temperature-sensitive cell division cycle (*cdc*) mutants of *S. pombe* have been isolated, some of which, like the genes encoding the M phase-specific protein kinase cdc2 (p34^{cdc2}) and its regulatory cyclin B subunit cdc13 (p63^{cdc13}), are extremely well characterized, others much less so. A classic temperature-sensitive cell cycle mutant goes through a normal division cycle at 25°C (the permissive temperature) but is arrested in cell cycle progress at 36°C (the restrictive temperature). Arrested cells continue to grow and, depending on the point in the cell cycle at which the thermolabile gene product is required, become highly elongated. A rarer class of mutations causes the cell to accelerate through the cell cycle and divide at roughly half the wild-type size. Mutants having specific defects in various aspects of mitosis and cytokinesis are also available (Yanagida, 1989).

A detailed account of fission yeast biology, genetics, and molecular biology has been published (Nasim *et al.*, 1989). The reviews of Russell and Nurse (1986), Mitchison (1990), and Nurse (1990b) also provide useful background reading. Methods for fission yeast cell and molecular biology can be found in

Moreno *et al.* (1991) and Alfa *et al.* (1993) and a detailed background on the cytology and ultrastructure of this organism may be found in Robinow and Hyams (1989) and Kanbe *et al.* (1989).

II. Immunofluorescence Microscopy

The localization of proteins in yeast cells by indirect immunofluorescence methods owes a great deal to the pioneering efforts of Adams and Pringle (1984) and Kilmartin and Adams (1984) in *S. cerevisiae*. A comprehensive review of antibody and other cytological methods in *S. cervisiae* has appeared (Pringle *et al.*, 1991) and contains much information that can be applied to *S. pombe*. Given reasonable luck, a decent antibody, and a good-quality microscope, there is no reason why any relatively abundant protein should not be detected by these techniques. The major differences between yeast and the mammalian cells for which immunofluorescence methods were originally developed is that yeast cells are not flat, they do not grow conveniently attached to a glass coverslip, and, most significantly, they are surrounded by a cell wall that must be substantially removed to allow entry of large antibody molecules. Cell wall digestion is a critical step because the enzymes employed contain proteases and are potentially highly destructive to cell structure.

The chapter first describes three immunofluorescence methods that were initially developed to localize cytoskeletal proteins in both *S. pombe* (Hagan and Hyams, 1988; Alfa *et al.*, 1989; Alfa *et al.*, 1993), and spindle pole body (Alfa *et al.*, 1990). The protocols have evolved a certain extent since their first publication (Hagan and Hyams, 1988), and no doubt they will continue to do so. They at least represent a reasonable starting point for any immunofluorescence study in this organism. The three methods involve fixation in (1) formaldehyde alone, (2) formaldehyde plus glutaraldehyde, and (3) methanol. The latter is far less labor intensive than the others but results in a certain degree of cell shrinkage and extraction of some of the proteins. It gives good preservation of spindle microtubules in *S. pombe* but cytoplasmic microtubules are relatively poorly preserved. It also gives excellent preservation of a number of cell cycle proteins. Microtubules are best preserved by both aldehydes in combination (Fig. 1). Formaldehyde alone appears to give excellent resolution of actin. Clearly, when attempting to localize an antigen for the first time it is necessary to consider a number of variations on these basic themes.

The state of the cells used will of course depend on the nature of the experiment. In most cases it is desirable to start with a culture in the midexponential phase of growth (a cell concentration of 5×10^6 cells/ml), although all three methods should work equally well for cells arrested at different phases of the cell cycle or in stationary phase. We will assume that a 50-ml culture is used in each case. Details for storing and growing fission yeast strains can be found in Alfa *et al.* (1993).

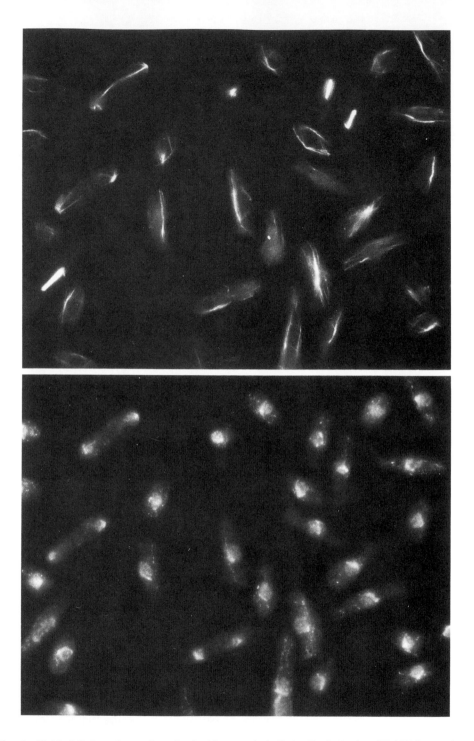

Fig. 1 Field of *S. japonicus* cells stained with an antitubulin antibody (top) and DAPI (bottom). [From Alfa, C. E., and Hyams, J. S. (1990). *J. Cell Sci.* **96,** 71–77, with permission.]

A. Formaldehyde Fixation Method: Actin Staining

Day 1

1. Count the cells of a 50-ml overnight culture by means of a hemacytometer or Coulter counter (Hialeah, FL) to check that they are actively growing. The presence of about 10% of cells with septa is indicative of an actively growing culture.

2. Prepare 10 ml of *fresh* 30% formaldehyde solution (see Section III).

3. *Fixation:* Fix the cells by addition of 6.67 ml (1 : 7.5 by volume) of 30% formaldehyde to the growing culture with agitation (final concentration, 4%). Replace the culture in the water bath and continue shaking for 60 min.

4. Harvest the cells by centrifugation [50-ml Falcon (Becton Dickinson, Oxnard, CA) tubes, benchtop centrifuge, 1000 g, 5 min].

5. Resuspend the pellet in 10 ml of PEM (see Section III,A).

6. Repeat steps 4 and 5 twice.

7. *Cell wall digestion:* Resuspend the cells to a density of $5 \times 10^7 - 1 \times 10^8$ cells/ml in PEMS (see Section III,A) containing cell wall-degradative enzymes at the following concentrations: 1 mg of Novozym 234/ml (Novo Biolabs, Denmark) and 0.3 mg of Zymolase 20T/ml (Seikagaku Co., Japan). Incubate the cells at 37°C. Observe the cells every 1 to 2 min by phase-contrast microscopy. Loss of cell wall material can be observed by a decrease in refractivity; digested cells essentially appear darker. When ~10% of the cells in the population show this appearance, *quickly* proceed to step 8 (below).

Notes on Cell Wall Digestion

The time taken to reach 10% digestion will vary with the activity of the particular batches of enzyme and the cell density. Once it has been determined for a particular case it should be fairly reproducible. It is well worth having a "dummy run" to check a new batch of enzyme. If the cells protoplast too rapidly, adjust the enzyme concentration. The reason for stopping when only 10% of the cells appear protoplasted is that this process continues to some extent during the subsequent centrifugation and wash steps.

8. Wash three times, each time with 10 ml of PEMS.

9. Resuspend the washed pellet in 10 ml of PEM containing 1% Triton X-100 and incubate for 30 sec.

10. Wash three times, each time with 10 ml of PEM.

11. Resuspend the pellet in 1 ml of PEMBAL (see Section III,A) and transfer to an Eppendorf tube a volume that will give a cell pellet of 20–30 μl when centrifuged in a microfuge operated at top speed for 2 min at room temperature (RT). Incubate for at least 30 min with continuous inversion. (The lysine in the buffer at this step binds unreacted aldehyde groups and reduces nonspecific background staining considerably.)

PEMBAL CONTAINS AZIDE, WHICH IS HIGHLY TOXIC.

12. Centrifuge and resuspend the pellet in 50–100 μl of anti-actin antibody (anti-chicken gizzard actin, mouse monoclonal N350; Amersham, Arlington Heights, IL) diluted 1/100 in PEMBAL.

13. Incubate overnight (12–16 hr) on a rotary inverter.

Day 2

14. Wash the cells three times, each time with 1 ml of PEMBAL.

15. Resuspend the pellet in 50–100 μl of rhodamine-conjugated rabbit anti-mouse secondary antibody (ICN Biomedicals, Irvine, CA) diluted 1/50 in PEMBAL.

16. Incubate overnight on a rotary inverter *in the dark* (wrap the tubes in foil).

Day 3

17. Wash three times, each time with 1 ml of PEMBAL.

18. Resuspend the pellet in 100–250 μl of PEMBAL to form a thin slurry and apply 50 μl to 13-mm diameter round coverslips (see Section III).

19. Dry the cells onto the coverslips by using a hairdryer and invert the coverslips onto ~2 μl of Elvanol mounting medium containing 4′,6-diamidino-2-phenylindole (DAPI) and antifade (Section III,A). Seal with nail varnish and observe. These mounts are not permanent.

B. Phalloidin Staining of *Schizosaccharomyces pombe*

An alternative method of actin staining is provided by rhodamine–phalloidin. This is quicker and simpler than the antibody method described above because it does not involve the removal of the cell wall. Two points should be borne in mind when using this method. First, phalloidin does not bind to G-actin. Second, rhodamine–phalloidin binds 10 times more efficiently to F-actin than does fluorescein–phalloidin.

1. Fix cells by addition of a 1/7.5 vol of freshly made 30% formaldehyde (see Section III,A) in PM buffer (Section V,A) to the growing culture with shaking. Agitate the cells for 60 min.

2. Pellet the cells (1000 g, 5 min) and wash three times (10 ml each) in PM buffer.

3. Permeabilize the cells by resuspension in 10 ml of PM containing 1% Triton X-100 and incubate for 30 sec. Pellet the cells.

4. Wash three times, each time with 10 ml of PM.

5. Resuspend the pellet in 1 ml of PM. Transfer a volume that will give a 20- to 30-μl pellet on centrifugation in an Eppendorf microfuge at top speed for 2 min at RT and resuspend in 100 μl of rhodamine-conjugated phalloidin (20 μg/μl), to form a slurry. To visualize both F-actin and the cell wall simultaneously, resuspend the pellet in 50 μl of rhodamine–phalloidin plus 50 μl of Calcofluor White M2R (sigma, St. Louis, MO) (5 mg/ml) (see Section IV).

PHALLOIDIN IS EXTREMELY TOXIC.

6. Incubate for 1 hr on a rotary inverter, in the dark.

7. Apply 50 μl of cell suspension to 13-mm diameter, washed, polylysine-coated coverslips and withdraw the excess (Section III). If the cell suspension is too thick for mounting, dilute with PM.

8. Dry the cell monlayers with a hairdryer.

9. Invert each coverslip onto a small drop ($\sim$2 μl) of PM containing 1 μg p-phenylene-diamine/ml and 1 μg DAPI/ml (Section III).

10. Seal the coverslips with nail varnish and observe. These mounts are not permanent.

C. Formaldehyde–Glutaraldehyde Fixation Method: Microtubule Staining

Day 1

1. Prepare 10 ml of *fresh* 30% formaldehyde solution (Section III).

2. Fix the cells by addition (to a 50-ml culture) of 5 ml of 30% formaldehyde, with agitation, followed 30–60 sec later by 0.44 ml of 25% glutaraldehyde [Sigma (St. Louis, MO) electron microscopy (EM) grade; final concentration, 0.2%(v/v)]. Agitate the cells for a further 90 min in a water bath.

3. Pellet the cells (50-ml Falcon tubes, benchtop centrifuge, 1000 g, 5 min at RT).

4. Wash the cells three times, each time with 10 ml of PEM.

5. Remove the cell walls by resuspending the cells at a density of 5×10^7–1×10^8 cells/ml in PEMS containing the following cocktail of cell wall-degradative enzymes: 1 mg of Novozym 234/ml and 0.3 mg of Zymolyase 20T/ml.

6. Incubate the cells at 37°c until $\sim$80% are judged, by a decrease in refractivity under phase-contrast optics, to have lost their cell walls. (*Note:* This cell wall digestion procedure differs from that described in Section II,A, step 1.)

7. Wash three times, each time with 10 ml of PEMS.

8. Resuspend in 10 ml of 1% Triton X-100 in PEM. Incubate for 30 sec and pellet.

9. Wash three times, each time with 10 ml of PEM.

10. Resuspend the cells in 10 ml of fresh sodium borohydride (1 mg/ml in PEM). Incubate for 5 min and pellet.

11. Repeat step 10 twice more.

12. Wash three times, each time with 10 ml of PEM.

13. Resuspend in 1 ml of PEMBAL and transfer an appropriate volume to an Eppendorf tube (see step 11, above). Incubate for at least 30 min with continuous inversion.

14. Spin in a microfuge (full speed for 10 sec) and resuspend each pellet in 50–100 μl of anti-tubulin primary antibody (see below).

15. Incubate overnight on a rotary inverter.

Day 2

16. Wash each sample three times, each time with 1 ml of PEMBAL.

17. Resuspend each pellet in 50–100 μl of rhodamine-conjugated rabbit anti-rat secondary antibody (ICN Biomedicals) diluted 1/50 in PEMBAL.

18. Incubate overnight on a rotary inverter.

Notes on the use of antitubulin antibodies for both immunofluorescence and immunoblotting in *S. pombe* may be found in Hagan and Hyams (1988) and in Alfa and Hyams (1990a). We have noted a wide variation in different batches of antibodies obtained commercially, some of which give excellent immunofluorescence staining of *S. cerevisiae* microtubules but fail to stain *S. pombe*. As a rule of thumb, antibodies raised against lower eukaryote tubulin are more likely to be successful than those raised against mammalian tubulin. Note particularly that *S. pombe* tubulin is neither detyrosinated nor acetylated (Alfa and Hyams, 1990b), so antibodies specific for these posttranlationally modified forms cannot be used.

Antibodies that should work are as follows:

YOL 1/34 (Kilmartin *et al.*, 1982)   Available commercially from Sero Tec.
YL1/2 (Kilmartin *et al.*, 1982)   (Kidlington, Oxfordshire, England)
TAT1 (Woods *et al.*, 1989)
KMX-1 (Birkett *et al.*, 1985)

Day 3

19. Wash three times, each time with 1 ml of PEMBAL.

20. Resuspend the pellet in 100–250 μl of PEMBAL to form a thin slurry and apply 50 μl to 13-mm diameter coverslips (Section III).

21. Dry the coverslips with a hairdryer and invert each onto ~2 μl of Elvanol–DAPI–antifade. Seal with nail varnish and observe. These mounts are not permanent.

D. Methanol Fixation Method: p34^{cdc2} Staining

Day 1

1. Rapidly filter the cells of a 50-ml overnight culture through a chilled 2.5-cm Millipore (Bedford, MA) membrane filter disk with a Whatman (Clifton, NJ) filter apparatus fitted with a 2.5-cm scintered glass plate. Fix the cells with ~10 ml of methanol precooled to −20°C (Falcon tube) as follows: Place the disk inside a Falcon tube (do not immerse in methanol) and rapidly wash off the cells with −20°C methanol, using a Pasteur pipette. Remove the disk. (*Note:* It is important that the methanol should remain as cold as possible.)

2. Leave the cells in methanol at −20°C for 8 min.

3. Remove the disk (taking care that no cells remain on it) and pellet the cells (1000 *g*, 5 min at 4°C).

4. Wash three times, each time with 10 ml of PEM.

5. Protoplast the cells by resuspending at a concentration of 1×10^7–1×10^8 cells/ml in PEMS containing 1 mg of Novozym 234/ml and 0.3 mg of Zymolyase 20T/ml. Incubate at 37°C until ~10% of the cells show a decrease in refractivity under phase-contrast optics.

6. Wash three times, each time with 10 ml of PEMS.

7. Resuspend the cells in 1 ml of PEMBAL and transfer an appropriate volume to an Eppendorf tube such that the subsequent pellet will be 20–30 μl. Incubate for at least 30 min on a rotary inverter.

8. Spin and resuspend in anti-p34^{cdc2} antibody (50–100 μl) diluted 1/50 in PEMBAL. This will form a thick cell suspension.

9. Incubate for 12–16 hr on a rotary inverter.

Day 2

10. Wash three times, each time with 10 ml of PEMBAL.

11. Resuspend the pellet in 50–100 μl of rhodamine-conjugated anti-rabbit secondary antibody (ICN Biomedicals) diluted 1–50 in PEMBAL.

12. Incubate for 8–16 hr on a rotary inverter in the dark.

Day 3

13. Wash three times, each time with 10 ml of PEMBAL.

14. Resuspend the pellet in 50–100 μl of PEMBAL to form a thick cell suspension (~5×10^7 cells/ml) and apply 50 μl to 13-mm diameter coverslips (Section III).

15. Dry the coverslips with a hairdryer and invert each onto ~2 μl of Elvanol–DAPI–antifade. Seal with nail varnish and observe.

Notes on Reducing Background Staining

In our experience, antibody-stained cells frequently have a high "background" of nonspecific staining. This is particularly true when the primary antibody is a polyclonal although this may be reduced by affinity purification of the serum. Any further background staining can often be reduced by preadsorbing the secondary antibody against yeast protein as follows (the same procedure can be used to reduce background in Western blots).

1. Break the yeast cells by vortexing in the presence of 425- to 600-μm glass beads and spin in an Eppendorf microfuge at top speed for 5 min at RT.

2. Remove the supernatant and boil the pellet in 10 vol of PBS containing 1% bovine serum albumin (BSA), 5% fetal calf serum (FCS), 1 m*M* PMSF for 5 min. Freeze in aliquots.

3. Dilute fluorochrome-conjugated secondary antibody 1 : 5 in the usual antibody buffer.

4. Add to an equal volume of absorbent and incubate for 30 min at room temperature.

5. Centrifuge in an Eppendorf microfuge (top speed, 5 min, RT).

6. The supernatant is secondary antibody diluted 1 : 10. Freeze in suitable aliquots.

Notes on Double Staining

In theory, the simultaneous localization of two antigens in yeast should present no more difficulty than in any other organism, provided monospecific antibodies raised in different hosts are available. In practice, this rarely proves to be the case. To colocalize two antigens we use methanol fixation, as this retains antigenicity of most proteins (Alfa *et al.*, 1990). The best results are obtained by sequential overnight incubations in antibody as follows: (1) primary 1, (2) secondary 1, rhodamine conjugated, (3) primary 2, and (4) secondary 2, fluorescein conjugated. With such a scheme it is likely that secondary 2 will also bind to some degree to primary 1. To abolish this you can preadsorb secondary 2 against primary 1 (Harlow and Lane, 1988) or block cells with excess unconjugated secondary 1 between steps (2) and (3). Alternatively, fixing cells with methanol at −20°C for 5 min between steps may reduce cross-reactivity.

III. Solutions and Notes for Immunofluorescence Microscopy

A. Solution for Immunofluorescence Microscopy

Formaldehyde (30%, make fresh): Add 3 g of *p*-formaldehyde to ~8 ml of PEM and mix well. Heat to 70°C in a water bath (*do not boil!*) and add ~1 ml of

5 M NaOH until the solution clears. Make up to 10 ml with PEM. Cool to the temperature of the growing culture

PEM: 100 mM Na-PIPES (piperazine-N,N'-bis(2-ethanesulfonic acid) (pH 6.9), 1 mM ethylene glycol-bis(β-aminoethyl ether)-N,N,N',N'-tetraacetic acid (EGTA), 1 mM MgSO$_4$

PEMS: PEM, 1 M sorbitol

PEMBAL: PEM, 0.1 M L-lysine, 1% BSA (globulin free), 0.1% sodium azide; *note:* azide is extremely toxic

DAPI (4',6-diamidino-2-phenylindole): 1 mg/ml in distilled H$_2$O

antifade (*p*-phenylenediamine): 10 mg/ml in PBS, pH 8.2

DAPI–antifade stock: Add 1 μl of DAPI stock to 99 μl of antifade stock dispensed into 10-μl aliquots and store at $-20°$C

Elvanol: Add 20 g of gelvatol (20/30 polyvinyl alcohol Aldrich 18463-2) to 80 ml of PBS, pH 8.2; heat to 70°C; add 40 ml of glycerol. Will keep on the bench for several weeks

We routinely mount in Elvanol containing DAPI–antifade prepared as follows: just before use thaw a 10-μl aliquot of DAPI–antifade and gently mix with 90 μl of Elvanol (avoid bubbles). Keep in the dark (wrap the tube in foil). For mounting cells following phalloidin staining add 10 μl of DAPI–antifade to 90 μl of PM buffer. *Note:* Elvanol cannot be used in this case.

B. Notes on Mounting Cells

Sticking cells firmly to the coverslip is essential for photography, which often requires long exposures (see below).

1. At the end of the protocol, resuspend the pellet in sufficient buffer to give a thick cell suspension (for an approximately 20- to 30-μl pellet in an Eppendorf tube, $\sim$100–250 μl is usually sufficient).

2. Apply 50 μl of the suspension to a coverslip (treated as below) held at an angle to allow most of the drop to run to the bottom. Withdraw the excess suspension and dry the cell monolayer with a hairdryer.

3. Pipette a small drop ($\sim$2 μl) of mounting medium onto a slide and gently place the coverslip onto it. Allow the drop to spread but do not apply pressure to the coverslip. Seal the edge of the coverslip with nail varnish.

4. Allow 2–5 min for the nail varnish to dry, then view. Such mounts are not permanent.

Preparation of Coverslips

Coverslips (and slides if necessary) should be washed free of both grease and dust as follows.

1. Immerse sequentially in the following (use forceps): (1) 5% Teepol (Teepol Products, Surrey, England), (2) distilled water, (3) distilled water, and (4) acetone. Blot dry on filter paper and coat with poly-L-lysine. (The latter creates a positively charged surface to which cells adhere.)

2. Cover the washed coverslips with a 0.1% aqueous solution of poly-L-lysine ($M_r > 70,000$, Milipore filtered, stored in refrigerator). Remove the excess and allow to dry.

IV. Fluorescence Microscopy

Fluorescent probes that specifically recognize a particular cellular structure or organelle are a familiar tool to most cell biologists. Some examples that work particularly well in *S. pombe* are given below.

A. DAPI

The fluorescent DNA stain 4′,6-diamidino-2-phenylindole (DAPI) (D-1388; Sigma) can be used both on fixed cells (e.g., in conjunction with antibody staining) or on live cultures that will continue to grow in the presence of a low concentration of DAPI (Toda *et al.*, 1981). DAPI-stained cells can also be analyzed by flow cytometry.

1. Pellet 1 ml of cells from a growing liquid culture and wash twice in 1 ml of PEM buffer. Resuspend the washed pellet in 50–100 μl of PEM.

2. Air dry cell the suspension onto 13-mm diameter coverslips.

3. To a frozen vial of DAPI–antifade (10 μl) add 90 μl of Elvanol, thaw, and mix gently to avoid air bubbles. Keep the tube wrapped in foil.

4. Invert each coverslip onto ~2 μl of mounting medium on a glass microscope slide. Seal the coverslips with nail varnish.

5. View by epifluorescence, using the appropriate filter set.

B. Calcofluor

A number of "brighteners," developed initially for the washing powder industry, bind with great efficiency to the yeast cell wall. This is particularly useful in *S. pombe,* in which the cell wall architecture provides a marker of progress through the cell cycle (Streiblova *et al.*, 1984). Compounds such as calcofluor with M2R (F-6959, "fluorescent brightener"; Sigma) have also proved to be extremely valuable in the characterization of septation mutants that were isolated in the original screen for temperature-sensitive cell cycle mutants (Streiblova *et al.*, 1984; Marks *et al.*, 1987).

1. Pellet a 1-ml aliquot of cells and wash twice with 1 ml of PEM.

2. Resuspend the pellet (or part) in 100 μl of PEM containing 5 mg of calcofluor/ml, to give a thick "sludge."

3. Incubate for 30 min in the dark.

4. Wash the cells three times, each time with 1 ml of PEM.

5. Mount and view cells as for DAPI.

C. Rhodamine 123

The existence of a proton gradient across the mitochondrial inner membrane results in the active uptake of a number of positively charged lipophilic molecules. The most widely used is rhodamine 123 (8004; Sigma) (Chen, 1988), largely because of its low toxicity.

1. Remove 1 ml of cells from an exponentially growing culture and place in an Eppendorf tube.

2. Add 10 μl of frozen rhodamine 123 stock solution (1.5 mg/ml in distilled H_2O) to cells (final concentration, 15 μg/ml).

3. Incubate for 60 min on an inverter at room temperature.

4. Pellet the cells in a microfuge (1 min, full speed).

5. Resuspend in 100 μl of medium and mount the cells.

D. CDCFDA

Vital staining of vacuoles in yeast can be achieved in a number of ways. One such way uses the fluorescein-based compound 5(6)-carboxy-2',7'-dichlorofluorescein diacetate (CDCFDA) (C-369; Molecular Probes, Eugene, OR), which stains vacuoles bright yellow as long as the pH of the medium is below 7.0. Labeling is probably the result of removal of acetate ester groups by vacuolar esterases.

1. Pellet 1 ml of each of the exponentially growing cultures provided.

2. Resuspend in 1 ml of low-pH YE medium (pH 3–4.5; Alfa et al., 1993).

3. Thaw a vial of CDCFDA stock [1 mM in dimethyl sulfoxide (DMSO)] and add 5 μl to each tube. Incubate on an inverter for 30 min at 30°C in the dark.

4. Wash in 1 ml of low-pH YE.

5. Resuspend in 100 μl of low-pH YE and mount the cells (see Section III,B). View immediately with a fluorescein filter.

V. Solutions and Notes for Fluorescence Microscopy

A. Solutions for Fluorescence Microscopy

Calcofluor: 5 mg/ml in PEM; make fresh and spin down undissolved material in a microfuge (5 min, full speed)

Rhodamine 123: 1.5 mg/ml (100×) stock in distilled H_2O; store at $-20°C$

CDCFDA: 1 mM (200×) stock in DMSO; store at $-20°C$

PM buffer: 35 mM KH_2PO_4 (pH 6.8), 0.5 mM $MgSO_4$; make fresh

Phalloidin: 20 μg/ml in PM; dilute from frozen stock on day of use

B. Microscopy and Photography

After all the efforts to obtain a well-stained preparation, the final frustration is not being able to get a good photographic image. Our laboratory has obtained excellent photographic results with two Zeiss (Thornwood, NY) microscopes, the Photomicroscope III fitted with a ×63 1.4 NA objective and an Axiophot equipped with a ×100 1.3 NA objective. Because the cells are small, these high magnifications are necessary. This means a dimmer image and relatively long exposure times for photography, so the higher the numerical aperture of the objective the better. Even so, for microtubule staining, for instance, we have to go to exposures in the range of 90 sec to 2 min, even with a fairly fast film such as Ilford FP4, Kodak TMAX or, for color, Kodak Ektachrome (all 400 ASA). Even after determining by an exposure series the correct exposure time, it is well worth taking a range of exposures that bracket this figure. Take lots of photographs!

VI. Immunogold Labeling of Yeast Ultrathin Sections

The small size of the fission yeast cell and the limit of resolution of the light microscope both restrict the amount of structural data that can be obtained by optical methods. To fully define the distribution of a given protein it is therefore necessary to take advantage of the increased resolving power of the electron microscope by using immunogold labeling of ultrathin sections. The application of gold labeling to yeast cells is well documented (Clark and Abelson, 1987; Wright *et al.*, 1988; Payne *et al.*, 1988; Veenhuis and Goodman, 1990; Karwan *et al.*, 1990; Potashkin *et al.*, 1990). However, these immunoelectron microscopic studies used conventional electron microscopy, that is, fixation with glutaraldehyde, formaldehyde, or a mixture of both, followed by dehydration and resin embedding. The cell wall is generally removed after fixation to aid the penetration of the resin.

Freeze-substitution electron microscopy involves immobilizing the cellular components in microseconds by freezing. It has several advantages over conventional electron microscopy. Fixatives are used at a low concentration or not at all and it is not necessary to remove cell walls. In order to have cells worthy of examination the water in the cell must freeze in a vitreous state. Once frozen, it is substituted, usually with an organic fluid such as acetone, ethanol, methanol, or tetrahydrofuran. The substituent may also contain a chemical fixative, for example, osmium tetroxide or glutaraldehyde, to further fix the cell components. The cells are then brought to room temperature and embedded conventionally for sectioning. If osmium tetroxide is deleted from the fixation schedule, or at least substantially reduced, specimens can be processed for immunocytochemistry.

Freezing of the specimen is the most crucial part of the freeze-substitution procedure. Freezing can be achieved with a variety of techniques. The method described below involves slamming the specimen at a rapid speed against a copper block cooled with liquid nitrogen or with liquid helium.

A. Freeze–Substitution for Immunocytochemistry

1. Harvest the cells by centrifugation at 1700 g on a Beckman GS-6 centrifuge 5 min at 4°C and wash once with 0.1 M potassium dihydrogen orthophosphate (pH 7.0).

2. Resuspend the pellet in an equal volume of 2% gelatin in phosphate-buffered saline (PBS, pH 7.4).

3. Place a droplet of suspension within an M3 nylon washer (o.d. 7 mm, i.d. 3 mm; RS Components, Corby, England) such that the meniscus protrudes just above the surface of the washer.

4. Rapidly freeze the washer containing the cell suspension by impact onto a liquid nitrogen-cooled copper block, using the Reichert MM80 metal mirror impact freezer attached to a Reichert KF80 main unit and set at the following values: speed 4, force 2, and thickness 2.

5. Transfer the frozen cells under liquid nitrogen to the chamber of a Balzers FSU 010 substitution unit, precooled to −80°C.

6. Cells to be used for immunocytochemistry are substituted in acetone [high-performance liquid chromatography (HPLC) grade] or acetone containing 0.25 to 0.50% glutaraldehyde at −80°C for 48 hr, −20°C for 2 hr, 4°C for 2 hr, and at room temperature for 30 min.

Note: For normal cytological examination greater contrast can be obtained by including 2% osmium tetroxide and 0.05% uranyl acetate in all steps (Fig. 2). This substitution medium is prepared as follows: (1) Dissolve uranyl acetate to a concentration of 0.05% in HPLC-grade acetone in a polypropylene container

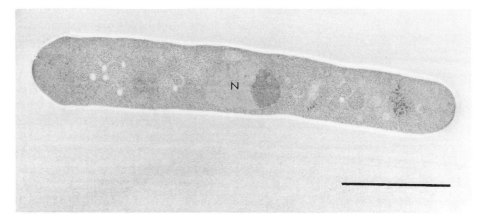

Fig. 2 *Schizosaccharomyces pombe* mutant *cdc10*. The cell division cycle mutant *cdc10* freeze-substituted with 2% osmium tetroxide and 0.05% uranyl acetate in acetone. N, Nucleus. (Bar: 5 μm.)

with a snap-on polyethylene lid (Falcon). This container and a similar one containing pure osmium tetroxide are cooled to −80°C; (2) pour the uranyl acetate–acetone solution into the osmium tetroxide container. The osmium tetroxide dissolves in about 1 hr at −80°C.

7. Infiltrate the samples with LRW resin (London Resin Company, London, England) in gelatin capsules (size 00; Agar Scientific, Essex, England) at 4°C for 2 days with several changes of resin.

8. Polymerize at 50°C for 24 hr.

Notes on Resins

Other resins can be used for infiltration, for example, Epon 812, Lowicryl HM20, and Lowicryl K4M (Hayat, 1986).

Notes on Sectioning

Using this method, frozen cells with a well-preserved ultrastructure appear in only the outer 10–15 μm of the sample. The definition of "well frozen" is subjective but in *S. pombe* the appearance of the vacuoles is a good guide. If these appear uniformly filled with a slightly granular content then we assume that the cells are well frozen. In badly frozen vacuoles the contents appear more clumped.

Sections are collected on 300- or 400-mesh nickel grids (Agar Scientific). Nickel grids are used to prevent any interaction of the colloidal gold particles of the secondary antibody with the grid.

Notes on Preparation of Solutions of Immunogold Labeling

Buffers used in the immunogold labeling procedure should be prepared from chemical stocks of reagent grade or better and filtered through a $0.20\text{-}\mu m$ pore size filter to remove small particulates that would interfere with colloidal gold visualization. The heavy metal stains used in the final steps of the immunogold labeling procedure are not only filtered but are also microcentrifuged for 10 min, just prior to use. The final centrifugation further removes aggregates and stain precipitates that may form after the stains have been prepared.

Notes on Antibodies

A major problem with polyclonal sera is nonspecific labeling of the cell wall (Fig. 3A and B). This can be overcome by preadsorbing the antibody with whole yeast cells as follows: harvest the cells by centrifugation and wash once with PBS. Add sodium azide to 10 mM and let stand for 5 min. Wash the cells twice with PBS. Dilute the antiserum 1:5 with PBS to 100 μl and add to 50 μl of washed yeast cells. Leave overnight at 4°C, then centrifuge at 500 g for 15 min at 4°C.

B. Immunogold Labeling Procedure

All steps are carried out on strips of Nescofilm (Nippon Shoji Kaisha, Ltd., Tokyo, Japan), which can be covered with a clear plastic lid between incubations. Drying of the sections or evaporation of the reagents between the incubations must be avoided, otherwise artifactual adsorption and clustering of the colloidal gold may occur. Too vigorous washing of sections should be avoided as this can displace gold particles and can produce nonuniform labeling.

Incubations are carried out by floating grids on a 50-μl drop of the different reagents so as to expose only one side of the section. Exposing both sides of the grids to the antibodies increases background staining.

1. Quench the sections in blocking solution [0.5% BSA, 2.5% fetal calf serum, 2.5% normal goat serum in Tris-buffered saline (TBS, pH 7.4), containing 0.5% Tween-20 (TBS-T)] for 40 min at room temperature, to block nonspecific binding sites.

2. Incubate the sections in primary antibody diluted in blocking solution for 2 to 3 hr. Each antibody must be titrated to determine the dilution that gives the best results. In general, higher concentrations of antibody give the highest backgrounds.

3. Wash the sections for 30 min in blocking solution, then eight times (5 min each) in 0.1% BSA in TBS-T.

4. Incubate the sections in secondary anti-IgG antibody–colloidal gold conjugate (GAR$_{15}$; BioCell Research, Cardiff) diluted 1:50 in 1.0% BSA in TBS-T.

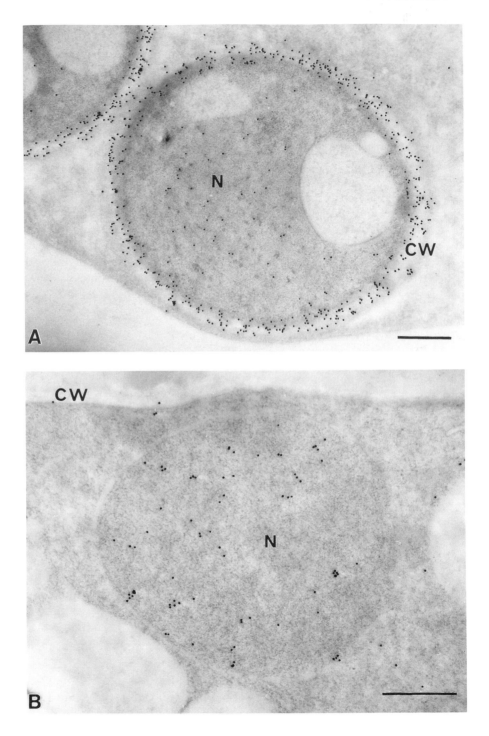

5. Wash the sections eight times (5 min each) in 0.1% BSA in TBS-T and twice (5 min each) in distilled water.

6. Stain the sections with 2% aqueous uranyl acetate for 15 min in a covered petri dish. Wash the grids with distilled water and stain with lead citrate (Reynolds, 1963) for 5 min. Wash the grids with distilled water and air dry.

7. Examine in a transmission electron microscope at 60 kV. The grids can be stored indefinitely at room temperature.

Notes on Controls

An essential feature of immunogold labeling is the demonstration of specificity. To check that nonspecific binding of either the primary or secondary antibody is not occurring it is necessary to carry out a number of controls.

1. Where possible, the immunolabeling of a specimen can be compared with that of a similar specimen lacking the antigen in question (and preferably only that antigen).

2. Treat the specimen with preimmune IgG in place of the primary antibody, followed by the conjugated secondary antibody.

3. Treat the specimen with the same primary and secondary antibody reagents as in the specific labeling experiment, but in the presence of an excess of a soluble homogeneous form of the antigen.

4. Treat the specimen with an unrelated antibody to see if nonspecific labeling of antibodies is occurring on any part of the specimen. For example, the carbohydrate moieties in the fungal cell wall causes nonspecific binding of antibodies (Srebotnik *et al.*, 1989).

Acknowledgments

We thank Kathryn Ayscough (Imperial Cancer Research Fund, London) and Iain Hagan (University of Kyoto) for their comments and assistance with some of the above methods.

References

Adachi, Y., and Yanagida, M. (1989). Higher order chromosome structure is affected by cold-sensitive mutations in *Schizosaccharomyces pombe* gene *crm 1*[+] which encodes a 115 Kd protein preferentially localized in the nucleus and its periphery. *J. Cell Biol.* **108**, 1195–1207.

Adams, A. E. M., and Pringle, J. R. (1984). Relationship of actin and tubulin distribution to bud-growth in wild-type and morphogenetic *Saccharomyces cerevisiae*. *J. Cell Biol.* **98**, 934–945.

Fig. 3 (A) Nonspecific labeling of the yeast cell wall. The degree of nonspecific labeling can be observed within the yeast cell wall (CW) by using a polyclonal antiserum to p63^{cdc13} (cyclin) that had not been preadsorbed with washed yeast cells. N, Nucleus. (Bar: 0.5 μm.) (B) Localization of p63^{cdc13} (cyclin) in *S. pombe*. Labeling of the yeast cell nucleus (N) with antiserum to p63^{cdc13} that had been preadsorbed with washed yeast cells. No nonspecific labeling of the yeast cell wall (CW) is visible. (Bar: 0.5 μm.)

Alfa, C. E., and Hyams, J. S. (1990a). Microtubules in the fission yeast *Schizosaccharomyces pombe* contain only the tyrosinated form of α-tubulin. *Cell Motil. Cytoskeleton* **18**, 86–93.

Alfa, C. E., and Hyams, J. S. (1990b). Distribution of actin and tubulin through the cell division cycle of the fission yeast *Schizosaccharomyces japonicus var. versatilis:* A comparison with *Schizosaccharomyces pombe. J. Cell Sci.* **96**, 71–77.

Alfa, C. E., Booher, R., Beach, D., and Hyams, J. S. (1989). Fission yeast cyclin: Subcellular localization and cell cycle regulation. *J. Cell Sci., Suppl.* **12**, 9–19.

Alfa, C. E., Ducommun, B., Beach, D., and Hyams, J. S. (1990). Distinct nuclear and spindle pole body populations of cyclin-cdc2 in fission yeast. *Nature (London)* **347**, 680–682.

Alfa, C. E., Gallagher, I. M., and Hyams, J. S. (1992). Subcellular localization of the p34^{cdc2}/ p36^{cdc13} protein kinase in fission yeast. *Cold Spring Harbor Symp. Quant. Biol.* LVI, 489–494.

Birkett, C. R., Foster, K., and Gull, K. (1985). Use of monoclonal antibodies to analyse the expression of a multitubulin family. *FEBS Lett.* **187**, 211–218.

Booher, R., Alfa, C. E., Hyams, J. S., and Beach, D. (1989). The fission yeast cdc2/cdc13/suc1 protein kinase: Regulation of catalytic activity and nuclear localization. *Cell* **58**, 485–497.

Chappell, T. G., and Warren, G. (1989). A galactosyl transferase from the fission yeast *Schizosaccharomyces pombe. J. Cell Biol.* **109**, 2693–2702.

Chen, L. B. (1988). Mitochondrial membrane potential in living cells. *Annu Rev. Cell Biol.* **4**, 155–181.

Clark, M. W., and Abelson, J. (1987). The subnuclear localization of tRNA ligase in yeast. *J. Cell Biol.* **105**, 1515–1526.

Gallagher, I. M., and Evans, C. S. (1990). Immunogold-cytochemical labelling of β-glucosidase in the white-rot fungus *Coriolus versicolor. Appl. Microbiol. Biotechnol.* **32**, 588–593.

Hagan, I. M., and Hyams, J. S. (1988). The use of cell division cycle mutants to investigate the control of microtubule distribution in the fission yeast *Schizosaccharomyces pombe. J. Cell Sci.* **89**, 343–357.

Hagan, I. M., Riddle, P. N., and Hyams, J. S. (1990). Intramitotic controls in the fission yeast *Schizosaccharomyces pombe:* The effect of cell size on spindle length and the timing of mitotic events. *J. Cell Biol.* **110**, 1617–1621.

Harlow, E., and Lane, D., eds. (1988). "Antibodies: A Laboratory Manual." Cold Spring Harbor Lab. Press, Cold Spring Harbor, New York.

Hayat, M. A. (1986). "Basic Techniques for Transmission Electron Microscopy." Academic Press, London.

Hirano, T., Konoha, G., Toda, T., and Yanagida, M. (1989). Essential roles for the RNA polymerase I largest subunit and DNA topoisomerases in the formation of fission yeast nucleolus. *J. Cell Biol.* **108**, 243–253.

Hiraoka, Y., Toda, T., and Yanagida, M. (1984). The NDA 3 gene of fission yeast encodes β-tubulin. A cold sensitive nda 3 mutation reversibly blocks spindle formation and chromosome movement in mitosis. *Cell* **39**, 349–358.

Hurt, E. C., McDowall, A., and Schimmang, T. (1988). Nucleolar and nuclear envelope proteins of the yeast *Saccharomyces cerevisiae. Eur. J. Cell Biol.* **46**, 554–563.

Kanbe, T., Kobayashi, I., and Tanaka, K. (1989). Dynamics of cytoplasmic organelles in the cell cycle of the fission yeast *Schizosaccharomyces pombe:* Three-dimensional reconstruction from serial sections. *J. Cell Sci.* **94**, 647–656.

Kanbe, T., Hiraoka, Y., Tanaka, K., and Yanagida, M. (1990). The transition of cells of the fission yeast β-tubulin mutant *nda3-311* as seen by freeze-substitution electron microscopy. Requirement of functional tubulin for spindle pole body separation. *J. Cell Sci.* **96**, 275–282.

Karwan, R. M., Laroche, T., Wintersberger, U., Gasser, S. M., and Binder, M. (1990). Ribonuclease H(70) is a component of the yeast nuclear scaffold. *J. Cell Sci.* **96**, 451–459.

Kilmartin, J. V., and Adams, A. E. M. (1984). Structural rearrangements of tubulin and actin during the cell cycle of the yeast *Saccharomyces. J. Cell Biol.* **98**, 922–933.

Kilmartin, J. V., Wright, B., and Milstein, C. (1982). Rat monoclonal antitubulin antibodies derived by using a new non-secreting cell line. *J. Cell Biol.* **93**, 576–582.

Marks, J., Hagan, I. M., and Hyams, J. S. (1987). Spatial association of F-actin with growth polarity and septation in the fission yeast *Schizosaccharomyces pombe*. *Spec. Publ. Soc. Gen. Microbiol.* No. 23, 119–135.

McCully, E. K., and Robinow, C. F. (1971). Mitosis in the fission yeast *Schizosaccharomyces pombe:* A comparative study with light and electron microscopy. *J. Cell Sci.* **9,** 475–507.

Mitchison, J. M. (1990). My favorite cell: The fission yeast *Schizosaccharomyces pombe*. *BioEssays* **12,** 189–191.

Moreno, S., Klar, A., and Nurse, P. (1991). Molecular genetic analysis of fission yeast *Schizosaccharomyces pombe*. *In* "Guide to Yeast Genetics and Molecular Biology" (C. Gutherie and G. Fink, eds.), Methods in Enzymology, Vol. 194, pp. 795–823. Academic Press, San Diego.

Nasmin, A., Young, P., and Johnson, B. J., eds. (1989). "Molecular Biology of the Fission Yeast." Academic Press, San Diego.

Nugent, J. H. A., Alfa, C. E., Young, T., and Hyams, J. S. (1991). Conserved structural motifs in cyclins identified by sequence analysis. *J. Cell Sci.* (in press).

Nurse, P. (1990a). Universal control mechanism regulating onset of M-phase. *Nature (London)* **334,** 503–508.

Nurse, P. (1990b). Fission yeast comes of age. *Cell* **61,** 755–756.

Okhura, H., Adachi, Y., Kinoshita, N., Niwa, O., Toda, T., and Yanagida, M. (1988). Cold-sensitive and caffeine-supersensitive mutants of the *Schizosaccharomyces pombe dis* genes implicated in sister chromatid separation during mitosis. *EMBO J.* **7,** 1465–1473.

Payne, G. S., Baker, D., van Tuinen, E., and Schekman, R. (1988). Protein transport to the vacuole and receptor-mediated endocytosis by clathrin heavy chain deficient yeast. *J. Cell Biol.* **106,** 1453–1461.

Potashkin, J. A., Derby, R. J., and Spector, D. L. (1990). Differential distribution of factors involved in pre-mRNA processing in the yeast cell nucleus. *Mol. Cell. Biol.* **10,** 3524–3534.

Pringle, J. R., Adams, A. E., Drubin, D., and Haarer, B. K. (1991). Immunofluorescence methods for yeast. *In* "Guide to Yeast Genetics and Molecular Biology" (C. Gutherie and G. Fink, eds.), Methods in Enzymology, Vol. 194, pp. 565–608. Academic Press, San Diego.

Reynolds, L. S. (1963). The use of lead citrate at high pH as an electron opaque stain in electron microscopy. *J. Cell Biol.* **17,** 208–212.

Robinow, C. F. (1981). A view through the microscope. *In* "Current Developments in Yeast Research" (G. G. Stewart and I. Russell, eds.), Pergamon, Oxford.

Robinow, C. F., and Hyams, J. S. (1989). General cytology of fission yeast. *In* "Molecular Biology and Morphogenisis of Fission Yeast" (A. Nassim, P. Young, and B. F. Johnson, eds.), pp. 273–330. Academic Press, San Diego.

Russell, P., and Nurse, P. (1986). *Schizosaccharomyces pombe* and *Saccharomyces cerevisiae:* A look at yeasts divided. *Cell* **45,** 781–782.

Schimmang, T., Tollervey, D., Kern, H., Frank, R., and Hurt, E. C. (1989). A yeast nucleolar protein related to mammalian fibrillarin is associated with small nucleolar RNA and is essential for viability. *EMBO J.* **8,** 4015–4024.

Srebotnik, E., Messner, K., Foisner, R., and Petterson, B. (1989). Ultrastructural localization of ligninase of *Phanerochaete chrysosporium* by immunogold labelling. *Curr. Microbiol.* **16,** 221–227.

Streiblova, E., Hasek, J., and Jelke, E. (1984). Septum patterns in ts mutants of *Schizosaccharomyces pombe* defective genes CDC 3, CDC 4, CDC 8, and CDC 12. *J. Cell Sci.* **69,** 47–65.

Tanaka, K., and Kanbe, T. (1986). Mitosis in fission yeast *Schizosaccharomyces pombe* as revealed by freeze-substitution electron microscopy. *J. Cell Sci.* **80,** 253–268.

Toda, T., Yamamoto, M., and Yanagida, M. (1981). Sequential alterations in the nuclear chromatin region during mitosis of the fission yeast *Schizosaccharomyces pombe:* Video fluorescence microscopy of synchronously growing wild type and cold sensite cdc mutants by using a DNA-binding fluorescent probe. *J. Cell Sci.* **52,** 271–287.

Umesono, K., Hirano, Y., and Yanagida, M. (1983). Visualization of chromosomes in mitotically arrested cells of the fission yeast *Schizosaccharomyces pombe*. *Curr. Genet.* **7,** 123–128.

Veenhuis, M., and Goodman, J. E. (1990). Peroxisomal assembly: Membrane proliferation precedes the induction of the abundant matrix proteins in the methylotrophic yeast *Candida boidinii*. *J. Cell Sci.* **96,** 583–590.

Woods, A., Sherwin, T., Sasse, R., MacRae, T. H., Bains, A. J., and Gull, K. (1989). Definition of individual components within the cytoskeleton of *Trypanosoma brucei* by a library of monoclonal antibodies. *J. Cell Sci.* **93,** 491–500.

Wright, R., Basson, M., D'Ari, L., and Rine, J. (1988). Increased amounts of HMG-CoA reductase induce ''karmellae'': A proliferation of stacked membrane pairs surrounding the yeast nucleus. *J. Cell Biol.* **107,** 101–114.

Yanagida, M. (1989). Gene products required for chromosome separation. *J. Cell Sci., Suppl.* **12,** 213–229.

CHAPTER 12

Nonfluorescent Immunolocalization of Antigens in Mitotic Sea Urchin Blastomeres

Brent D. Wright and Jonathan M. Scholey

Section of Molecular and Cellular Biology
Division of Biological Sciences
University of California, Davis
Davis, California 95616 and
Department of Cellular and Structural Biology
University of Colorado Health Science Center
Denver, Colorado 80262

METHODS IN CELL BIOLOGY, VOL. 37

I. Introduction

Immunocytochemistry is a powerful and productive tool for analysis of the molecular architecture of cells (Osborn and Weber, 1982; Harlow and Lane, 1988). Sea urchin eggs and blastomeres are particularly useful cells for identifying and dissecting the roles of various intracellular proteins, particularly those involved in mitosis and embryogenesis (Wright and Scholey, 1992; Schroeder, 1986). Here we discuss the rationale, methodology, and some results of applying peroxidase–anti-peroxidase (PAP) staining for immunolocalization of antigens in early sea urchin blastomeres, focusing exclusively on the methods currently used and results obtained in our laboratory during our studies on the localization of the microtubule (MT)-based motor protein kinesin.[1]

II. Immunolocalization and Sea Urchin Embryos

A. Sea Urchin Embryos

Sea urchin eggs and early embryos have proved to be a useful system for biochemical and cytological analysis of the processes of cell division and embryonic development (for brief review see Wright and Scholey, 1992; for methods see Schroeder, 1986). The large volumes of gametes routinely obtained produce sufficient quantities of rapidly and synchronously dividing embryos for stage-specific isolation, biochemical analysis, and immunolocalization of cellular proteins. We have been using this system to investigate the role of cytoskeletal motors, particularly the kinesin superfamily of MT-based motors, in dividing sea urchin blastomeres.

B. Immunolocalization

Immunofluorescence microscopy is commonly employed for the visualization of cellular antigens (see, e.g., Osborn and Weber, 1982; Harlow and Lane, 1988), yet there are inherent limitations to this technique. First, auto-fluorescence of some cell types, particularly in sea urchin embryos, limits the

---

[1] We note that the PAP-staining method described here originated in the McIntosh laboratory (B. Neighbors *et al.*, unpublished observations), and has evolved to its current state over several years.

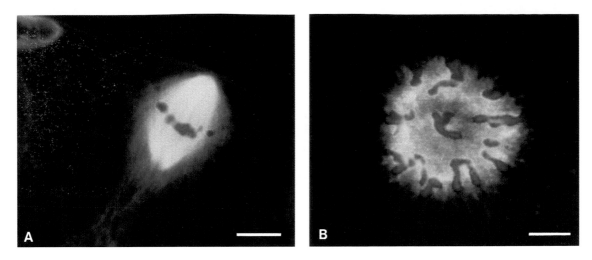

Color Plate 1   Dual fluorescence of microtubules and chromosomes in methanol-fixed *Xenopus* eggs. (A) Longitudinal section of a second meiotic spindle (the first polar body is apparent in the upper left corner) in an unfertilized *Xenopus* egg fixed with methanol. Upon closer examination, microtubules can be seen to be fragmented and collapsed into bundles. Cytoplasmic microtubules are very poorly preserved under these conditions. (B) Cross-section of the metaphase plate of a similar meiotic spindle. Eggs were fixed and stained with antitubulin (green) and propidium iodide (red) as described in Chapter 9, Table III, and were examined using the dual wavelength filter set (Texas red: fluorescein) provided with the MRC-600. Scale bar is 10 μm in A, and 5 μm in B. Reprinted from Gard, 1992. (For more information see Chapter 9.)

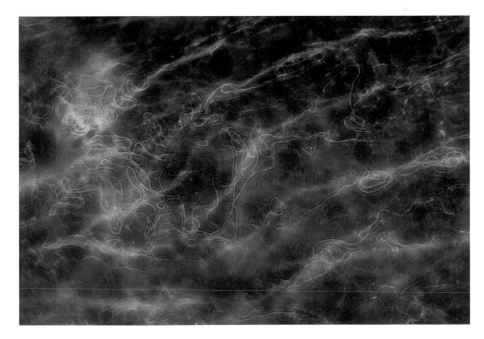

Color Plate 2   Double label immunofluorescence microscopy of calmodulin and tubulin in the green alga *Ernodesmis verticillata*. Rabbit anticalmodulin and mouse monoclonal antitubulin were followed by fluorescein-tagged goat anti-rabbit IgG and rhodamine-tagged goat anti-mouse IgG and were photographed via a double exposure. Courtesy of Russell H. Goddard and John W. La Claire II, University of Texas, Austin. (For more information see Chapter 10.)

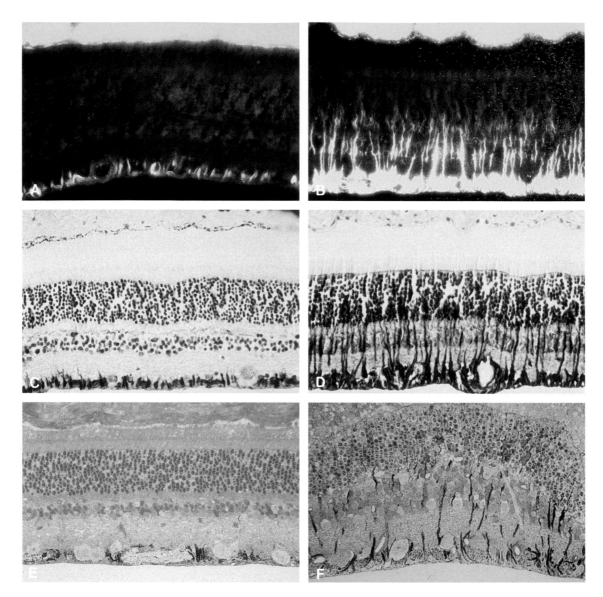

Color Plate 3   Three different immunocytochemical detection methods. A comparison is made using an antibody against GFAP on normal (A,C,E) and 3-day-detached cat retinal sections (B,D,F). (A,B) *Immunofluorescence technique on paraffin-embedded tissue.* With the immunofluorescence technique, the FITC-tagged secondary antibody appears as a yellow-green color. (The yellow at the top of the sections is autofluorescence from a specialized reflecting layer in the cat retina, the tapetum.) (C,D) *Immunoperoxidase technique on paraffin-embedded tissue.* The anti-GFAP was detected with a peroxidase-conjugated secondary antibody which results in a brown reaction product. These sections have also been counterstained with hematoxylin which stains the nuclei blue. (E,F) *Immunogold silver-enhanced technique on LR White-embedded tissue.* The gold-conjugated secondary antibody is enhanced with silver which produces a black reaction product. These sections have been counterstained with basic fuchsin. The paraffin-embedded tissue (A,B,C,D) results in a higher labeling density but less tissue resolution as compared to the tissue embedded in LR White resin (E,F). A comparison of results from all three techniques in normal retina, to the retinas that were detached for 3 days, shows how the distribution of GFAP dramatically increases after the detachment. The use of electron microscopic immunocytochemistry shows that the label is associated with intermediate filaments in a specific class of retinal glial cells, the Müller cells (see Chapter 15, Fig. 3). Magnification ×218.

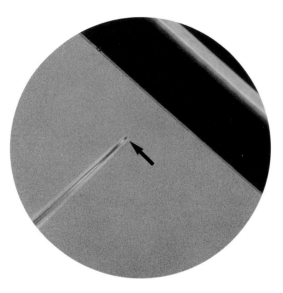

Color Plate 4   Range of colors displayed by dispersed light at the microcapillary tip. The tip portion of a microcapillary pulled into an injection needle as seen under the microscope. The black and yellow striped object in the upper right is part of the wall of a microcapillary shaft that is used to break off the black tip of the needle (arrow). Magnification: ×400. (For more information see Chapter 17.)

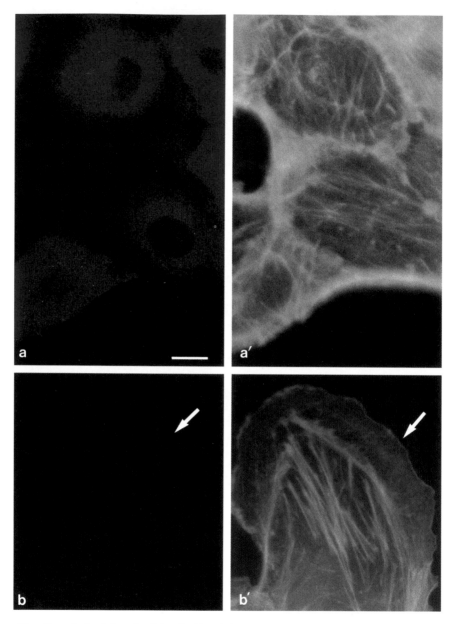

Color Plate 5   The effect of microinjected antivinculin (1 mg/ml) on the microfilament organization in an epithelioid sheet. The double label images show the distribution of the injected antibody (a, revealed by Rh-staining with a second antibody) in several individual cells of such a sheet, and the actin-containing microfilament bundles (a′, b′, stained with FITC-phalloidin). The injected cells show a severe disruption of the radial bundles concomitant with a reorganization of the actin filaments into small star-like structures and a disorganization of the peripheral belts which show multiple disruptions (a′). In contrast, a noninjected cell present in the same sheet (b, arrow) shows prominent radial stress fibers (b′). (For more information see Chapter 17.)

applicability of fluorescent staining. Second, elevated background levels from information out of the focal plane, especially in large cells such as sea urchin blastomeres, limits our ability to resolve fine details of intracellular structure. Newer techniques such as those involving digital image processing and laser scanning confocal microscopy can eliminate much of this confounding background, but require sophisticated skills and expensive equipment for routine cytochemical analysis. Third, the ability to detect relatively nonabundant antigens is limited by the low number of specifically localized molecules, the number of epitopes on each molecule recognized by the antibody probe, and the intensity of fluorescent labeling of the probe. Fourth, the tendency of fluorescent stains to photobleach with prolonged or intensive illumination, especially as used for confocal imaging, reduces the useable life of these samples.

Immunoperoxidase staining of sea urchin embryos with a tertiary PAP system (see Fig. 1) avoids a number of the problems associated with immunofluorescence (see Fig. 2 for example). The peroxidase reaction produces a colored product that is resistant to photobleaching. This stain is visible by simple bright-field illumination, thus decreasing background by avoiding cell autofluorescence and increasing specific information by utilizing a greater depth of field. The detection of a weak signal (such as a nonabundant antigen) is also enhanced by tertiary labeling, using three sequential antibody probes, each of which geometrically increases specific signal, and a final signal amplification produced by a controllable, chromogenic, enzymatic reaction to generate the desired degree of staining instead of relying on the amount of fluorescent stain conjugated to an antibody (see Fig. 1B). A major drawback of peroxidase staining is the inability to perform double-labeling experiments. However, doing simultaneous localizations in parallel batches can be nearly as informative. Using this staining method we have characterized the localization of sea urchin kinesin (Fig. 2) and its association with other cytoplasmic components (Figs. 3 and 4) through division and during early sea urchin embryogenesis (Fig. 5). This technique should prove useful for similarly localizing and characterizing additional cellular antigens in sea urchin embryos as well as in other cell systems.

III. Reagents and Buffers

A. Culturing Sea Urchin Embryos

1. Prepare sea-water (SW) solutions of appropriate salinity and temperature (see Schroeder, 1986, for formulas) for the species of sea urchin selected (see Section V,A below). Natural sea water should be sterile filtered ($0.22-\mu$m pore size; Millipore, Bedford, MA) (MFSW), and calcium-free sea water artificially made from salts (CFSW). All SW solutions should have antibiotics added (10 U of penicillin G/ml and 10 μg of streptomycin/ml, from a 1000 × stock; Sigma Chemical Co., St. Louis, MO) to prevent bacterial growth and prolong the

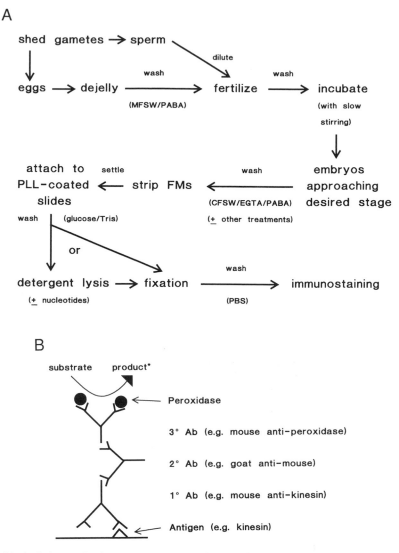

Fig. 1 Methods for production, treatment, mounting, and fixation of sea urchin embryos (A) and the ''sandwich'' technique for tertiary peroxidase–anti-peroxidase (PAP) immunostaining (B). (A) provides a simplified flow diagram of the steps required from shedding of sea urchin gametes through mounted and fixed, stage-specific, sea urchin embryos ready to incubate with antibodies (Abs) for PAP immunolocalization of the desired antigen (see Section V for detailed methods). (B) is a schematic representation of a completed tertiary antibody labeling sandwich providing three levels of antibody labeling amplification as well as a final level of signal amplification provided by the controlled enzymatic production of a colored (*) product (see Section VI for detailed methods).

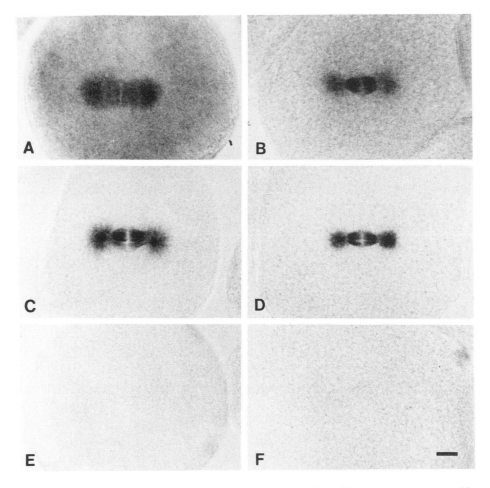

Fig. 2 Anti-kinesin, anti-tubulin, and control antibody staining of first-metaphase sea urchin blastomeres. Cells were fixed without (A, C, and E) or with Nonidet P-40 (NP-40) lysis (B, D, and F), then probed with anti-kinesin (A and B), anti-tubulin (C and D), or nonspecific control (E and F) monoclonal antibodies. Note that detergent lysis removes most of the cytoplasmic anti-kinesin staining but enhances its fibrous character in the mitotic apparatus (MA). (Bar: 10 μm). [From Wright *et al.* (1991). *J. Cell Biol.* **113**, 817–833, Fig. 6.]

experimental life of eggs/embryos, and should be thoroughly aerated shortly before use.

2. Prepare the following solution:

p-Aminobenzoic acid (PABA): 1 *M* in distilled H_2O, pH 8.1
Sterile-filtered sea water (MFSW), 10 m*M* PABA, pH 8.1

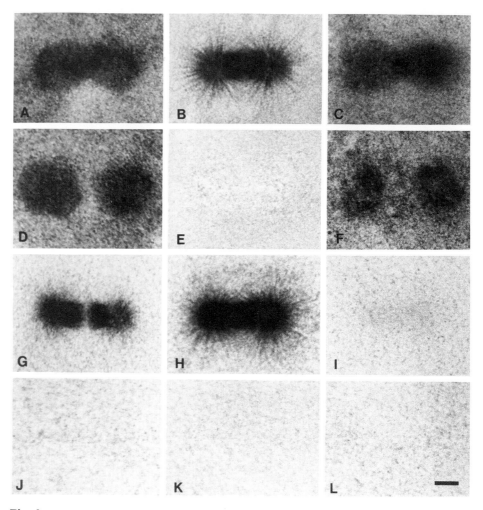

Fig. 3 Effects of microtubule depolymerization and/or detergent lysis on MA staining for kinesin, tubulin, and calsequestrin. First-metaphase embryos were fixed (A–C), nocodazole treated and fixed (D–F), detergent lysed and fixed (G–I), or nocodazole treated, and detergent lysed, and fixed (J–L), then probed with monoclonal anti-kinesin (A, D, G, and J), anti-tubulin (B, E, H, and K), or anti-calsequestrin (C, F, I, and L). Kinesin (A) and calsequestrin (C) are both concentrated into the MA, although there clearly is diffuse cytoplasmic staining as well. The pattern of staining for kinesin (D) and calsequestrin (F) is altered by MT depolymerization, but remains clearly visible in the region of the MT-depleted (E) MA. Detergent extraction of membranes removes calsequestrin as expected (I), whereas anti-kinesin staining of the MA becomes more "fibrous" (G). However, sequentially depolymerizing MTs and detergent extracting cells before fixation results in the loss of kinesin (J) and calsequestrin (L) as well as tubulin (K). (Bar: 10 μm) [From Wright *et al.* (1991). *J. Cell Biol.* **113**, 817–833, Fig. 7.]

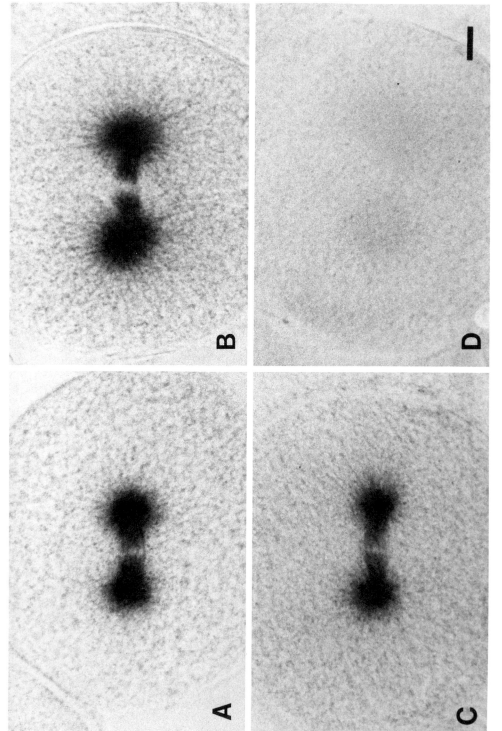

Fig. 4 Nucleotide sensitivity of kinesin staining of MAs in detergent-extracted cells. First-metaphase embryos were detergent lysed in the presence of 10 m*M* MgSO₄ (A), 10 U of Mg·apyrase/ml (B), 10m*M* Mg·AMP-PCP (C), or 10 m*M* Mg·ATP (D). Compared to controls (A), Mg·AMP-PCP had no effect on MA staining somewhat. Mg·ATP, the optimal nucleotide substrate for kinesin, totally abolished kinesin localization to the MA (D). The addition of nucleotides had no effect on anti-tubulin staining (not shown). (Bar: 10 µm.) [From Wright *et al.*, (1991). *J. Cell Biol.* **113**, 817–833, Fig. 8.]

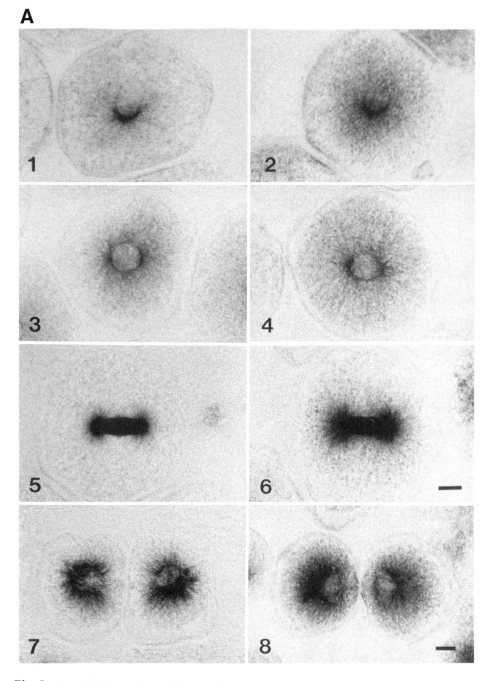

Fig. 5 (A and B) Comparison of kinesin (right) and tubulin (left) localization during cleavage stage development. Aliquots of synchronously developing embryos at the following stages were stained with anti-kinesin (even numbers) or anti-tubulin (odd numbers): sperm aster (1 and 2), diaster (3 and 4), first metaphase (5 and 6), second interphase (7 and 8), second metaphase (9 and 10), third

B

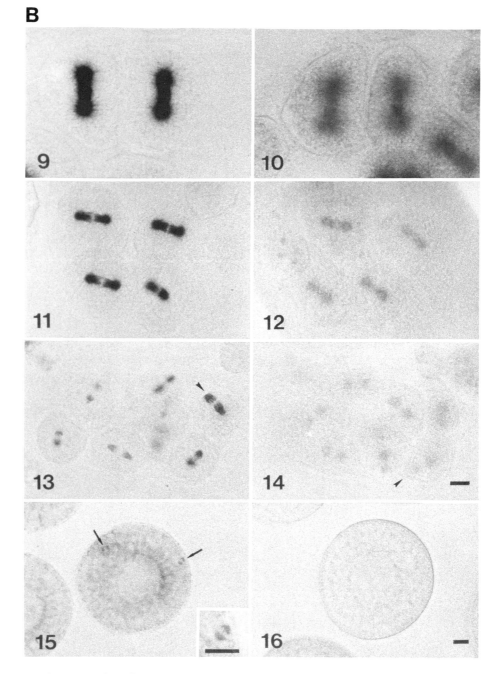

metaphase (11 and 12), fourth metaphase (13 and 14), and hatching blastulas (15 and 16). All but the blastulas were lysed prior to fixation. Note the gradual loss of intensity of anti-kinesin staining and the presence of acentric spindles in the asymmetric fourth division (13 and 14; arrowheads). By the blastula stage, diminutive spindles are clearly present (15, arrows and inset) that do not seem to stain for kinesin (16). (Bar: 10 μm.) [From Wright *et al.* (1991) *J. Cell Biol.* **113**, 817–833, Fig. 9.]

Calcium-free sea water (CFSW), 5 mM ethylene glycol–bis (β-aminoethyl ether)-N,N,N',N'-tetra acetic acid (EGTA), 10 mM PABA, pH 8.1

B. Embryo Immobilization and Fixation

Poly-L-lysine (PLL) solution: 1.2 mg/ml in distilled H$_2$O (M_r 300,000; Sigma)

Detergent lysis buffer (LB): 25 mM piperazine-N,N'-bis (2-ethanesulfonic acid) (PIPES) (free acid), 1 mM MgSO$_4$, 6 mM EGTA, 20% (v/v) glycerol, 1% (v/v) Nonidet P-40 (NP-40) detergent, (pH 6.9; adjust pH with KOH), plus protease inhibitors [1 mM dithiothreitol (DTT), 1 mg of $N\alpha$-p-tosyl-L-arginine methyl ester hydrochloride (TAME-HCl)/ml, 100 μM phenylmethylsulfonyl fluoride (PMSF), 100 μg of soybean trypsin inhibitor (SBTI)/ml, 2 μg of aprotinin/ml, 1 μg of leupeptin/ml, 1 μg of pepstatin/ml]

Glucose–Tris wash: 1 M glucose, 10 mM Tris buffer, pH 8.0

Methanol–EGTA fixative: 90% (v/v) methanol, 50 mM EGTA, pH 6.0, −20°C

C. Immunostaining and Mounting of Embryos

Phosphate-buffered saline (PBS): 135 mM NaCl, 5 mM KCl, 0.5 mM MgSO$_4$, 10 mM sodium phosphate buffer, pH 7.4

Blocking solution (PBS-NGS): PBS, 10% (v/v) normal goat serum (NGS; ICN Biomedicals, Inc., Costa Mesa, CA)

Tris-buffered saline (TBS): 150 mM NaCl, 10 mM Tris buffer, pH 7.5

Developing solution: TBS, 20% (v/v) methanol, 0.5 mg of 4-chloro-1-napthol/ml, 0.025% H$_2$O$_2$

Mounting medium: PBS, 90% (v/v) glycerol

IV. Antibody Preparation

A. Primary Antibodies

The production, characterization, selection, and purification of primary antibodies for immunocytochemistry have been documented exhaustively in [2], this volume and elsewhere (see Harlow and Lane, 1988, for detailed methods) and only those points considered relevant to this work will be briefly discussed here.

The foremost issue involved in choice of a primary antibody is the optimization of a specific antibody–antigen binding reaction. Assuming equivalent antibody affinities, the basic decision to be made (limited by availability) is whether to use monoclonal or polyclonal antibodies. Monoclonal antibodies are often highly antigen specific but usually bind to only one epitope per antigen molecule, thus reducing immunolabeling. Polyclonal antibodies often recognize and bind to numerous epitopes per molecule, thus increasing the stoichiometry of antibody–antigen labeling, but usually require extensive purification to achieve sufficient antigen specificity. For monoclonal antibodies, either batch ammonium sulfate or column purification (e.g., protein A for IgGs or gel filtration for IgMs) is acceptable. If possible, polyclonal antibodies should be affinity purified directly against the specific antigen of interest (Olmsted, 1981; Wright *et al.,* 1991; Harlow and Lane, 1988; article 5, this volume). In either case, the specificity of these antibodies *must* be characterized, at least by immunoblotting, to provide useful and interpretable immunolocalization results. Positive and negative staining controls in *every* experiment are also essential. Anti-tubulin [e.g., mouse monoclonal anti-β-tubulin (M. Klymkowski, University of Colorado, Boulder, CO) or rabbit polyclonal anti-tubulin (ICN Biomedicals, Inc.)] and nonspecific IgG [e.g., mouse monoclonal MOPC 21 (Sigma) or protein A-purified *preimmune* rabbit IgG, from the *same* animal as the immune serum] are good choices for control primary antibodies.

B. Peroxidase–Anti-peroxidase Secondary and Tertiary Antibodies

These antibodies are commercially available from a number of sources (e.g., ICN Biomedicals, Inc.), but should be of the highest quality available and should match the selected primary antibody(s) (e.g., see Fig. 1B). Thus, if the primary antibody is a mouse monoclonal (e.g., IgG) the secondary antibody should be an affinity-purified goat anti-mouse fraction directed against immunoglobulins of the same class as the primary (e.g., AP-GAM IgG). A secondary antibody to both heavy and light chains (H + L) will produce increased labeling of the primary antibody over one that is heavy chain specific (e.g., γ or Fc specific). The tertiary antibody (like the primary) should then be a mouse monoclonal peroxidase–anti-peroxidase complex (PAP). Similarly, a rabbit polyclonal primary antibody requires an affinity-purified goat anti-rabbit secondary (e.g., AP-GAR IgG, H + L) and an affinity-purified rabbit PAP tertiary antibody.

V. Preparation of Sea Urchin Embryos

Methods for handling of sea urchins, obtaining and storing gametes, fertilization, embryonic culture, and fixation have been discussed extensively elsewhere (see, e.g., Wright *et al.,* 1991; see Schroeder, 1986, for detailed methods). Here we will briefly cover the methods routinely used in our laboratory (see Fig. 1A, for summary).

A. Species Selection

Numerous sea urchin species are acceptable for immunolocalization and should be chosen based on antigen immunocross-reactivity with the primary antibodies chosen, availability, optimal temperature range, and egg clarity and size. We routinely use *Lytechinus variegatus* (24°C), *Lytechinus pictus*, and *Strongylocentrotus franciscanus* (14°C), which have larger (100- to 125-μm diameter) and relatively transparent eggs. *Strongylocentrotus purpuratus* (14°C) urchins are also useable, although their smaller (70-μm diameter) and more opaque, yolk-laden eggs make intracellular observation and embryonic staging more difficult.

B. Gamete Collection and Storage

1. Induce gamete shedding by an intracoelomic injection of 0.56 M KCl. Collect eggs by inverting females over breakers of MFSW at the appropriate temperature for the species selected. Collect sperm "dry" by inverting males over plastic petri dishes, then rapidly transferring the undiluted semen to sealed plastic test tubes and storing on ice until use (up to 3 days later).

2. Dejelly the eggs by passing them seven times through a prewetted 150-μm pore size Nitex screen (Small Parts, Inc., Miami, FL), then wash by settling, aspiration, and gentle resuspension three times (>20 vol each) in MFSW at the appropriate temperature. A useful apparatus for aspiration of solutions can be made by attaching a glass Pasteur pipette with a length of rubber tubing to a vacuum flask. Eggs can be stored up to 24–36 hr in MFSW containing antibiotics when settled as monolayers in large, loosely covered, flat-bottomed glass dishes and stored at the appropriate temperature.

C. Fertilization, Softening of Fertilization Membranes, and Embryo Culture

At fertilization sea urchin embryos elevate fertilization membranes (FMs), which harden as they "age" by an ovoperoxidase-dependent thiol cross-linking reaction (Showman and Foerder, 1979). These FMs interfere with the ability of embryos to attach directly to PLL-coated slides, preventing efficient embryo immobilization. Fertilization membranes are also relatively impermeable and can inhibit rapid detergent lysis, fixation, and complete antibody permeation.

1. Before fertilization, wash dejellied eggs into MFSW containing 10 mM PABA (a peroxidase inhibitor) to prevent subsequent hardening of FMs (Leslie *et al.*, 1987).

2. Dilute and activate sperm by adding one drop of concentrated semen to 50 ml of MFSW–PABA, suspending the sperm with a Pasteur pipette.

3. Fertilize the eggs by adding approximately 1 ml of this diluted sperm suspension to 500 ml of diluted (10–20 vol) dejellied eggs, and swirling the egg–sperm mixture to rapidly and thoroughly mix the gametes. To optimize monospermic fertilization, the appropriate egg-to-sperm ratio can be determined in advance by titering the sperm with small batches of eggs and monitoring them for percentage fertilization and normal first cleavage.

4. After the fertilized eggs have settled, remove excess sperm by aspirating the overlying solution and resuspending the embryos in fresh, aerated MFSW–PABA.

5. Keep the embryos suspended by gentle stirring at their appropriate temperature. This provides uniform oxygenation, allowing synchronous division and development.

D. Monitoring Development and Collection of Stage-Specific Embryos

1. Stage the developing embryos by periodically removing aliquots and observing with a dissecting microscope. Their specific stage in the cell cycle can be determined with polarization optics to observe birefringent MT-containing cytoplasmic figures.

2. Prepare PLL-coated microscope slides while the embryos are developing to the desired stage. We have found it convenient to use epoxy-coated multiwell slides (10 well; Carlson Scientific, Inc., Peotone, IL), to allow several antibody incubations (e.g., an antibody dilution series) simultaneously on a single slide. Slides are coated, one at a time, in the following manner.

a. Ionize and heat the surface of each slide by passing it through the flame of a Bunsen burner, then place it horizontally (face up) on a cooling rack.

b. Carefully apply a single drop of PLL solution to each well of the warm slide. Care should be taken to avoid spilling the PLL onto the epoxy mask, which would allow wetting on this surface, resulting in mixing of the individual drops during the embryo attachment and antibody incubation steps.

c. After all 10 wells have been wetted with PLL, remove the solution by aspiration. The wells should dry almost immediately by evaporation, leaving a uniformly thin coating of PLL on the glass surface in each well.

d. Chill and store the coated slides at the correct temperature for the embryos used. A covered glass petri dish is a convenient way of handling several slides at once. Fresh PLL-coated slides should be made each day.

3. Beginning at a stage that allows enough time for handling and preparation of embryos for fixation before reaching the desired stage (about 15 min), wash the embryos into CFSW containing 5 mM EGTA and 10 mM PABA by settling, aspiration, and resuspension.

4. Strip softened FMs from embryos by passing rapidly one to seven times through a prewetted Nitex screen of mesh size slightly larger than the egg diameter (150 μm for *L. variegatus, L. pictus,* and *S. franciscanus;* 105 μm for *S. purpuratus*), monitoring FM removal by microscopic observation of aliquots.

5. Concentrate FM-stripped embryos into a small volume of CFSW–EGTA–PABA (<1 vol), either by settling, or *gentle* centrifugation, and aspiration, for attachment to preprepared, PLL-coated microscope slides.

E. Embryo Immobilization, Detergent Lysis, and Fixation

1. Aliquot appropriately staged, demembranated and concentrated embryos dropwise into each well of the PLL-coated slides.

2. Allow the embryos to attach to the slides for 1–3 min (remember that they are still continuing to develop). Now all of the steps involving treating these embryos may be easily accomplished by having solutions in Coplin jars (slotted vertical slide staining dishes) and gently transferring the slides, with a pair of forceps, from one jar to another.

3. Quickly rinse the slides by gently dipping them into glucose–Tris solution at room temperature to remove SW salts and unattached embryos.

4. Immediately fix, or detergent lyse and fix, the embryos.

a. Slides of rinsed embryos can be rapidly fixed and permeabilized by transferring them to methanol–EGTA at −20°C (Harris, 1962).

b. Alternatively, embryos can be membrane extracted by immersing in detergent lysis buffer (Balczon and Schatten, 1983; Wright *et al.*, 1991) for 15 min at room temperature. This treatment dissolves lipids in an MT stabilization buffer, resulting in removal of soluble cytoplasmic antigen as well as cytosolic yolk and membranes, allowing higher resolution visualization of antigen specifically associated with the stabilized, detergent-extracted cytoskeleton. These detergent lysed embryos are then directly transferred to −20°C methanol–EGTA fixative.

5. Store slides of embryos in methanol–EGTA fixative at −20°C for 1–72 hr. Prolonged fixation does not alter staining of stable structures (e.g., MTs) but may result in eventual extraction of more soluble antigens (e.g., kinesin).

VI. Immunostaining

A. Antibody Labeling

All antibody solutions should be made up in PBS with 10% normal goat serum as a blocking agent to reduce nonspecific binding. Optimal dilutions for primary antibodies depend on concentration and affinity of specific antibody as well as abundance of antigen, and should be determined empirically by dilution series (see Fig. 1B, for diagrammatic example).

1. Remove slides of fixed or lysed/fixed cells from methanol–EGTA fixative and wash three times (10 min each) in PBS at room temperature.

2. Incubate the cells with primary antibody.

a. Place the slides face up in the bottom of a humidification chamber (e.g., a covered glass petri dish containing a wetted filter paper).

b. Carefully dry the epoxy mask by aspiration while leaving the cells moist.

c. Pipette about 15 μl of primary antibody into each well and spread the drop, using the pipette tip but not touching the cells, to cover the well evenly.

d. Cover the humidification chamber and incubate for 1–3 hr at 37°C, 5–8 hr at room temperature, or overnight at 4°C.

e. After incubation, aspirate the antibody off the cells and rinse by dipping the slide into a *large* volume of PBS at room temperature and immediately transferring to the first wash.

3. Repeat washing and incubation as above for both secondary (used at a 1:20 dilution) and tertiary (used at 1:40) antibodies.

4. After tertiary binding, wash the cells three times (10 min each) in PBS, then three times (5 min each) in TBS, and then develop.

B. Color Development

1. Develop the stain by placing the slides in glass petri dishes containing peroxidase developer and monitoring color generation under a dissecting microscope. Development times, typically from 1 to 15 min, will vary with different primary antibodies.

2. Once the cells are sufficiently stained, stop the reaction by quenching in cold distilled H_2O. Changing this wash several times helps prevent developer "run-on" from overstaining the cells.

C. Mounting and Photography

1. Remove the slides from the distilled H_2O wash, briefly touch the edge of each slide to a paper towel to wick off excess fluid, and place face up on a piece of filter paper.

2. Pipette a sufficient amount of glycerol–PBS into the middle of the slide to cover the cells.

3. Apply a #1, 22 × 50 mm coverslip (Corning, Inc., Corning, NY) over the cells, starting at one edge and tipping onto the fluid to avoid bubbles.

4. Flatten the cells slightly for better viewing and absorb excess fluid by pressing the coverslipped slide *gently* between filter papers.

5. Seal the edges with clear nail polish. Mounted slides should be stored in the dark to prevent slow photodegradation of the stain.

6. Single cells can be viewed with bright-field optics, using a ×40 objective.

7. Reasonably detailed photographs can be taken by standard meter readings and photographic techniques.

VII. Alternative Techniques

1. Difficulty in stripping FMs from embryos may indicate insufficient softening of FMs. Transferring fertilized embryos into CFSW–EGTA–PABA before culturing to the desired stage may prevent calcium-induced cross-linking and facilitate FM removal. However, calcium is also required for cell–cell adhesion, so culturing embryos past the first cell division in the absence of calcium may result in embryonic disaggregation. Using other peroxidase inhibitors, such as 10 mM 3-amino-1,2,4-triazole (Showman and Foerder, 1979), is also effective at blocking FM hardening.

2. Fixation of ciliated embryos in 50 mM EGTA will result in deciliation, which can be avoided by lowering the EGTA concentrations to 10 mM (Harris, 1962).

3. Cold methanol–EGTA fixative provides good structural and antigenic preservation overall, but some structures (e.g., endoplasmic reticulum or the actin cytoskeleton) may be distorted or disrupted. Thus, other fixatives may result in better staining for certain antigens. Some useful possibilities are as follows:

Paraformaldehyde (3%) in 90% methanol with 10 mM EGTA, pH 6.0, at −20°C

Paraformaldehyde (3%) plus 0.1% glutaraldehyde in CFSW at room temperature

Glutaraldehyde (and sometimes paraformaldehyde) fixation requires reduction of free aldehyde groups by incubating in sodium borohydride (0.5 mg/ml in PBS) for 15 min at room temperature. In addition, fixation without methanol may require subsequent permeabilization, to allow antibody permeation, by incubating in 0.1% Triton X-100 in PBS for 15 min at room temperature.

4. Although 4-chloro-1-naphthol yields the most photodense staining, other peroxidase substrates such as 3,3'-diaminobenzidine (see Harlow and Lane, 1988, for methods) produce different colored products that may enhance visualization of some staining patterns.

5. A final method of enhancing staining is to use a peroxidase-conjugated secondary antibody in addition to the PAP tertiary complex (Leslie *et al.*, 1987). This increases the amount of chromogenic enzyme bound to the antigen, but also tends to increase background staining.

6. Like detergent lysis experiments, perturbing embryos in other ways to observe antigen redistribution can offer insights into protein function (Wright *et al.*, 1991). Treating embryos for 15 min prior to either lysis or fixation with nocodazole (10 μg/ml) will disassemble cytoskeletal MTs (Fig. 3), cytochalasin B (10 μg/ml) may similarly disrupt the actin cytoskeleton (not shown), while

adding millimolar concentrations of nucleotides to the lysis buffer can perturb nucleotide-sensitive interactions (Fig. 4).

VIII. Results and Perspectives

Using PAP-staining (see Fig. 1), the MT-associated motor kinesin was first localized to the mitotic spindle of dividing sea urchin blastomeres (Scholey *et al.*, 1985; Leslie *et al.*, 1987). Subsequently, it was discovered that kinesin was a single member of a large superfamily of related proteins. Sequence similarity in certain regions of these molecules raises the possibility of immunocross-reactivity between distinct members of the superfamily. To distinguish which molecule was responsible for the observed spindle staining in sea urchin embryos, we created a panel of domain-specific antibodies to distinguish "conventional" kinesin from "kinesin-like" proteins (Ingold *et al.*, 1988; Wright *et al.*, 1991). PAP-staining sea urchin spindles with these antibodies clearly showed that it was conventional kinesin that associated with mitotic apparatus (MA) MTs (Fig. 2). We then used PAP immunolocalization in experimentally perturbed embryos to characterize kinesin association with other cytoplasmic components, demonstrating that sea urchin kinesin was apparently functioning as a membrane motor along spindle MTs in early embryos (Figs. 3 and 4). Similarly, PAP-staining showed that this pattern of spindle MT association changed during early embryogenesis (Fig. 5). We are now attempting to identify and purify new kinesin-like proteins from sea urchin eggs (Cole *et al.*, 1992). Using these proteins as antigens to produce new antibody probes for PAP immunolocalization studies in sea urchin embryos should help us to characterize and define the biological functions for these novel kinesin-like proteins. Finally, these techniques are promising for characterizing other sea urchin proteins and, with modifications, should prove equally useful in many other cell systems.

Acknowledgments

We would like to express grateful acknowledgment of the contributions of B. Neighbors, P. Grissom, and J. R. McIntosh to the early stages of this work. Mouse monoclonal anti-β-tubulin ascites was generously supplied by M. Klymkowsky (University of Colorado, Boulder, CO) and rabbit anti-sea urchin calsequestrin antiserum was generously provided by D. Begg (Harvard Medical School, Boston, MA) and B. Kaminer (Boston University Medical School, Boston, MA). Work in our laboratory was supported by grants from the NIH (#GM46376-01), ACS (#BE-46D), and March of Dimes Birth Defects Foundation (#1-1188) to J.M.S. B.D.W. was supported by a March of Dimes Birth Defects Foundation Graduate Research Training Fellowship (#18-89-44).

References

Balczon, R., and Schatten, G. (1983). Microtubule-containing detergent-extracted cytoskeletons in sea urchin eggs from fertilization through cell division. *Cell Motil.* **3**, 213–226.

Cole, D. G., Cande, W. Z., Baskin, R. J., Skoufias, D. A., Hogan, C. J., and Scholey, J. M. (1992). Isolation of a sea urchin egg kinesin-related protein using peptide antibodies. *J. Cell Sci.* **101,** 291–301.

Harlow, E., and Lane, D. (1988). "Antibodies: A Laboratory Manual." Cold Spring Harbor Lab. Press, Cold Spring Harbor, New York.

Harris, P. (1962). Some structural aspects of the mitotic apparatus in sea urchin embryos. *J. Cell Biol.* **14,** 475–487.

Ingold, A. L., Cohn, S. A., and Scholey, J. M. (1988). Inhibition of kinesin-driven microtubule motility by monoclonal antibodies to kinesin heavy chains. *J. Cell Biol.* **107,** 2657–2667.

Leslie, R. J., Hird, R. B., Wilson, L., McIntosh, J. R., and Scholey, J. M. (1987). Kinesin is associated with a nonmicrotubule component of sea urchin mitotic spindles. *Proc. Natl. Acad. Sci. U.S.A.* **84,** 2771–2775.

Olmsted, J. B. (1981). Affinity purification of antibodies from diazotized paper blots of heterogenous protein samples. *J. Biol. Chem.* **256,** 11955–11957.

Osborn, M., and Weber, K. (1982). Immunofluorescence and immunocytochemical procedures with affinity purified antibodies: Tubulin-containing structures. *Methods Cell Biol.* **24,** 97–132.

Scholey, J. M., Porter, M. E., Grissom, P. M., and McIntosh, J. R. (1985). Identification of kinesin in sea urchin eggs, and evidence for its localization in the mitotic spindle. *Nature (London)* **318,** 483–486.

Schroeder, T. E., ed. (1986). *Methods Cell Biol.* **27.**

Showman, R. M., and Foerder, C. A. (1979). Removal of the fertilization membrane of sea urchin embryos employing aminotriazole. *Exp. Cell Res.* **120,** 253–255.

Wright, B. D., and Scholey, J. M. (1992). Microtubule motors in the early sea urchin embryo. *Curr. Top. Dev. Biol.* **26,** 71–91.

Wright, B. D., Henson, J. H., Wedaman, K. P., Willy, P. J., Morand, J. N., and Scholey, J. M. (1991). Subcellular localization and sequence of sea urchin kinesin heavy chain: Evidence for its association with membranes in the mitotic apparatus and interphase cytoplasm. *J. Cell Biol.* **113,** 817–833.

CHAPTER 13

Ultrasmall Gold Probes: Characteristics and Use in Immuno(cyto)chemical Studies

Peter van de Plas and Jan L. M. Leunissen

AURION
Immuno Gold Reagents and Accessories
6702 AA Wageningen
The Netherlands

I. Introduction

A. Markers in Immunocytochemistry

Different immunocytochemical methods have been developed in order to visualize the distribution of immunoreactive sites. Markers are used either directly bound to the antigen-specific antibody (direct method) or linked to bridging molecules that bind specifically to the primary antibody (indirect method).

Markers such as enzyme conjugates or radioisotopes give a diffuse image of the label. The "diffuse" markers are generally sensitive because of inherent amplification properties. However, besides the risk of diffusion of the reaction product, fine structural details of the labeled sites are often obscured (Novikoff, 1980). The use of markers of particulate nature results in a higher resolution localization of the antigen. Ferritin and Imposil are components with an electron-dense iron core that give relatively high labeling densities (Tokuyasu and Singer, 1976; Geiger *et al.*, 1981). Their use is limited to electron microscopy (EM), in which the visibility of the iron core may be hampered by the electron density of the specimen. Furthermore, ferritin is a naturally occurring protein and is found occasionally as an endogenous component.

To avoid difficulties inherent to "diffuse" and iron-core based markers, colloidal gold markers were introduced in 1971 (Faulk and Taylor, 1971) and developed both for transmission electron microscopy (TEM) and scanning electron microscopy and for light microscopy (LM) and bioassays.

B. Colloidal Gold as Marker Particles

The success of gold particle-based immunodetection systems is determined by the following.

Gold sols can be prepared in a wide range of diameters, from 2 to 3 nm to 150 nm, with a narrow size distribution. This allows double labeling at the EM level (Geuze *et al.*, 1981; Leunissen and De Mey, 1989; Watkins *et al.*, 1990).

Macromolecules coupled to colloidal gold generally retain their bioactivity. The probes are stable over a long period of time.

Because of the Z number of the element gold and the accumulation of gold atoms in a colloidal particle the marker is highly electron dense and clearly visible even in heavy metal ion-contrasted specimens in transmission electron microscopy.

Colloidal gold emits secondary electrons and backscatters electrons and therefore it is also an appropriate marker in scanning electron microscopy (De Harven *et al.*, 1984; Gross and De Boni, 1990).

Large conglomerates of colloidal gold are intensely colored and as such visible in the light microscope and on blotting membranes (Brada and Roth, 1984; Moeremans *et al.*, 1984).

Gold particles reflect and depolarize light. These properties are used in epipolarization microscopy to increase detectability (De Waele *et al.*, 1988).

Colloidal gold particles act as catalytic nuclei for the deposition of silver (Danscher, 1981; Danscher and Rytter-Nörgaard, 1983). The sensitivity of detection in the light microscope and on blots can thus be substantially improved (Holgate *et al.*, 1983; Moeremans *et al.*, 1984). The immunogold silver staining (IGSS) technique is suited for double labeling at the EM level (Krenács *et al.*, 1990).

Colloidal gold is preferred to radioactive or enzyme markers because of its particulate nature, which results in improved resolution. In addition, gold probes are nonhazardous reagents.

C. Rationale of Gold Probe Preparation

There are several methods to produce monodisperse gold sols with particles of different size. They are based on the controlled reduction of an aqueous solution of tetrachloroauric acid. Particle size and size distribution are determined by the reducing agent, by varying the ratio between reducing agent and chloroauric acid and by temperature. The sodium citrate technique of Frens (1973) may be used to produce particles ranging in size from 15 to 150 nm. Reproducibility is best for the lower particle sizes. By slightly modifying this sodium citrate method (M. Moeremans, personal communication) solutions can be produced with an average particle diameter of 8 to 10 nm. White phosphorus is a stronger reductor than sodium citrate, therefore smaller particles can be produced. With the white phosphorus method described by Faulk and Taylor (1971), as modified by Slot and Geuze (1981), a mean particle size of 5 nm can be reached. For the preparation of gold colloids down to 2 nm sodium borohydride (Tschopp et al., 1982) and thiocyanate (Baschong and Lucocq, 1985) have been used.

Colloidal gold particles cary a net negative surface charge due to ions that are adsorbed on the surface of the particles. Positively charged proteins can therefore bind electrostatically. This phenomenon has been exploited to stain protein bands on blots (Moeremans et al., 1985) and to develop a sensitive protein assay (Ciesolka and Gabius, 1988). This electrostatic stabilization is less suitable for the production of immunogold probes, as these interactions will give rise to the formation of large protein–gold aggregates.

Next to its charge characteristics, colloidal gold particles also display hydrophobic properties. Hydrophobic interactions are maximally expressed at the isoelectric point of the macromolecule to be conjugated. To let the hydrophobic interactions prevail and to reduce the risk of electrostatic aggregation, proteins are coupled at a pH value that is equal to or slightly above their isoelectric point (Geoghean and Ackerman, 1981). When sufficient protein has been adsorbed to the surface of colloidal gold particles, the surface charge is shielded in such a way that electrostatic interactions will not give rise to particle aggregation.

Although not all of the hydrophobic binding sites may be saturated, it is generally accepted that a further increase of added protein results in the formation of additional layers of protein rather than more protein directly adsorbed to the gold particle surface (Goodman et al., 1981). The binding strength of the second and further layers is less and hence the stability of the probe decreases. The stability of ''monolayer'' probes is remarkable and is probably due to multiple point interactions of the protein. Secondary stabilizers, for example, bovine serum albumin (BSA) and Carbowax 20M (Fluka, Buchs, Switzerland) are used to cover potentially remaining bare surface areas on the gold particle.

Excess of free protein can be removed by centrifugation (Leunissen and De Mey, 1989). Several molecular species form active complexes with gold. In histochemical research immunoglobulins, protein A and lectins are most commonly used.

The different aspects of gold sols and probes and their preparation and use in immuno(cyto)chemical studies are covered in several reviews (Horisberger, 1981; De Mey, 1983; Beesley, 1985; Hayat, 1989; Verkleij and Leunissen, 1989).

II. Improvement of Detection System

When trying to improve the performance of colloidal gold probes, several considerations must be taken into account. This is understood when performance is expressed in terms of two types of sensitivity.

sensitivity of an immunogold reaction: The sensitivity of an immunogold reaction is generally expressed as the number of gold particles per unit of antigen. This depends on several factors, such as the nature and size of the detecting molecule used (e.g., protein A or immunoglobulins), the number of active binding sites per probe, and the size of the gold probe, which is in part determined by the diameter of the marker. Steric hindrance and penetration are also directly related to the probe size. It was observed by numerous authors that labeling efficiency increases with decreasing particle size (van Bergen en Henegouwen and Leunissen, 1986; Yokata, 1988).

sensitivity of detection: The sensitivity of detection is inversely related to the particle size: larger particles produce more contrast in the electron microsope, higher color intensity in the light microscope and on blots, and easier depolarization is observed in epipolarization microscopy.

In summary, the particle size should be as small as possible during incubation and be of a convenient (larger) size for observation. These requirements are not necessarily incompatible. Both are fulfilled when (1) small particles can be prepared and coupled in a stable way to macromolecules, (2) such complexes are at least as immunoreactive as larger sized probes, and (3) small gold particles can be homogeneously and efficiently enlarged by metal (silver) precipitation.

III. Characteristics of Ultrasmall Gold Probes

Baschong and Lucocq (1985) and Tschopp *et al.* (1982) describe procedures for the preparation of approximately 2- to 3-nm particles. We initially succeeded in accomplishing a further reduction of the size to approximately 1 nm and later down to an average diameter of 0.8 nm. Hainfeld (1987) reported the preparation of undecagold, noncolloidal organometallic complexes with gold clusters consisting of 11 gold atoms. The relevance of reduced particle diameters for high-resolution electron microscopy was clearly demonstrated. The applicability of undecagold for light microscopy and bioassays appears to be limited because silver enhancement of undecagold clusters is not often successful. This is most likely due to the fact that the gold core is encapsulated in an organochemical structure (J. Hainfeld, personal communication).

Contrary to the larger sized particles, *uncoated* ultrasmall colloidal gold particles are not stable and will aggregate in time. However, after adsorption to proteins the particles are stabilized and maintain their original size.

Gold probes based on ultrasmall particles are called *ultrasmall gold probes*. They have been commercially available since early 1989. Ultrasmall gold probes differ in their characteristics and performance when compared with probes built around larger particles. Whereas the "classic" probes can be considered as particles coated with proteins, ultrasmall gold probes can be considered as proteins coated with one or more gold particle(s).

These features are favorable, for example, in ligand coupling and receptor–ligand interaction studies. A reduction of the particle size down to 1 nm or smaller will lead to a dramatic decrease in the particle weight, which likely favors increased diffusion rates. The surface-to-volume ratio also increases favorably (Leunissen and van de Plas, 1993).

A reduction of the gold particle size not only leads to a reduced probe size: colloidal gold particles and gold probes behave as negatively charged particles/conjugates as well. Many proteins are negatively charged at physiological pH. Following aldehyde fixation the net negative charge (at the pH of incubation) will be even more pronounced. These characteristics may result in electrostatic repulsion between gold probes and aldehyde-fixed specimens, which in turn may generate low background staining but create relatively high-energy barriers that must be overcome in the approach and binding of gold probes to antigens. With ultrasmall gold probes charge-determined repulsion will be diminished.

Ultrasmall gold particles can be visualized directly, for example, with high-angle annular dark-field STEM (scanning TEM) (Humbel *et al.*, 1991; Otten *et al.*, 1992). A widely used method that manipulates the gold signal before visualization (with or without artificial contrast enhancement) is the immunogold silver staining (IGSS) technique (Danscher, 1981; Danscher and Rytter-Nörgaard, 1983; Holgate *et al.*, 1983). From theoretical studies and practical testing, a nucleus of only two gold atoms (or four silver atoms) has an electron cloud configuration that favors the catalyzed electron transfer from reducing agents (developer molecules) to silver ions (Hamilton and Logel, 1974). Metallic silver is deposited on the colloidal gold particles by physical development and the gold particles increase in size (making them easily detected by electron microscopy and epipolarization microscopy). Next to this increase in size a dark brown-to-black signal is generated with high contrast in the light microscope and on blots. Results obtained in this way are comparable to those found with immunofluorescence microscopy, with the advantage that fading of the signal with time does not occur. Danscher (1981) provided a useful recipe for the silver enhancement of colloidal gold. There have been several reviews dealing with a number of aspects and modifications of silver enhancement (Scopsi, 1989; Hacker, 1989). In combination with controlled silver enhancement, ultrasmall gold particles are excellent markers for electron microscopy, light microscopy, and immunoblotting. As such, ultrasmall gold probes are valuable tools for integrated immunolabeling studies.

The incubation protocol for the indirect immunodetection of antigens described in the following sections has been developed especially for ultrasmall gold probes but can be universally applied, independent of the particle size; however, for most EM applications silver enhancement is omitted when "classic" probes are used.

IV. Application of Ultrasmall Gold Probes

A. Reagents and Solutions

1. Buffer Solutions

Phosphate-buffed saline (PBS): 10 mM phosphate buffer, 150 mM NaCl, pH 7.6

Aldehyde-block buffer: PBS (pH 7.6) supplemented with 0.05 M glycine (Merck, Rahway, NJ) (LM/EM) or 0.05 M hydroxyl ammonium chloride (LM/EM), or 0.1% NaBH$_4$ (Merck) (LM), freshly prepared

Incubation buffer: PBS (pH 7.6) supplemented with 0.8% bovine serum albumin (BSA) (fraction V; Sigma, St. Louis, MO), 0.1% cold-water fish skin gelatin (Sigma), 20 mM NaN$_3$ (Merck); check the pH and adjust to 7.6 if necessary

Block buffer: Incubation buffer supplemented with 5% bovine serum albumin and 5% normal serum prepared from the same species as the ultrasmall gold-labeled secondary antibody; for immunoblotting the normal serum can be left out

Detection buffer: Incubation buffer supplemented with 1% normal serum and 2% glutaraldehyde (BDH, Poole, England) in PBS (pH 7.6)

2. Immunodetection Reagents

Specific primary antibody, preferably affinity purified

Ultrasmall gold-labeled reagents [Amersham (Arlington Heights, IL), Aurion (Wageningen, the Netherlands), BioCell (Cardiff, United Kingdom)]

3. Silver Enhancement Reagents

Ready-to-use reagents:
-Aurion R-gent (Aurion)

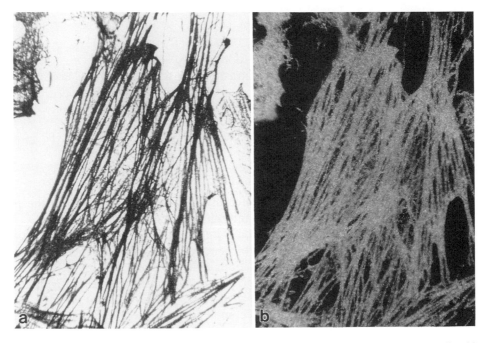

Fig. 1 Light microscopic detection of stress fibers in cultured fibroblasts with ultrasmall gold probes and silver enhancement. Coverslip cultures were fixed in 4% paraformaldehyde and permeabilized with 0.1% triton X-100. A polyclonal antibody to actin was used as primary antibody. (a) Bright-field mode. (b) Epipolarization image of the same specimen.

-Biocell silver enhancement kit (BioCell)
-IntenSE (Amersham)
-Gum arabic (for EM)
Silver enhancement according to Danscher (1981) (for EM):
Protecting colloid: Dissolve 100 g of gum arabic (Merck/Fluka) in 200 ml distilled water (this lasts several days). Filter the solution through layers of gauze. The solution can be aliquoted and stored frozen
Citrate buffer: Dissolve 2.55 g citric acid·H_2O [Merck/Fluka (Buchs, Switzerland)] and 2.35 g of sodium citrate·$2H_2O$ (Merck/Fluka) in distilled water to make 10 ml
Reducing agent: Dissolve 0.57 g of hydroquinone (Fluka) in 10 ml of distilled water. Prepare immediately before use and protect from light
Silver ion supply: Dissolve 0.073 g of silver lactate (Fluka) in 10 ml of distilled water. Prepare immediately before use and protect from light

4. Additional Reagents

For light microscopy:
Counterstaining; e.g., hematoxylin/eosin (Fluka)

Ethanol (Merck/Fluka)

Xylene (Merck/Fluka)

Pertex (Histolab, Göteborg, Sweden)

For electron microscopy:

Heavy metal contrasting solution; e.g., uranyl acetate/Reynold's lead citrate

Methylcellulose (Fluka) for ultrathin cryosections

B. Incubation Procedure

1. General Remarks

a. Light Microscopy

Living cells are preferably incubated at 0–4°C or in the presence of 0.05–0.02% NaN$_3$ to avoid internalization of immunoreagents.

For immunodetection of intracellular antigens, permeabilization steps may be required. The degree of penetration of immunoreagents depends on specimen characteristics and fixation on the one hand, and on incubation time on the other hand. For the detection of cytoskeletal antigens or antigens associated with the cytoskeleton the use of 0.1–0.5% Triton X-100 (Sigma) (applied only after aldehyde fixation) is recommended. Successful labeling of intracellular antigens in nerve tissue has been reported in the absence of detergent (van Lookeren Campagne, 1991).

Monolayers (or sections mounted) on coverslips can be easily incubated with six-well culture plates (Falcon®, Beckton-Dickinson, Etten-Leur, The Netherlands). Washing steps are carried out on a rocking table, using about 2 ml of washing medium per well.

Cell suspensions are gently pelleted after each incubation step. The pellets are resuspended in the medium used in the next step. Incubations are carried out on a rocking table.

Sections mounted on microscope slides are washed in 250-ml staining trays with a separate slide rack, on a magnetic stirrer.

The specimens must be fully covered by the immunoreagents (in practice, 50–200 μl is sufficient). Incubate in a humid chamber to prevent the specimens from drying. Incubations are carried out at room temperature. Incubation times can be shortened by using a microwave oven (Jackson et al., 1988).

b. Electron Microscopy

Preembedding immunolabeling: Use the marking set-up as described for light microscopy. If penetration problems occur, Triton X-100 may be applied (applied only after aldehyde fixation) but the ultrastructure may suffer from this treatment. For some specimens a compromise between ultrastructural preservation and penetration can be obtained with saponin (Sigma) or the use of a graded series of ethanol followed by 100% acetone and rehydration in the ethanol series.

If necessary, osmium tetroxide fixations may be used. Osmification must be performed before silver enhancement because osmium tetroxide removes metallic silver by oxidation.

Postembedding immunolabeling: The use of nickel grids covered with a carbon-coated Formvar or Parlodion (Pelco Int., Reddington, CA) film is recommended because silver enhancement procedures will be used. For most applications grids are floated on top of 50-μl drops of (immuno)reagents on a sheet of Parafilm.

c. Immunoblotting

Depending on the size of blots or strips, incubations can be carried out in sealed plastic bags, Petri dishes, or in disposable screw-cap sealed tubes. The volume of immunoreagents needed is between 1 and 5 ml. It is important to use a rocking table or tilting apparatus for all incubation steps except blocking.

2. Actual Procedure

Aldehyde-fixed specimens are incubated in 0.05 M glycine or hydroxyl ammonium chloride (LM/EM), or in 0.1% $NaBH_4$ (LM) for 15 min to inactivate residual aldehyde groups. Rinse in PBS twice (1 min each) before proceeding.

1. Incubate in block buffer at room temperature for 30 min. For blots, incubate in block buffer at 45°C for 30 min.

2. Incubate with primary antibody (1–5 μg/ml) in detection buffer for 30–60 min.

3. Remove unbound primary antibody by washing in incubation buffer three times (10 min each). The washing should be extended to six times (5 min each) for EM on-grid marking.

4. Incubate with the appropriate ultrasmall gold conjugate reagent, diluted according to manufacturer instructions, in detection buffer for 120 min. It is recommended that the optimal dilution for each new localization study be tested.

5. Remove excess gold probe by washing in incubation buffer four times (10 min each). The washing steps should be extended to eight times (5 min each) for EM on-grid marking.

6. Wash in PBS three times (5 min each).

7. Postfix in 2% glutaraldehyde in PBS for 15 min.

8. Wash in distilled water three times (5 min each).

9. Visualize the gold signal with silver enhancement as follows: For LM and immunoblotting, mix the enhancer and the initiator/developer in a 1 : 1 ratio immediately before use. The specimens should be fully covered by the mixture. Typical enhancement times are between 15 and 30 min (room temperature). For LM the on-going process can be monitored with an (inverted) light microscope with dimmed light conditions.

To obtain a homogenous enhancement for EM, gum arabic [50% (w/v) in distilled water] is added to the enhancement mixture in a 1:2 ratio. Typical enhancement time is between 10 and 20 min.

An alternative to EM visualization is the silver enhancement procedure described by Danscher (1981): Prepare the enhancement solution by mixing 6 ml of gum arabic, 1 ml of citrate buffer, and 1.5 ml of hydroquinone. Add 1.5 ml of silver lactate (the jar must be protected from light) and mix carefully. The enhancement solution should be used immediately, and the silver enhancement procedure should be shielded from direct light. After 20 min at room temperature the ultrasmall gold–silver particles will have reached a diameter between 5 and 10 nm.

10. Wash the specimens with an excess of distilled water three times (5 min each). For complete removal of the gum arabic more but shorter wash steps are necessary (e.g., 10 times, 2 min each).

11. For light microscopy, specimens can be counterstained in, for example, hematoxylin/eosin or basic fuchsin. They are dehydrated in a graded series of ethanol and xylene and embedded in a *water-incompatible* mounting medium (e.g., Pertex). This results in a permanent gold–silver signal.

For electron microscopy, uranyl acetate and lead citrate (Reynold's) can be used for heavy metal contrasting. Plastic sections are air dried. Ultrathin cryosections are protected from drying artifacts by mounting in a thin layer of methyl cellulose.

Immunoblots are air dried.

C. Troubleshooting

Background staining, a weak signal intensity, or no signal at all does sometimes occur in immunogold silver staining.

1. Prevention of Background Staining

To unravel the cause of background staining it is advisable to include at least the following controls in every new experimental set-up:

Control 1 (omission of the primary antibody): Specimens are incubated with the gold-labeled secondary antibody and silver enhanced.
Control 2 (omission of primary and secondary antibody): Specimens are only silver enhanced.

When both control 1 and control 2 are negative the primary antibody has caused an unintended staining pattern.

1. To evaluate performance and usability of every new primary antibody, test a dilution series.

2. In sera many clones of antibodies are present and some of them may be directed to other antigens. A test with preimmune serum may elucidate this. Affinity purify the serum or use a different, well-characterized serum or primary antibody.

3. Polyclonal antibodies recognize many different epitopes. Sometimes these epitopes are present in other compounds besides the one under investigation. In this case cross-adsorption will substantially improve the specificity. The remaining activity will be sufficient for a successful immunolabeling. If the same problem occurs with a monoclonal antibody it is necessary to use a clone recognizing a different epitope of the antigen to be detected.

When control 1 is positive, background staining is caused by the gold-labeled secondary antibody. The following suggestions may be helpful.

1. If possible avoid the use of positively charged adhesives such as poly-L-lysine. If charge-based interactions occur between the gold probe and the specimen itself (e.g., with collagen or histone proteins, which carry positive charges), negatively charged additives in the incubation and detection buffer can prevent this type of background. In practice 0.1–0.2% acetylated BSA (BSAc) (Aurion) in PBS (pH 7.6) is found to be most effective. In the immunoincubation procedure insert two washing steps with 0.1–0.2% BSAc in PBS before the incubation step with the ultrasmall gold reagent. Use this BSAc buffer also to dilute the ultrasmall gold probe and to remove unbound gold reagent.[1]

2. Increase the number of washing steps after the secondary antibody incubation.

3. When the positive signal is strong, it may be helpful to dilute the gold reagent further in order to improve the signal-to-noise ratio.

When background staining is present in control 2, it may be necessary to apply silver enhancement for longer than 25 min. If so, it is advisable to follow a two-step procedure to avoid autonucleation. After an initial enhancement period of 25 min, wash the specimen extensively in distilled water and proceed with the second enhancement step, using freshly mixed reagents.

2. Weak (or Absent) Signal Intensity

It is advisable to include a positive control, using a specimen in which the reactivity of the primary antibody is known to be positive and using the same

---

[1] More information on the subject of background prevention can be found in *Aurion Newsletter* No. 1 (October 1990).

secondary reagents. If this control gives satisfactory results it can be concluded that the procedure is reliable.

If staining of the positive control is also too weak, troubleshooting should be directed to: the activity of the immunogold/silver reagents, the reactivity of the primary antibody, and the fixation or embedding medium/procedure.

Activity of Silver Enhancement Reagents

The activity of silver enhancement agents can be easily tested. Apply 1 μl of a dilution of immunogold reagent in PBS (1/10, 1/100, and 1/1000) on a strip of nitrocellulose membrane. Allow the spots to dry and wash the strip briefly in distilled water. Proceed with silver enhancement. The last spot should be visible after 20 min of enhancement.

Bioactivity of Immunogold Probe

The bioactivity of the immunogold probe can be assessed in a dot-spot test (Moeremans *et al.*, 1984). One microliter of a serial dilution (250–0.1 μg/ml) of the primary antibody in PBS plus 50 μg of BSA/ml is spotted on a strip of nitrocellulose paper. The spots are allowed to dry. The strip is washed in PBS. Proceed as for immunoblotting. After silver enhancement 1 ng of primary antibody should be detected.

When the problem of low signal intensity is not caused by the immunogold silver reagents the reactivity of the primary antibody is suboptimal or the antigen was destroyed, in which case the fixation and/or embedding procedure must be adapted.

Effect of Chemical Fixation on Immunoreactivity

The effect of chemical fixation on immunoreactivity can be tested with the antigen-spot test (Moeremans *et al.*, 1984). In this way the effect of a fixation regime on preservation of antigenicity can be evaluated without the use of time-consuming preparation and microscopic techniques.

For electron microscopic applications the influences of the preparation technique and the immunodetection procedure should be evaluated by light microscopy, for example, by optimizing the labeling conditions on semithin cryosections (LM) and extrapolating these conditions to the electron microscopic level.

V. Discussion

The application of ultrasmall colloidal gold probes with particle diameters as low as 1 nm was reported for the first time by Leunissen *et al.* (1989) and van de Plas and Leunissen (1989) for the light microscopic detection of tubulin, by Moeremans *et al.* (1989) for immunoblotting, and by van Bergen en Henegouwen and van Lookeren Campagne (1989) for the subcellular detection of phosphoprotein B-50. Overall it was found that the use of ultrasmall gold probes and silver enhancement gave an increase in sensitivity of two to three times when compared to results obtained with immunogold probes built around 5- to 6-nm particles. Nielsen and Bastholm (1990) achieved similar results with ultrathin cryosections of the pituitary gland labeled for growth hormone and laminin.

Studies by Humbel *et al.* (1991) and Otten *et al.* (1993), using a field emission gun and high-angle annular dark-field scanning transmission electron microscopy, demonstrate the usefulness of immunolabeling with ultrasmall gold probes without silver enhancement. This combination offers the means for direct visualization and high-resolution detection.

However, silver enhancement is necessary to visualize ultrasmall gold probes in normal transmission and scanning electron microscopy and in light microscopy and bioassays such as immunoblotting.

The silver amplification method of Danscher (1981) gives the most homogeneous and efficient enhancement of ultrasmall gold particles (Stierhof *et al.*, 1991, and unpublished observations). This is of particular importance when ultrasmall gold probes are used in electron microscopy. A homogeneous silver enhancement of ultrasmall gold particles also opens the possibility of double-labeling experiments (Leunissen and van de Plas, 1993).

The gum arabic present in the silver enhancement mixture (Danscher, 1981) plays a major role in the enhancement characteristics. Addition of gum arabic to the commercially available silver enhancement reagents to a final concentration of 30–50% brings about results similar to those achieved by the low-pH Danscher method (Stierhof *et al.*, 1991, and unpublished observations).

The use of gum arabic is not required in applications in which homogeneity of the diameter of the ultrasmall gold/silver particles is not important (e.g., in light microscopy and immunoblotting).

The method published by Bienz *et al.* (1986), using a silver halogenide coating as the silver source, gives excellent and reproducible enhancement of 10-nm gold particles. We have no experience with this method for the enhancement of ultrasmall probes.

In using immunogold probes based on "large" gold particles on cryosections, it has been observed, both by stereo micrograph imaging (Van Bergen en Henegouwen and Leunissen, 1986) and by cutting transverse sections from plastic-embedded immunolabeled cryosections (Stierhof *et al.*, 1986), that even in ultrathin cryosections (where the biological material is fully hydrated during incubations) only a limited degree of penetration exists. From these studies it became evident that the "classic" gold label remains restricted to the exposed section surface.

The use of ultrasmall gold probes provides the possibility of detecting antigens inside cryosections, as we demonstrated by the labeling of tubulin in semi- and ultrathin cryosections of Chinese hamster ovary cells (Leunissen and van de Plas, 1993).

The features of ultrasmall gold probes provide a sound basis for preembedding electron microscopic studies. We compared the penetrative power and labeling efficiency of 5-nm gold probes versus ultrasmall gold probes for the intracellular labeling of tubulin in 1% glutaraldehyde-fixed whole-mount epithelial rat kangaroo cells at the light microscopic level, in order to evaluate permeabilizing agents and procedures that might be more suited to adequate ultrastructural preservation (van de Plas and Leunissen, 1989).

Extensive Triton X-100 treatment during fixation permeabilized the cell membrane and resulted in the extraction of the soluble fraction of the cell, including subcellular structures. Little difference was observed between the labeling densities of the two types of gold probes. Triton X-100 treatment after fixation did not produce the same effect on the subcellular components because these were, at least in part, prefixed. In such specimens primary antibodies and ultrasmall gold probes diffuse readily to their targets, whereas 5- to 6-nm probes appear to be hampered by the cytoplasmic matrix. This is even more pronounced when saponin (Willingham, 1983) is used as a permeabilizing agent.

When the cells were fixed without any permeabilization, intracellular labeling was found only when the ultrasmall probes were used. By using incubation periods of 1 hr for the primary antibody and 2 hr for the ultrasmall gold probe, intense labeling of microtubules was found in those cytoplasmic areas that border the cell membrane. Extending the incubation with the ultrasmall gold probe from 2 hr to overnight showed that cells were labeled throughout the cytoplasm.

With controlled fixation and extended incubation periods, possibilities exist for preembedding labeling of specimens without extensive permeabilization steps. This is also demonstrated by the work of van Lookeren Campagne (1991) in nerve cells, De Graaf et al. (1991) for labeling of the nuclear matrix, and by Yi et al. (1990) for the labeling of human alveolar macrophages.

The use of ultrasmall gold probes and silver enhancement leads to increased sensitivity in immunoblotting. In light and electron microscopy increased density of immunostaining and high-resolution immunolabeling can be obtained. Because ultrasmall gold probes can be used at all levels of immunostaining they can be employed in integrated labeling studies. These advantages, in combination with the improved penetration characteristics, make probes based on ultrasmall gold particles valuable tools in immuno(cyto)chemistry.

References

Baschong, W., and Lucocq, J. M. (1985). "Thiocyanate gold": Small (2–3 nm) colloidal gold for affinity cytochemical labeling in electron microscopy. *Histochemistry* **83,** 409–411.

Beesley, J. E. (1985). Colloidal gold: A new revolution in marking cytochemistry. *Proc. RMS* **20,** 187–197.

Bienz, K., Egger, D., and Pasamontes, L. (1986). Electron microscopic immunocytochemistry. Silver enhancement of colloidal gold marker allows double labeling with the same primary antibody. *J. Histochem. Cytochem.* **34,** 1337–1342.

Brada, D., and Roth, J. (1984). "Golden Blot." Detection of polyclonal and monoclonal antibodies bound to antigens on nitrocellulose by protein-A gold complexes. *Anal. Biochem.* **142,** 79–83.

Ciesolka, T., and Gabius, H. J. (1988). An 8- to 10-fold enhancement in sensitivity for quantitation of proteins by modified application of colloidal gold. *Anal. Biochem.* **168,** 280–283.

Danscher, G. (1981). Localization of gold in biological tissue. A photochemical method for light and electron microscopy. *Histochemistry* **71,** 81–88.

Danscher, G., and Rytter-Nörgaard, J. (1983). Light microscopic visualization of colloidal gold on resin-embedded tissue. *J. Histochem. Cytochem.* **31,** 1394–1398.

De Graaf, J., Van Bergen en Henegouwen, P. M. P., Meijne, A. M. L., Van Driel, R., and Verkleij, A. J. (1991). Ultrastructural localization of nuclear matrix proteins in HeLa cells using silver enhanced ultra small gold probes. *J. Histochem. Cytochem.* **39,** 1035–1046.

De Harven, E., Leung, R., and Christensen, H. (1984). A novel approach for scanning electron microscopy of colloidal gold labeled cell surfaces. *J. Cell Biol.* **99,** 53–57.

De Mey, J. (1983). Colloidal gold probes in immunocytochemistry. *In* "Immunocytochemistry: Practical Application in Pathology and Biology" (S. Polak and S. Van Noorden, eds.), pp. 82–112. Wright, London.

De Waele, M., Renmans, W., Segers, E., Jochmans, K., and Van Camp, B. (1988). Sensitive detection of immunogold-silver staining with darkfield and epi-polarization microscopy. *J. Histochem. Cytochem.* **36,** 679–683.

Faulk, W. P., and Taylor, G. M. (1971). An immunocolloid method for the electron microscope. *Immunochemistry* **8,** 1081–1083.

Frens, G. (1973). Controlled nucleation for the regulation of the particle size in monodisperse gold suspensions. *Nature (Phys. Sci.)* **241,** 20–22.

Geiger, B., Dutton, H., Tokuyasu, K. T., and Singer, S. J. (1981). Immunoelectron microscope studies of membrane filament interactions: Distribution of α-actinin, tropomyosin and vinculin in intestinal brush border and chicken gizzard smooth muscle cells. *J. Cell Biol.* **91,** 614–628.

Geoghean, W., and Ackerman, G. (1981). Adsorption of horseradish peroxidase, ovomucoid and anti immunoglobulin to colloidal gold for the indirect detection of concanavalin A, wheat germ agglutinin and goat anti-human immunoglobulin G on cell surfaces at the electron microscopic level, new method, theory and application. *J. Histochem. Cytochem.* **25,** 1187–1200.

Geuze, H. J., Slot, J. W., Van der Ley, P., Schuffer, R., and Griffith, J. (1981). Use of colloidal gold particles in double labeling immuno electron microscopy on ultrathin frozen sections. *J. Cell Biol.* **89,** 653–665.

Goodman, S. L., Hodges, G. M., Trejdosiewicz, L. K., and Livingstone, D. C. (1981). Colloidal gold markers and gold probes for routine application in microscopy. *J. Microsc. (Oxford)* **123,** 201–213.

Gross, D. U., and De Boni, U. (1990). Colloidal gold labeling of intracellular ligands in dorsal root sensory neurons, visualized by scanning electron microscopy. *J. Histochem. Cytochem.* **38,** 775–784.

Hacker, G. W. (1989). Silver enhanced colloidal gold for light microscopy. *In* "Colloidal Gold: Principles, Methods, and Applications" (M. Hayat, ed.), Vol. 1, pp. 297–321. Academic Press, San Diego.

Hainfeld, J. (1987). A small gold-conjugated antibody label: Improved resolution for electron microscopy. *Science* **236,** 450–453.

Hamilton, J. F., and Logel, P. G. (1974). The minimum size of silver and gold nuclei for silver physical development. *Photogr. Sci. Eng.* **18,** 507–511.

Hayat, M., ed. (1989). "Colloidal Gold: Principles, Methods, and Applications," Vols. 1 and 2. Academic Press, San Diego.

Holgate, C. S., Jackson, P., Cowen, P. N., and Bird, C. (1983). Immunogold-silver staining: New method of immunostaining with enhanced sensitivity. *J. Histochem. Cytochem.* **31,** 938–944.

Horisberger, M. (1981). Colloidal gold: A cytochemical marker for light and fluorescent microscopy and for transmission and scanning electron microscopy. *Scanning Electron Microsc.* **2,** 9–28.

Humbel, B. M., Stierhof, Y.-D., Hermann, R., Otten, M. T., and Schwarz, H. (1991). Ultra-small gold particles directly visualized on immuno gold labelled resin sections. *Proc. Dutch Soc. Electron Microsc.* p. 42.

Jackson, P., Lalani, E. N., and Goutsen, J. (1988). Microwave-stimulated immunogold silver staining. *Histochem. J.* **20,** 353–358.

Krenács, T., Krenács, L., Bozoky, B., and Ivanyi, B. (1990). Double and triple immunocytochemical labeling at the light microscope level in histopathology. *Histochem. J.* **22,** 530–536.

Leunissen, J. L. M., and De Mey, J. R. (1989). Preparation of gold probes. *In* "Immunogold Labeling in Cell Biology" (A. J. Verkleij and J. L. M. Leunissen, eds.), pp. 3–16. CRC Press, Boca Raton, Florida.

Leunissen, J. L. M., and van de Plas, P. F. E. M. (1993). Ultra small gold probes in cryoultramicrotomy. *In* "Immunoelectron Microscopy in Virus Diagnosis and Research" (B. Eaton and A. Hyatt, eds.), pp. 327–348. CRC Press, Boca Raton, Florida.

Leunissen, J. L. M., van de Plas, P. F. E. M., and Borghgraef, P. E. J. (1989). Auroprobe One: A new and universal ultra small gold particle based (immuno)detection system for high sensitivity and improved penetration. *In* "Aurofile" (Janssen Life Sciences, ed.), Vol. 2, pp. 1–2. Janssen Life Sci., Olen, Belgium.

Moeremans, M., Daneels, G., van Dijck, A., Langanger, G., and De Mey, J. (1984). Sensitive visualization of antigen–antibody reactions in dot and blot immuno overlay assays with the immunogold and immunogold/silver staining. *J. Immunol. Methods* **74,** 353–360.

Moeremans, M., Daneels, G., and De Mey, J. (1985). Sensitive colloidal gold or silver staining of protein blots on nitrocellulose membranes. *Anal. Biochem.* **145,** 315–321.

Moeremans, M., Daneels, G., De Raeymaeker, M., and Leunissen, J. L. M. (1989). Auroprobe One in immunoblotting. *In* "Aurofile" (Janssen Life Sciences, ed.), Vol. 2, pp. 3–4. Janssen Life Sci., Olen, Belgium.

Nielsen, M. H., and Bastholm, L. (1990). Improved immunolabeling of ultrathin cryosections using antibody conjugated with 1-nm gold particles. *Proc. Int. Congr. Electron Microsc., 12th, San Francisco* pp. 928–929.

Novikoff, A. (1980). DAB cytochemistry: Artifact problems in its current uses. *J. Histochem. Cytochem.* **28,** 1036–1038.

Otten, M. T., Stenzel, D. J., Cousens, D. R., Humbel, B. M., Leunissen, J. L. M., Stierhof, Y.-D., and Busing, W. M. (1992). High angular annular dark-field STEM imaging of immuno-gold labels. *Scanning* **14,** 282–289.

Scopsi, L. (1989). Silver-enhanced colloidal gold method. *In* "Colloidal Gold: Principles, Methods, and Applications" (M. Hayat, ed.), Vol. 1, pp. 251–295. Academic Press, San Diego.

Slot, J. W., and Geuze, H. J. (1981). Sizing of protein A-colloidal gold probes for immuno-electron microscopy. *J. Cell Biol.* **90,** 533–536.

Stierhof, Y.-D., Schwarz, H., and Hermann, F. (1986). Transverse sectioning of plastic embedded immuno-labeled cryosections: Morphology and permeability to protein A–gold complexes. *J. Ultrastruct. Mol. Struct. Res.* **97,** 187–196.

Stierhof, Y.-D., Humbel, B. M., and Schwarz, H. (1991). Silver enhancement of ultra-small gold markers on immunolabelled ultra-thin resin sections. *Proc. Dutch Soc. Electron Microsc.* p. 49.

Tokuyasu, K. T., and Singer, S. J. (1976). Improved procedures for immunoferritin labeling of ultrathin frozen sections. *J. Cell Biol.* **71,** 894–906.

Tschopp, J., Podack, E. R., and Muller-Eberhard, H. J. (1982). Ultrastructure of the membrane attack complex of complement. Detection of the tetramolecular C9-polymerizing complex C5b-8. *Proc. Natl. Acad. Sci. U.S.A.* **79,** 7474–7478.

van Bergen en Henegouwen, P., and Leunissen, J. (1986). Controlled growth of colloidal gold particles and implications for labelling efficiency. *Histochemistry* **85,** 81–87.

van Bergen en Henegouwen, P., and Van Lookeren Campagne, M. (1989). Subcellular localization of phosphoprotein B-50 in isolated presynaptic nerve terminals and in young adult rat brain using a silver-enhanced ultra-small gold probe. *In* "Aurofile" (Janssen Life Sciences, ed.), Vol. 2, pp. 6–8. Janssen Life Sci., Olen, Belgium.

van de Plas, P. F. E. M., and Leunissen, J. L. M. (1989). Immunocytochemical detection of tubulin in whole mount preparations of PtK2-cells: Improved penetration characteristics of Auroprobe One. *In* "Aurofile" (Janssen Life Sciences, ed.), Vol. 2, pp. 3–4. Janssen Life Sci., Olen, Belgium.

van Lookeren Campagne, M. (1991). Redistribution of the growth-associated protein B-50 during neuronal development and maturation. Ph.D. Thesis, Univ. of Utrecht, Utrecht.

Verkleij, A. J., and Leunissen, J. L. M. (1989). "Immuno-Gold Labeling in Cell Biology." CRC Press, Boca Raton, Florida.

Watkins, S. C., Raso, V., and Slayter, H. S. (1990). Immunoelectron-microscopic studies of human platelet thrombospondin, Von Willebrand factor and fibrinogen redistribution during clot formation. *Histochem. J.* **22,** 507–518.

Willingham, M. (1983). An alternative fixation-processing method for preembedding ultrastructural immunocytochemistry of cytoplasmic antigens, the GBS glutaraldehyde–borohydride–saponin procedure. *J. Histochem. Cytochem.* **31,** 791–798.

Yi, H., Pueringer, R. J., Moore, K. C., and Hunninghake, G. W. (1990). Immunolocalization of human alveolar macrophage prostagladin H synthetase using 1-nm gold probe and silver enhancement techniques. *Proc. Int. Congr. Electron Microsc., 12th, San Francisco,* pp. 362–363.

Yokata, S. (1988). Effect of particle size on labeling density for catalase in protein A-gold immunocytochemistry. *J. Histochem. Cytochem.* **36,** 107–109.

CHAPTER 14

Immunogold Electron Microscopy: Mapping Tubulin Isotypes on Neurite Microtubules

Harish C. Joshi

Department of Anatomy and Cell Biology
Emory University School of Medicine
Atlanta, Georgia 30322

I. Introduction

The modern cell biologist is challenged to describe cellular structure and function at the resolution of molecules. The impact of exploiting antibodies to discover, isolate, and characterize the molecular components of cells is evident throughout this volume of *Methods in Cell Biology*. This article describes immunogold electron microscopy, a powerful ultrastructural approach, to visually map the location of cytoskeletal molecules in their natural intracellular environment(s). In this technique, gold particles of distinct size are linked to an antibody, and the antibody–gold conjugate is allowed to react with cells or tissues containing the specific antigenic epitopes. The location of the antigen is then detected with an electron microscope by visualizing the gold particles.

The foundation for immunoelectron microscopy was laid in 1959 when Singer first coupled an antibody to ferritin, an electron-dense crystalline protein (Singer, 1959). This iron-rich protein conferred high electron density on the antibody molecules without inactivating antigen binding. Thus the antibody–protein conjugates were visible with the electron microscope. A decade later, Faulk and Taylor (1971) made the next advance in this powerful technique by successful adsorption of colloidal gold particles to functionally active antibodies. Gold particles offered striking contrast and hence could be used after conventional staining techniques of transmission electron microscopy. In addition, through the ability of gold particles to generate secondary electrons, gold-tagged antibodies also proved useful as suitable markers in scanning electron microscopy (Horisberger, 1979).

The recognition of the potential of this technique in molecular cell biology led many workers to invest effort toward improving the quality of the reagents. The goals were twofold: (1) to devise standardized procedures that yield colloidal dispersions of uniformly sized gold particles, and (2) to determine conditions that enable efficient coupling of gold particles to the antibody of choice without compromising antigen binding. Standardized procedures have now emerged that succeed in achieving both of these goals. Briefly, colloidal gold is formed by releasing gold atoms from tetrachloroauric acid solution by controlled chemical reduction of it with either phosphorus-saturated ether (Horisberger, 1979), sodium ascorbate (Horisberger, 1979), or sodium citrate (Frens, 1973). The atomic gold aggregates then provide nuclei for further crystal growth. Appropriate choice of the reducing agent and careful manipulation of the reduction conditions allow both controlled nucleation and growth of gold microcrystals. These techniques result in reproducible colloidal gold suspensions of uniform average particle diameter with a small coefficient of variance (<15%) (Leunissen and De Mey, 1989). The narrow size distribution thus achieved allows simultaneous visualization of more than one antigen in the same intracellular structure if different populations of distinctly sized gold particles are used (a procedure known as double-label immunogold electron microscopy). The second goal of coupling gold particles to functionally active antibodies was accomplished by maximizing the hydrophobic interactions of the antibody at a reaction pH adjusted near the pI of the antibody (De Mey, 1983). These procedures are extensively reviewed elsewhere (De Mey, *et al.*, 1986; Leunissen and De Mey, 1989). Thus, the ease of preparing gold particles of narrow size distribution, their efficient coupling to antibody molecules, and their high contrast on stained biological material combine to provide a means to build a molecular map of cellular structures at the resolution of macromolecules.

II. Scope and Application

The purpose of this article is to introduce practical aspects of advances in immunogold electron microscopy and describe its usefulness in localizing individual cytoskeletal molecules on subcellular structures. Details concerning the preparation of colloidal gold and antibody coupling are not within the scope of this article but are treated comprehensively elsewhere (Leunissen and De Mey, 1989). Described here are the general principles of the protocols, followed by a specific example of mapping different tubulin gene products on single microtubules. The issue addressed in this example is to determine if animal cells use different tubulin polypeptide isotypes to construct cytoplasmic microtubules of different function.

The experimental paradigm employs a clonal cell line, rat PC12, that extends neurites in response to nerve growth factor (NGF). In the absence of NGF, PC12 cells grow mitotically as round, chromaffin-like cells. In response to NGF, PC12 cells withdraw from the cell cycle and differentiate into sympathetic neuron-like cells. In so doing, these cells transform their mitotic and cytoplasmic microtubule arrays into microtubules whose functions include neurite outgrowth, establishment and maintenance of asymmetric morphology, and fast axonal transport of vesicles through the cytoplasm. Specific β-tubulin genes expressed during differentiation in rat PC12 cells can be categorized into "induced" and "constitutive" β-tubulin isotypes (Joshi and Cleveland, 1989). A question that arises is whether "induced" isotypes construct functionally specialized subsets or segments of axonal microtubules. Immunogold electron microscopic methods presented here reveal a nonuniform "patchy" distribution of an "induced" neuron-specific tubulin isotype (type III) whose localization contrasts with the uniformity of constitutively expressed tubulin isotype (type IV). The functional significance this biochemical heterogeneity of axonal microtubules along microtubule lengths remains unresolved. Nonetheless, this example from our study underscores the utility of double-label immunogold electron microscopy for the detection and resolution of individual gene products within a cellular organelle assembled from similar, but not identical, polypeptides.

III. Labeling Protocols (Preembedding)

Immunogold localizations can be performed in three fundamentally different ways: (1) prior to embedding cells in plastic (preembedding), (2) on thin sections of plastic-embedded cells (postembedding), and (3) on the unembedded frozen sections or after freeze substitution. Two methods, postembedding and freeze substitution, are discussed in two separate articles of this volume (see articles 15 and 16). The preembedding method will be described here, as applied in our studies of neuronal microtubules. In the preembedding procedure, cells that are lightly fixed and permeabilized are incubated with the antibody reagents. This procedure is suitable for microtubule labeling because semi-quantitative, three-dimensional information about tubulin distribution is readily obtained (Geuens et al., 1986). The following protocols for staining neuronal microtubules are modified from the procedures developed for other cultured cell types by M. DeBrabander and others at Janssen Pharmecutica (Beerse, Belgium) (see, e.g., De Mey et al., 1986; Geuens et al., 1986). Details of the protocols we employed are listed following (also see Fig. 2).

A. Tubulin Antibodies and Characterization of Their Specificity

We have used two rabbit polyclonal antibodies, specific to type III or type IV β tubulins (originally described in Lopata and Cleveland, 1987), and a mouse monoclonal antibody that binds to both of these polypeptides (DM1B; Amersham Corp., Arlington Heights, IL). The tubulin isotype-specific polyclonal antibodies were raised against 9- to 17-amino acid synthetic peptides corresponding to consensus sequences for the carboxy-terminal domains of vertebrate β-tubulin isotypic classes III and IV (Fig. 1A). Antibodies were affinity purified according to Lopata and Cleveland (1987). To determine if type III and type IV antibodies are in fact specific to their respective tubulin isotypes, we needed bacterially expressed isotype-defining domains of β tubulin. We linked cDNA sequences encoding entire carboxy-terminal domains of type III and type IV β tubulins, in frame, to the amino-terminal 32-kDa of inducible protein trpE. The expression of the chimeric protein induced by indole acrylic acid is shown in Fig. 1B. The specificities of all three antibodies used in our study are determined by immunoblotting (Fig. 1B, ii–iv). We then determined the specificity of the antibody in PC12 cells by immunoblotting total protein extracts of these cells at different stages of their NGF-induced morphogenesis (Fig. 1C–E). We found that the antibodies specifically bind β tubulin throughout the differentiation of these round, mitotically active cells (on day 0 of NGF exposure) to a postmitotic cell with highly elaborate neuronal morphology (on day 6 of NGF exposure) (Fig. 1C–E).

B. Cell Culture and Their Differentiation

PC12 cells are maintained in culture and can be induced to differentiate either on glass coverslips (20 × 20 mm; VWR Scientific, San Francisco, CA), or plastic petri dishes as described earlier (Greene and Tischler, 1976; Joshi and Cleveland, 1989). There are distinct advantages of growing cells on glass coverslips or on plastic petri dishes. Glass coverslips offer optimal optical surfaces for sophisticated light microscopy and the ability to process this same coverslip with its attached cells for electron microscopy. Cells grown on plastic petri dishes offer ease of handling for electron microscopy but preclude some light microscopic analyses, for example, Nomarski optics (although Hoffman contrast microscopy is feasible). Both methods are described here. Nevertheless, when possible, the use of plastic petri dishes is recommended.

1. Cells on Plastic

Coat Permanox dishes (Lab-Tek division, Miles Laboratories, Inc., Naperville, IL) with 0.1% poly-L-lysine (Mazia *et al.*, 1975) by pipetting 1 ml of sterile solution onto the plating surface, cover, and leave for 30 min at room temperature. Wash three times with 5 ml of distilled sterile water, 30 sec each. To induce differentiation of PC12 cells, triturate cells from the surface of the stock culture dish and plate (3×10^4 cells/cm^2) onto the poly-L-lysine-coated dish. Add 100 ng of NGF/ml to the culture medium and change the medium every 48 hr for 6 days. (Permanox dishes are used because they are resistant to epoxy resins.)

2. Cells on Glass Coverslips

Spray the glass coverslips uniformly with the dry lubricant MS-123 release agent (Miller Stephenson Chemical Company, Inc., Danbury, CN), wipe dry with a Kim-wipe, and sterilize by placing the coverslips under ultraviolet (UV) light for 30 min. Place each coverslip (sprayed side up) into individual plastic petri dishes, coat the coverslips with poly-L-lysine, and plate the cells for differentiation as mentioned above.

All subsequent steps, including fixation, washing, incubation, and embedding can be performed on cells still in the original culture dish. At the completion of each step, solutions (e.g., fixatives, antibody solution, or washes) from culture dishes can be conveniently aspirated with a Pasteur pipette, or a plastic pipette tip connected to a vacuum trap. The aspirator pipette, when mounted on a stand, is so convenient that it has been called a "third hand" by De Mey *et al.*, (1986).

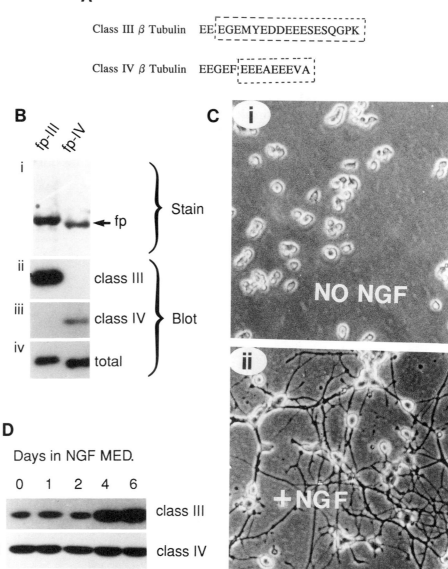

A

Class III β Tubulin   EE EGEMYEDDEEESESQGPK

Class IV β Tubulin   EEGEF EEEAEEEVA

B

fp-III fp-IV

i ← fp   } Stain

ii class III

iii class IV   } Blot

iv total

C

i   NO NGF

ii   +NGF

D

Days in NGF MED.

0 1 2 4 6

i   class III

ii   class IV

Fig. 1 Demonstration of specificities of the peptide-derived isotype-specific antibodies, both by bacterially expressed proteins and at different stages of NGF-induced neuronal differentiation of rat PC12 cells. (A) Carboxy termini of two evolutionarily conserved classes of β-tubulin polypeptides. Carboxy-terminal sequences of type III and type IV β tubulins are shown with the single-letter amino acid code. The boxed oligopeptide regions represent synthetic peptides used for generating isotype-specific antibodies. (After Sullivan and Cleveland, 1986.) The detailed procedures for

C. Fixation and Permeabilization of Cells

Cells are fixed and permeabilized at room temperature (25 ° C), using 2.0 ml for each change of solution.

1. Rinse twice, quickly, with phosphate-buffered saline (PBS) prewarmed at 37°C, to remove the culture medium.

2. Incubate for 1 min at room temperature with freshly prepared 1% Triton X-100 and 0.5% glutaraldehyde in buffer I [Hanks' buffered salt solution containing 5 mM piperazine-N,N'-bis(2-ethane sulfonic acid) (PIPES), 2 mM MgCl$_2$, and 2 mM ethylene glycol-bis(β-aminoethyl ether-N,N,N',N'-tetraacetic acid (EGTA) (pH 6), Geuens *et al.*, 1986].

3. Fix the cells again, for 1 min, with 0.5% glutaraldehyde in buffer I (in the absence of Triton X-100).

4. Wash the cells for 15 min with 0.5% Triton X-100 alone in buffer I (this step removes loose debris).

5. Reduce the unreacted aldehyde groups by incubating for 15 min with freshly prepared 0.5 mg/ml NaBH$_4$ in buffer I.

6. Remove the residual borohydride with TBS (10 mM Tris, 140 mM NaCl, pH 7.4) for 5 min.

D. Blocking and Immunostaining (Indirect Immunolabeling Procedure)

The volumes of the (valuable) antibody reagents for subsequent immunostaining steps can be minimized effectively by the following procedures. For cells grown on glass coverslips, remove the coverslips with a pair of forceps from the original dish and place, cell side up, into a fresh, dry plastic petri dish. Do not let

---

coupling the oligopeptides to keyhole limpet hemocyanin (KLH), immunizations, and affinity purification of resulting antibodies are described in Lopata and Cleveland (1987). (B) Specificity of affinity-purified antibodies to each of type III and type IV mammalian β-tubulins (i) Coomassie blue-stained SDS-polyacrylamide gel of bacterial lysates containing highly induced, cloned fusion proteins composed of 32-kDa bacterial trpE protein linked to carboxy-terminal (ca. 100) β-tubulin residues (ii and iii) Immunoblots of bacterial fusion proteins shown in (i), using peptide-derived isotype-specific polyclonal antibodies, type III (ii) and type IV (iii). (iv) Immunoblots of identical protein gels with a monoclonal antibody that recognizes both type III and type IV β tubulins. (C) Nerve growth factor (NGF)-induced differentiation of rat PC12 cells. Uninduced PC12 cells have a round morphology similar to chromaffin-like cells (i), which extend long cylindrical neurites on exposure to 100 ng of NGF/ml for 6 days in culture (ii). (D) Accumulation of two β-tubulin isotypes during NGF-induced differentiation of PC12 cells. (i and ii) Immunoblot analysis of type III (i) and type IV (ii) β tubulin accumulation during neurite outgrowth. Fifty micrograms of total cell protein from PC12 cells that had previously been treated with 100 ng of NGF/ml for 0, 1, 2, 4, and 6 days was electrophoresed on two parallel gels. Autoradiograms are of immunoblots with type III and type IV antibodies. (From H. C. Joshi and D. W. Cleveland, unpublished observations; parts of this work adapted from Joshi and Cleveland, 1989.)

Characterization of antibody/antigen-binding specificities

⬇

Cell culture plastic, glass, polylysine

⬇

Fixation and permeabilization

⬇

Reduction of unreacted aldehyde groups

⬇

Blocking
BSA fraction V (globulin free)
Serum of host species

⬇

Staining with primary antibody

⬇

Staining with secondary antibody

⬇

Postfixation

⬇

Processing for EM

⬇

Flat embedding, LM observation, marking, mounting

⬇

Sectioning

⬇

Poststaining (contrasting)

⬇

Micrographs, analysis

Fig. 2   Flow sheet showing a general scheme of preembedding immunogold labeling protocols.

the cells dry out at any point in this procedure. All incubations are performed by pipetting 70 μl of solution onto the coverslip. For cells grown on a plastic surface, scratch a circle (1.5- to 2.0-cm diameter) with a diamond marker pen (VWR Scientific) to delineate a small staining area on the surface of the dish. Wipe the cells outside the circle with a piece of folded Kim-wipe held by a pair of forceps. Do not let the cells within the circle dry out at any time. All the following incubations are performed by pipetting 70 μl of solution onto the circle.

1. Incubate the cells for 15 min at 37°C in blocking solution. [a 1:20 dilution of nonimmune goat serum, 0.2% bovine serum albumin (BSA) in 1× TBS (pH 7.4)]. This blocking solution is for the experiments requiring secondary antibodies made in goat. Do not use this blocking solution if gold-conjugated protein A will be used as a probe. Rather, use 2% fraction V BSA (this is free of globulins; Sigma, St. Louis, MO).

2. Incubate in the appropriate primary antibody diluted in blocking solution, for 3 hr at 37°C. For double-labeling experiments, we have used two primary antibodies simultaneously or successively without any significant differences in labeling efficiency.

3. Wash three times (10 min each) with 0.2% BSA in TBS (pH 7.4) (washing steps are done on a rocking platform).

4. Block as in step 1 with 5% goat serum and 0.2% BSA in TBS (pH 8.2).

5. Incubate, for 3 hr at 37 °C, with gold-labeled secondary antibodies. [We have used 5- and 10-nm gold-conjugated goat–anti-rabbit IgG (GAR5 and GAR10, respectively; Janssen Pharmecutica) diluted 1:3 in 0.2% BSA in TBS (pH 8.2)].

6. Wash five times (10 min each) with 0.2% BSA in TBS (pH 8.2). Proceed with the following postfixation and embedding steps. [Staining can be monitored under bright-field or differential contrast interference microscopy (DIC) (stain appears pale to red under bright-field microscopy and grayish black in DIC).]

E. Postfixation

1. Rinse cells with buffer I, and refix them for 30 min with 1% glutaraldehyde in buffer I.

2. Wash twice with buffer I (5 min each).

3. Osmicate by incubating with 2% OsO_4 for 30 min.

4. Impregnate with 0.5% uranyl acetate for 30 min.

5. Dehydrate with an ethanol series of 50, 60, 70, 80, 90, and 95% ethanol, and finally twice with 100% ethanol (5 min each), and embed in Epon as follows.

F. Embedding

The protocol for embedding depends on whether the cells were processed on a glass or plastic surface.

1. Cells on Glass Coverslips

Glass coverslips with their attached cells are flat-embedded in a silicon mold. Epon drops are placed in the middle of the mold (Electron Microscopy Sciences, Fort Washington, PA) and the coverslip is placed, cell side down, on the drops of Epon such that there are no air bubbles left between the cell surface of the coverslip and the Epon in the mold. Be careful not to allow Epon on the top surface of the coverslip. Cure the Epon for 30 hr at 50°C. Separate the thin wafer of the Epon polymer-containing cells from the glass coverslips by alternate treatment with liquid N_2 and boiling water. The embedded cells can then be examined by phase optics in the Epon wafer to select appropriate areas.

2. Cells on a Plastic Surface

Remove 100% ethanol from the petri dish and add a thin layer of Epon (a 0.3- to 0.4-cm thick layer) in the dish. The Epon is then cured in the dish for 30 hr at 50°C. The polymerized Epon is separated from the plastic by first breaking away the rims of the petri dish with the help of metal pliers. This yields a disk of plastic attached to a disk of polymerized Epon. Next, hold the disk vertically on the bench top and tap the top edge firmly with a hammer. This will begin the detachment of the two disks at the edge (usually at the point of contact with the hammer). The place where the two disks are detached will be clearly visible due to the insertion of air between the Epon disk and the original plastic disk. Pry the two disks apart by wedging a razor blade between the two disks at the point of initial detachment. The embedded cells can then be visualized in the polymerized Epon disk by phase optics. The area of interest can be identified and circled with a diamond mounted eccentrically onto an objective (Diamond Marker Objective, Zeiss).

The encircled areas of the Epon wafer are cut out and mounted, with unpolymerized Epon, onto small Epon cylinders prepolymerized in gelatin capsules (EM Sciences, Fort Washington, PA). These "sandwiches" are then baked at 50°C for 24 hr. Serial thin sections, of silver to gray interference color, are picked up onto a Polyvar-coated, carbon-stabilized electron microscopy (EM) grid. These EM grids are stained successively with lead citrate and uranyl acetate in the following manner.

1. Spread a piece of Parafilm on a flat surface.
2. Place 50-μl drops of lead citrate and uranyl acetate onto the Parafilm.
3. Float the EM grid, section side down, on a lead citrate drop for 1 min.
4. Wash (from a syringe) with a few drops of distilled water while holding the EM grid with a fine pair of forceps (water drops will trickle down the forceps tip, to the grid, and fall off).
5. Drain the residual water onto fiber-free filter paper by gently touching the paper to the grid edge.
6. Now stain with uranyl acetate for 1 min and repeat the wash step.

Grids are now ready for transmission electron microscopy. [This preembedding procedure satisfactorily labels cellular microtubules, both in the soma and in the neurites (see Results and Perspectives).]

G. Labeling of *in Vitro*-Polymerized Microtubules

Tubulin, purified from microtubule-associated proteins by phosphocellulose chromatography, is polymerized in a microcentrifuge tube (2 mg/ml) in 100 mM Na-PIPES (pH 6.94), 1 mM MgCl$_2$, 1 mM GTP, and 10 μg of taxol/ml at 37°C. Polymers were fixed in 0.5% glutaraldehyde and diluted 50× in TBS. A drop of the dilute microtubule suspension is applied to plastic-coated, carbon-stabilized nickel grid. The excess liquid is drained from the grid by gently touching the grid edge to a fiber-free filter paper. The EM grid with attached microtubules is processed for immunostaining by transferring it successively onto drops of the following solution placed on Parafilm in a humidified chamber.

1. Sterile water for 5 min
2. Blocking solution for 15 min
3. Primary antibody solution for 1 hr
4. Blocking solution, twice, 15 min each
5. Secondary antibody solution for 1 hr
6. Blocking solution twice, 15 min each
7. TBS for 1 min
8. Glutaraldehyde (1%) for 15 min, and the EM grid is washed with water

After immunolabeling, the grids are negatively stained for observation in the microscope.

IV. Results and Perspectives

We sought to determine how PC12 cells use, during their differentiation, the constitutively expressed type IV β tubulin to construct neurite microtubules as compared to an induced neuronal isotype, type III β tubulin. These preembedding immunogold studies have revealed, for the first time, subtle differences in the tubulin isotype composition of axonal microtubules. Examples of both single-label and double-label immunogold staining of microtubule populations of differentiated PC12 neurites are shown in Fig. 3. The uniform distribution of 5-nm gold particles along the entire lengths of neurite microtubules represents a (more or less) homogeneous incorporation of the constitutively expressed type IV β-tubulin isotype (Fig. 3A). In contrast, the neuron-specific β-tubulin type III that is induced as a function of the differentiation is localized only to short segments that alternate with stretches of unstained segments along the length of single microtubules. In contrast to a random assembly of type IV β tubulin, type III tubulin is therefore recruited into short segments of generally long axonal microtubules (Fig. 3B). Using the polyclonal type IV tubulin antibody (large, 10-nm gold particles) and a monoclonal antibody that binds both type III and

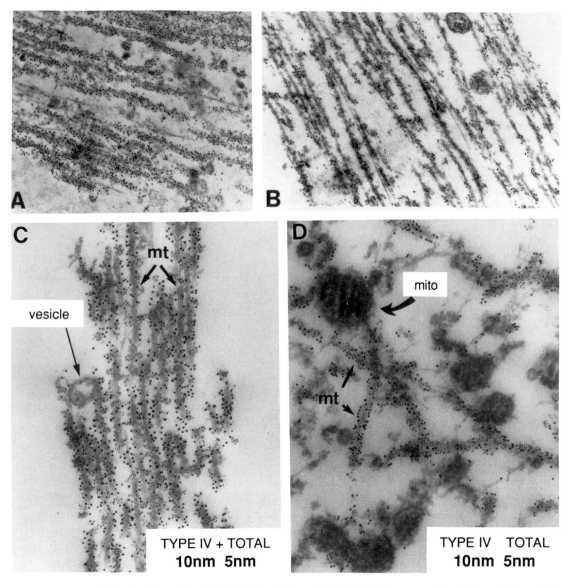

Fig. 3 Examples of both single-label and double-label immunogold staining of microtubule populations of differentiated PC12 neurites. (A and B) Transmission electron micrographs of the neurite microtubule arrays along their long axis. (A) Immunogold localization of type IV β tubulin, whose expression is independent of the differentiation. The uniform distribution of 5-nm gold particles along the entire lengths of neurite microtubules represents a more or less homogeneous incorporation of this constitutively expressed β-tubulin isotype. Compare this pattern of gold particle distribution with that of type III (B). (B) Immunogold localization of the neuron-specific β-tubulin type III (a differentiation-induced tubulin isotype). Note that short stained segments

type IV β-tubulin isotypes (small, 5-nm gold particles), we were able to determine the β-tubulin isotype composition of microtubules in both the neurite (Fig. 3C) and the soma (Fig. 3D).

To determine if type IV tubulin is polymerized randomly along the lengths of *in vitro*-polymerized microtubules, analogous to its *in vivo* distribution, we double-labeled microtubules with isotype-specific type IV tubulin polyclonal antibody (detected by small, 5-nm gold particles), and a mouse monoclonal antibody that binds with all known β-tubulin isotypes. These results are shown in Fig. 4. This double-label immunogold staining of microtubules polymerized *in vitro* from purified brain tubulin shows clearly that type IV tubulin is incorporated randomly, as *in vivo*, along the lengths of *in vitro*-polymerized microtubules.

The differential tubulin isotype composition of axonal microtubules along their length may, in fact, explain the noted heterogeneous stability properties of axonal microtubules (Baas and Black, 1990; Joshi *et al.*, 1986). The heterogeneous subunit composition within the same microtubule may be responsible for the differences in the stability and polymerization dynamics observed in the contiguous segments of the individual long microtubules. To explore the role of tubulin isotype heterogeneity along these polymers, one needs first to build a molecular map of individual microtubules along their entire lengths. A problem encountered in these studies is that often individual microtubules leave and reenter the plane of a thin section. One way to circumvent this problem is to reconstruct the entire array of neurite microtubules from the serial thin sections (Fig. 5). Microtubules and neurite outlines are traced on transparent polyester plastic sheets from the electron micrographs of consecutive serial sections. Frequent lateral blebs of the axolemma are used as registration markers to align and superimpose the traces of consecutive serial sections. The final alignment is achieved by moving the traces within a range of two microtubule diameters to maximize the number of matches between microtubule ends in two consecutive

---

(patches), along the length of single microtubules, alternate with stretches of unstained segments. In contrast to the uniform distribution of type IV β tubulin, type III tubulin is therefore recruited into short segments of generally long axonal microtubules. (C and D) Simultaneous localization, both in the neurite (C) and the soma (D), by double-label immunogold staining, with a polyclonal type IV tubulin antibody (large, 10-nm gold particles) and a monoclonal antibody that binds both type III and type IV β-tubulin isotypes (small, 5-nm gold particles). For detecting the relative distribution of each tubulin isotype we used a preembedding immunogold labeling procedure on cells that are fixed and permeabilized simultaneously with 0.5% glutaraldehyde and 1% Triton X-100 (see text) for 1 min. This treatment did not disrupt the morphology of intracellular vesicles [e.g., arrowed vesicle in (C)], and yet allowed gold–antibody conjugates free access to the intracellular microtubules. mt, Microtubule, mito, mitochondria. (From H. C. Joshi *et al.*, unpublished observations.) (The primary antibody dilutions used in these examples are 1 : 10 for type III, 1 : 40 for type IV, and 1 : 200 for common monoclonal.)

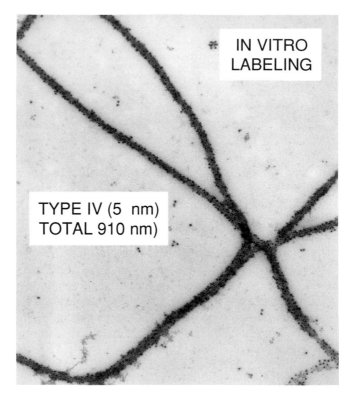

Fig. 4   Double-label immunogold staining of negatively stained microtubules, polymerized *in vitro* from purified brain tubulin. To determine if type IV tubulin is incorporated randomly, as *in vivo,* along the lengths of *in vitro*-polymerized microtubules, we double-labeled microtubules with isotype-specific type IV tubulin polyclonal antibody (small, 5-nm gold particles) and a mouse monoclonal antibody that binds with all known β-tubulin isotypes (large, 10-nm gold particles). Note uniform intermingled distribution of both small (5 nm, type IV), and large (10 nm, total) gold particles along microtubule lengths. Brain tubulin, purified from other microtubule-associated proteins by phosphocellulose chromatography, is polymerized in a microcentrifuge tube (2 mg/ml) in 100 mM Na-PIPES (pH 6.94), 1 mM MgCl$_2$, 1 mM GTP, and 10 μg of taxol/ml at 37°C. Polymers were fixed in 0.5% glutaraldehyde and diluted 50× in TBS. A drop of the dilute microtubule suspension is applied to a plastic-coated, carbon-stabilized nickel grid. The excess liquid is drained from the grid by gently touching the grid edge to fiber-free filter paper. After immunolabeling the grids are negatively stained for observation under the electron microscope. (From H. C. Joshi *et al.,* unpublished observations.)

sections. Various computer-based programs are now available for automated counting and further statistical analyses of the data (for example, Bioquant, R & M Biometrics, Inc. TN).

With what spatial precision can the immunogold EM technique map macromolecular structures within cells? The answer depends on the type of reagents used. The most common isotypes of circulating antibody molecules, IgG and

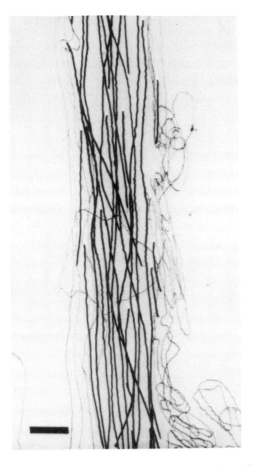

Fig. 5 Serial section reconstruction of PC12 neurite microtubules. Two-dimensional reconstruction of a 6.0-μm segment of a differentiated PC12 neurite. The fainter, peripheral lines represent the boundaries of the neurite on serial sections. Microtubules are represented by the bold lines within the neurite boundary. Twenty-two microtubules were identified in this region. The alignment was achieved by maximizing the number of matches between microtubule ends within two microtubule diameters. (Bar: 0.5 μm.) (After Joshi *et al.*, 1986.)

IgM, constitute the reagents for immunocytochemistry. An IgG molecule is shaped like the letter Y, each arm of which is roughly 8 nm long. Therefore the longest distance by which the antigen could be separated in space from the gold particle bound to the Fc region of the antibody is 16 nm, when the primary antibody itself is labeled (direct labeling experiment). This distance could extend up to 30 nm in an indirect immunolabeling experiment, in which binding of the primary antibody is detected by a gold-conjugated secondary antibody that specifically binds the unlabeled primary antibody. Singer *et al.* (1982) caution that often, in practice, one could obtain a much better resolution because of the ordered structural arrangement of antigens in the sections of biological materials. IgM molecules, being pentameric, are rather large in size and their use as intact molecules should be avoided for immunogold electron microscopy. Procedures are available to prepare small active fragments, F(ab'), by controlled proteolytic digestions of both IgG and IgM molecules (Lin and Putnam, 1978; Poljak *et al.*, 1973; Padlan *et al.*, 1973; Nisonoff and Rivers, 1961). In conjunction with the reactive antibody fragments, newly developed 1- to 2-nm gold particles (J. L. M. Leunissen, personal communication), will improve the resolution of immunogold EM tremendously. New insights into the structures previously not amenable to this technique, have begun to emerge. Earnshaw has used 1-nm gold-conjugated antibodies to detect a polypeptide tucked within mammalian kinetochores, dense structures at constricted region of chromosomes (Cooke *et al.*, 1990).

V. Controls and Critical Considerations

Two types of controls are essential to correctly interpret the results obtained from these immunogold labeling experiments. These include (1) controls for the specificity of antibodies used and (2) controls for the penetrance of the gold–antibody adducts.

A. Controls for Antibody Specificity

The binding specificity of an antibody is demonstrated most convincingly by observing the gradual disappearance of bound antibody when the binding reaction is performed in the presence of increasing concentrations of competing antigen. The availability of sufficient amounts of the pure antigen is a prerequisite for this control experiment. Therefore this is feasible only in the case of antibodies elicited by synthetic peptides, biochemically purified antigens, or a few soluble, bacterially expressed fusion-protein antigens. If the pure antigen is not available in the soluble form, for example, in the case of insoluble bacterially expressed fusion polypeptides, one could use the insoluble antigen polypeptide to deplete specific antibodies from a mixture of polyclonal antibodies by immunoadsorption. Immunoabsorbed antibodies can then be used in place of the polyclonal antibodies containing the specific antibody. In the event the pure antigen is not available one must settle for the following alternative control. A parallel preparation of the sample can be stained with the antibodies (usually IgG) from the preimmune animals in place of the specific primary antibodies obtained after the immunization. All subsequent staining steps in the preimmune control specimen should be identical to that of the experimental specimen. Furthermore, it is most comforting to see more than one independent antibody produce identical immunolocalization results. Whenever possible, therefore, one should attempt the localization experiment with additional independent polyclonal antibodies or independent monoclonal antibodies that bind different epitopes on the antigen molecule.

B. Controls for Penetrance of Gold–Antibody Adducts

In the preembedding immunogold EM procedure, the antibody molecules with their gold cargo must diffuse to often-congested intracellular locations. Therefore insufficient extraction of cells could result in a nonuniform staining pattern. To determine if the intracellular structures of interest are accessible to the secondary antibody–gold conjugate, do the following control: omit the primary antibody incubation step and incubate the cells with gold-labeled secondary antibody. Omit the washing step and process for EM. The absence of gold particles in intracellular spaces will indicate an extraction problem. Alternatively, some double-immunolabeling experiments can be internally controlled. In our study aimed at determining the distribution of two different tubulin isotypes along axonal microtubules, we used two distinctly sized gold particles (5 and 10 nm) to reveal simultaneously the specific distribution of type III and type IV β-tubulin subunits (by isotype-specific rabbit polyclonal antibodies), and the general distribution of both subtypes (by a mouse monoclonal β-tubulin antibody) along the same microtubule. By using an alternative combination of gold-labeled secondary antibodies, we could verify the results by reversing marker gold particle sizes.

C. Criteria for Making Decision for Preembedding or Postembedding Procedures

How does one choose between preembedding, postembedding, or frozen section (without embedding) immunogold localization techniques? Each of these procedures has distinct advantages and disadvantages. Although successfully used for cytoskeletal structures, the preembedding procedures require partial lysis of cellular membranes, precluding their use for the localization of membraneous components or soluble cytoplasmic antigens (De Mey *et al.*, 1986; Langanger and De Mey, 1989). The postembedding procedure offers the advantage of well-preserved ultrastructure, but is limited by two factors: stearic constraints posed on antigen accessibility as well as antigen modifications (masked antigens) can lead to poor antibody binding. Nevertheless this procedure has been a valuable tool in obtaining qualitative information about antigen localizations on the exposed surfaces of the antigens. Introduction of hydrophilic resins and etching techniques that unmask antigens has now improved the efficacy of this procedure (for review, see Singer *et al.*, 1982). Although the third procedure of staining unembedded frozen thin sections (ultracryomicrotomy; e.g., Leunissen and Verkleij, 1989) is most attractive, its use in processing cultured cells is still in experimental stages and currently is technically challenging (at least for some). Hence, the choice of the correct technique depends on the system and the kind of question asked.

D. Standardizing Fixation/Permeabilization

This is the most crucial step in this technique because the two aims of successful fixation contradict one another. First, preservation of cellular structure and the associated antigen is generally achieved by extensive chemical cross-linking of cellular molecules. However, the second aim, permeabilization of cells to allow antibody diffusion and the retention of antigen structure, is made difficult by chemical cross-linking. Optimal conditions are thus determined empirically. Glutaraldehyde, a bifunctional aldehyde, is the most commonly used fixative that reacts with secondary amino groups of basic amino acid residues within cellular proteins. Some epitopes, particularly ones that contain basic amino acids within or adjacent to them, lose their ability to bind antibodies efficiently. Therefore a careful choice of fixative and fixation conditions is a prerequisite for a successful immunogold experiment. Although many cellular structures can be reasonably well preserved by formaldehyde, with only minimal effect on the antibody-binding activity of cellular antigens (Kyte, 1976a,b), microtubules are best preserved by glutaraldehyde. A low concentration of glutaraldehyde (0.1–0.5% in phosphate-buffered saline, pH 7.4) mixed with 2–4% formaldehyde is also recommended by several authors as an efficient fixative for immunogold staining. The most effective concentration of the fixative and the time of fixation should be determined in advance. Monolayers of cultured cells behave well in all the fixation conditions recommended here. The quick screening of fixation conditions can first be done by immunofluorescence light microscopy.

E. Consideration of Antibody Properties

The immunogold localization, like other immunological procedures, is dependent on antigen–antibody interaction. Several aspects of antibody/antigen characteristics have a bearing on these experiments. The foremost requirement for a useful localization experiment is the absolute binding specificity of the antibody. Specificity of the antibodies must be determined for each cell type used in the localization experiment. Appropriate controls are necessary to interpret the localization results (see Section VI for specificity controls). Although high-affinity purified antibodies are most suitable, the utility of each antibody for the immunolocalization experiments should be empirically determined. The stability of the antigen–antibody complex (avidity) is dependent on the intrinsic affinity of the antibody, the valency of the antibody/antigen, and the spatial arrangement of the antigen/antibody molecules (Harlow and Lane, 1988). Sometimes a low-affinity antibody may produce optimal results due a high local concentration of the antigen molecules, attached in a distinct spatial arrangement, on a solid support (fixed in place).

F. Other Gold-Conjugated Secondary Reagents

Various high-quality secondary reagents attached to gold particles of defined size are commercially available. Among these are Protein A and protein G, bacterial proteins that bind with high affinity to antibodies (particularly IgG) from many mammals, and streptavidin, a 15-kDa bacterial protein that binds tightly to a coenzyme, biotin, which in turn can be attached to antibodies of choice. These secondary reagents can sometimes be used to amplify a rather faint signal.

VI. Troubleshooting Guide

Problem	Probable cause and the remedy
Glass coverslip does not detach from Epon polymer	Epon overflow on top of glass coverslips, insufficient polymerization of Epon: Use minimal amount of Epon, check the ratio of resin ingredients
	Improper application of release agent to the glass coverslip
No staining	Inappropriate matching of secondary antibodies: Check the primary antibody isotype and species (rabbit IgG, mouse IgM, etc.) and use secondary antibody of appropriate specificity (anti-rabbit IgG, anti-mouse IgM, etc.)
	Dead primary antibody: use freshly purified, highest quality antibody
	Antibody dilution too high: Determine appropriate antibody dilution empirically
	Wrong binding pH: Check pH of solutions
	Destabilization of gold–antibody conjugate: Wash away free dissociated antibody from the gold conjugates (centrifugation), use fresh secondary antibody
	Masked antigen: Manipulate fixation conditions (see Section V), perform antibody labeling prior to fixation (e.g., for cytoskeletal elements)
Irregular patches of stained areas	Insufficient permeabilization: Increase detergent concentration, time of permeabilization, prolong time/temperature of antibody incubations

(continues)

(Continued)

Problem	Probable cause and the remedy
High background on the specimen	Insufficient quenching of aldehyde groups: Use fresh sodium borohydride to reduce unreacted aldehyde groups
	Air drying of cells: Do not allow cells to dry at any step during staining
	Insufficient washing: Increase number of washing steps, include 0.1% Triton X-100 in the wash.
	Insufficient blocking: Increase blocking time, increase protein concentration in the blocking solution. (do not use normal serum as a blocking agent if protein A is used instead of secondary antibody; use fraction V BSA instead)
	Wrong reaction pH: Check pH
	Primary antibody concentration too high: Titrate it
	Destabilization of antigen–antibody complex: Fix cells well after the completion of antibody staining
High background around the sample on the plastic	Blocking error
	Salt concentration too high
Gold lumps	Primary antibody aggregates: Centrifuge antibody solution prior to use
	Secondary antibody aggregates: Use a fresh batch of good-quality secondary antibody
	Primary antibody concentration too high: Titrate antibody concentration to determine optimal dilution

VII. Conclusion

This article provides principles and a brief rationale of the immunogold staining technique and its practical application for mapping primary isotypes of β tubulin on neuronal microtubules, and thus the study of the utilization of tubulin gene family members during neuronal differentiation. The decisive functional dissection of multiple gene families, such as that of tubulins, is far from being complete. We hope that these examples from our ongoing work convey the power of immunogold electron microscopic techniques to establish a link between gene families, their encoded polypeptides, and the intracellular structures that are specialized to perform unique functions within differentiated eukaryotic cells.

=========== **VIII. Appendix**

A. Reagents

Glutaraldehyde (should be free of polymers): 50% solution in sealed ampoules from Polysciences (Warrington, PA), stored frozen at $-20°C$. Open the ampoule and dilute into the recommended buffer just before use

Formaldehyde (4%): Do not use commercially available 37% formaldehyde solution (use high-quality paraformaldehyde). Make solutions fresh, as follows:

1. Heat 50 ml of water to 60°C in a fume hood
2. Add 4 g of paraformaldehyde and 0.1 ml of 1 N NaOH
3. Let the clear solution cool to room temperature
4. Add an equal volume of 2× PBS

Poly-L-lysine (average M, 400,000): 1 mg/ml in distilled water; store frozen at $-20°C$

Sodium borohydride ($NaHBO_4$): Store desiccated, should be dissolved just prior to use

Epon 812, nadic methyl anhydride (NMA), dodecenyl succinic anhydride (DDSA), 2,4,6,tri(dimethylaminomethyl) phenol (DMP 30) (Ladd Research Industries, Inc., Burlington, VT): Mix according to Geiselman and Burke (1973) and store frozen in plastic syringes

Uranyl acetate (2%): Dissolve 1 g of uranyl acetate in 50 ml water, filter once, and add 1 drop (ca. 30 μl) of Triton X-100. (*Caution:* Remember that uranyl acetate is radioactive)

Lead Citrate: dissolve one pellet of NaOH in 50 ml of water and add 0.25 g of lead citrate. Mix thoroughly and add 1 drop (30 μl) of Triton X-100

Gold-conjugated reagents (antibodies, protein A, and streptavidin): Purchase from Janssen Pharmecutica (supplied in the United States by Amersham)

B. Suppliers

Electron Microscopy Sciences, 321 Morris Road, P.O. Box 251, Fort Washington, PA 19034

Ladd Research Industries, Inc., P.O. Box 1005, Burlington, VT 05402

Lux, Lab-Tek division, Miles Laboratories, Inc., Naperville, IL 60540

Miller Stephenson Chemical Company, Inc., Danbury, CN 06810

Polysciences, Inc., 400 Valley Road, Warrington, PA 18976

Sigma Chemical Co., P.O. Box 14508, St. Louis, MO 63178

Acknowledgments

I thank Dr. Marc DeBrabander of Janssen Pharmecutica (Beerse, Belgium) for his hospitality during my visit to his laboratory and for inculcating my interest in immunogold electron microscopy; all the members of the DeBrabander laboratory for their support and encouragement; Dr. Don Cleveland, Dr. Steven Heidemann, Dr. Jan Leunissen, Mr. S. Geuens, Dr. Peter Baas, Dr. Winfield Sale, and Dr. Stephen Bonasera for discussions; Dr. Don Cleveland and Margaret Lopata for providing tubulin antibodies; Dr. Doug Murphy for providing purified chicken brain tubulin; and Ms. Elizabeth Chávez for her help in the preparation of the manuscript. This work is supported partially by grants to H.C.J. from the American Cancer Society, the National Institutes of Health.

References

Baas, P. W., and Black, M. M. (1990). Individual microtubules in the axon consist of domains that differ in both composition and stability. *J. Cell Biol.* **111,** 495.

Bendayan, M. (1984). Protein-A gold electron microscopic immunocytochemistry, methods, applications and limitations. *J. Electron Microsc. Tech.* **1,** 243.

Bennet, H. (1984). Decoration with myosin subfragment-1 disrupts contacts between microfilament and the cell membrane in isolated *Dictyostelium* cortices. *J. Cell Biol.* **99,** 1434.

Chang, J. P. (1971). A new technique for separation of coverglass substrate from epoxy-embedded specimens for electron microscopy. *J. Ultrastruct. Res.* **37,** 370.

Cooke, C. A., Bernat, R. L., and Earnshaw, W. C. (1990). CENP-B: A major human centromere protein located beneath the kinetochore. *J. Cell Biol.* **110,** 1475.

De Mey, J. (1983). Colloidal gold probes in immunocytochemistry. *In* "Immunocytochemistry" (J. M. Polak and S. van Noorden, eds.), p. 82. J. G. Wright, Bristol.

De Mey, J., Langanger, G., Geuens, G., Nuydens, R., and DeBrabander, M. (1986). Preembedding for localization by electron microscopy of cytoskeletal antigens in cultured cell monolayers using gold-labeled antibodies. *In* "Structural and Contractile Proteins, Part C: The Contractile Apparatus and the Cytoskeleton" (R. Vallee, ed.), Methods in Enzymology, Vol. 134, p. 592. Academic Press, Orlando, Florida.

Faulk, W. P., and Taylor, G. (1971). An immunocolloid method for the electron microscope. *Immunocytochemistry* **8,** 1081.

Frens, G. (1973). Controlled nucleation for the regulation of the particle size in monodisperse gold suspensions. *Nature Phys. Sci.* **241,** 20.

Geiselman, C. W., and Burke, C. N. (1973). Exact anhydride: Epoxy percentages for araldite and araldite–Epon embedding. *J. Ultrastruct. Res.* **43,** 220.

Geuens, G., Gundersen, G. G., Nuydens, R., Cornelissen, F., Bulinski, J. C., and DeBrabander, M. (1986). Ultrastructural colocalization of tyrosinated and detyrosinated α-tubulin in interphase and mitotic cells. *J. Cell Biol.* **103,** 1883.

Greene, L. A., and Tischler, A. S. (1976). Establishment of a noradrenergic clonal cell line of adrenal pheochromocytoma cells which respond to nerve growth factor. *Proc. Natl. Acad. Sci. U.S.A.* **73,** 2424.

Harlow, E., and Lane, D. (1988). "Antibodies: A Laboratory Manual." Cold Spring Harbor Lab. Press, Cold Spring Harbor, New York.

Horisberger, M. (1979). Evaluation of colloidal gold as a cytochemical marker for transmission and scanning electron microscopy. *Biol. Cell.* **36,** 253–258.

Joshi, H. C., and Cleveland, D. W. (1989). Differential utilization of β-tubulin isotypes in differentiating neurites. *J. Cell Biol.* **109,** 663.

Joshi, H. C., Baas, P., Chu, D. T., and Heidemann, S. R. (1986). The cytoskeleton of neurites after microtubule depolymerization. *Exp. Cell Res.* **163,** 233.

Kyte, J. (1976a). Immunoferritin determination of the distribution of $(Na^+ + K^+)$ ATPase over the plasma membranes of renal convoluted tubules. II. Proximal segment. *J. Cell Biol.* **68,** 309.

Kyte, J. (1976b). Immunoferritin determination of the distribution of $(Na^+ + K^+)$ ATPase over the plasma membranes of renal convoluted tubules. I. Distal segment. *J. Cell Biol.* **68**, 287.

Langanger, G., and De Mey, J. (1989). Detection of cytoskeletal proteins in cultured cells at the ultrastructural level. *In* "Immuno-Gold Labeling in Cell Biology" (A. J. Verkleij and J. L. M. Leunissen, eds.), pp. 335–351. CRC Press, Boca Raton, Florida.

Leunissen, J. L. M., and De Mey, J. R. (1989). Preparation of gold probes. *In* "Immuno-Gold Labeling in Cell Biology" (A. J. Verkleij and J. L. M. Leunissen, eds.), p. 3. CRC Press, Boca Raton, Florida.

Leunissen, J. L. M., and Verkleij, A. J. (1989). Cryo-ultramicrotomy and immuno-gold labeling. *In* "Immuno-Gold Labeling in Cell Biology" (A. J. Verkleij and J. L. M. Leunissen, eds.), p. 95. CRC Press, Boca Raton, Florida.

Lin, L. C., and Putnam, F. W. (1978). Cold pepsin digestion: A novel method to produce the Fv fragment from immunoglobulin M. *Proc. Natl. Acad. Sci. U.S.A.* **75**, 2649.

Lopata, M. A., and Cleveland, D. W. (1987). *In vivo* microtubules are copolymers of available β-tubulin isotypes: Localization of each of six vertebrate β-tubulin isotypes using polyclonal antibodies elicited by synthetic peptide antigens. *J. Cell Biol.* **105**, 1707.

Mazia, D., Schatten, J., and Sale, W. S. (1975). Adhesion of cells to surfaces coated with polylysine. *J. Cell Biol.* **66**, 198.

Nicklas, R. B., Kubai, D. F., and Hays, T. S. (1982). Spindle microtubules and their mechanical association after micromanipulation in anaphase. *J. Cell Biol.* **95**, 91.

Nisonoff, A., and Rivers, M. N. (1961). Recombination of a mixture of univalent antibody fragments of different specificity. *Arch. Biochem. Biophys.* **93**, 460.

Padlan, E. A., Segal, D. M., Spande, T. F., Davies, D. R., Rudikoff, S., and Pottor, M. (1973). Structure at 4.5Å resolution at a phosphorylcholine-binding Fab. *Nature (London), New Biol.* **245**, 165.

Poljak, R. J., Amzel, L. M., Avey, H. P., Chen, B. L., Phizackerley, R. P., and Saul, F. (1973). Three dimensional stucture of the Fab' fragment of a human immunoglobulin at 2.8Å resolution. *Proc. Natl. Acad. Sci. U.S.A.* **70**, 3305.

Singer, S. J. (1959). Preparation of an electron dense antibody conjugate. *Nature (London)* **183**, 1523.

Singer, S. J., Tokuyasu, T. Y., Dutton, A. H., and Chen, W.-T. (1982). High-resolution immuno-electron microscopy of cell and tissue ultrastructure. *In* "Electron Microscopy in Biology" (J. D. Griffith, ed.), Vol. 2, pp. 55–106. Wiley, New York.

Sullivan, K. F., and Cleveland, D. W. (1986). Identification of conserved isotype-defining variable region sequences for four vertebrate β-tubulin polypeptide classes. *Proc. Natl. Acad. Sci. U.S.A.* **83**, 4327.

Varndell, I. M., and Polak, J. M. (1984). Double immunostaining procedures: Techniques and applications. *In* "Immunolabeling for Electron Microscopy" (J. M. Polak and I. M. Varndell, eds.), p. 155. Elsevier, Amsterdam.

CHAPTER 15

Postembedding Immunocytochemical Techniques for Light and Electron Microscopy

Page A. Erickson,* Geoffrey P. Lewis, and Steven K. Fisher

Neuroscience Research Institute and
Department of Biological Sciences
University of California, Santa Barbara
Santa Barbara, California 93106

I. Introduction

Postembedding immunocytochemistry takes advantage of the fact that many proteins still react with antibodies directed against them in tissue that has been chemically fixed and embedded for microscopy. The tissue may by embedded in a resin (e.g., LR White, London Resin Co., London, England) that allows thin sections (1 μm thick) to be cut for light microscopy and ultrathin sections (70 nm thick) to be cut for electron microscopy. An antigen is localized to a specific tissue or a specific cellular or subcellular location by applying a specific antibody (the primary antibody) to the tissue section. The primary antibody is then detected by a secondary antibody conjugated to some molecule that allows its

* Present Address: BioTek Solutions, Inc., Santa Barbara, CA. 93117

METHODS IN CELL BIOLOGY, VOL. 37

visualization by light or electron microscopy. The postembedding techniques differ from the preembedding techniques, in which the antibodies are applied to tissues (or cells in culture for another example) prior to embedment and sectioning. There are advantages and disadvantages to both techniques and these should be considered before choosing one. The postembedding techniques have a major advantage by virtue of being able to apply different antibodies to different (e.g., serial) sections. Within the limitations listed below one could, theoretically, use an infinite variety of antibodies on any given tissue as long as sections are available and the antibodies in question recognize antigens in fixed and embedded tissues. In this article we will demonstrate the embedding of tissue in one of the newer resins (LR White) that can be used for both light and electron microscopy. In this case, the resin remains with the tissue sections and only antigenic sites on the surface of the sections are available for reaction with the antibody. Resolution and tissue integrity are usually quite good but the number of sites available for antibody binding (hence labeling density or signal) are usually low. In other variants, the tissue is embedded in a medium that is removed after the sections are cut and prior to antibody labeling (e.g., paraffin, Paraplast Xtra; Sherwood Medical, St. Louis, MO). In this case labeling density is usually higher than is achieved by the first technique, because antibodies penetrate and bind to antigenic sites throughout the full thickness of the section, but resolution and structural preservation are usually less. Postembedding techniques can also be used with frozen tissue in which no fixative is used (other than the freezing itself) and the tissue remains unembedded or embedded in a water-soluble medium such as OCT compound (e.g., Tissue-Tek; Miles, Inc., Elkhart, IN). In general a technique that produces the highest signal has the lowest resolution.

For the postembedding techniques, tissue preservation must often be sacrificed in order to retain adequate antigenicity, that is, the ability of the primary antibody to recognize and bind to its specific antigenic site. Because this article considers both light and electron microscopic immunocytochemistry, the bias will be toward retaining maximum tissue preservation while still allowing the detection of a broad range of antigens. The inability to use osmium tetroxide as a secondary fixative, after primary fixation with aldehydes, as is commonly done in conventional tissue processing for electron microscopy, is a limitation of the technique. Osmium tetroxide provides better tissue preservation and contrast, and although it has been used for some immunolabeling experiments (Pelletier *et al.,* 1981; Schwendemann *et al.,* 1982; Smith and Keefer, 1982) its use as a fixative for immunocytochemistry is limited because it bonds covalently to proteins and alters their conformation (Lenard and Singer, 1968), usually to the ". . . extinction of antigenicity in the great majority of proteins" (Pearse, 1980).

The immersion of aldehyde-fixed tissue in a solution of uranyl acetate, either before or during dehydration, is also used to increase tissue contrast and is often referred to as *en bloc* staining. Uranyl acetate, used in this manner, also has fixative properties (Silva *et al.*, 1968, 1971; Terzakis, 1968). However, the fixative action of uranyl acetate appears to be electrostatic (Hyatt, 1981), rather than covalent, thus reducing the problem of conformational changes. Thus we commonly use uranyl acetate as a secondary fixative after aldehyde fixation to obtain better tissue preservation and still allow the immunolabeling of a wide variety of proteins (Erickson *et al.*, 1987; Lewis *et al.*, 1989). This procedure has the additional advantage of quantitatively increasing the immunolabeling of a number of proteins when compared to immunolabeling after simple aldehyde fixation (Erickson *et al.*, 1987).

There are other important variables in tissue processing that can also affect the quality of fixation and immunolabeling, including the following.

1. The tissue preservation technique, which may involve freezing or the use of a coagulative (e.g., acetone or ethanol), or a chemical fixative
2. Temperature of solutions during fixation
3. Length of fixation
4. Type, pH, and osmolality of buffer(s)
5. Dehydrating agent(s)
6. Concentrations and temperatures of dehydrating agent(s) during dehydration
7. Choice of embedding medium

Because the conformation of proteins, and thus the ability of antibodies to recognize them, can be altered by virtually any of these variables the choice of tissue preparation procedures is largely empirical. However, we have included fixation protocols that we routinely use that work well for a broad range of proteins (cytoplasmic, cytoskeletal, membrane-bound and extracellular matrix) in mammalian tissues and at least some cells in culture. If these tissue-processing protocols give inadequate immunolabeling, consider decreasing the glutaraldehyde concentration or omitting it altogether (e.g., as is recommended in the paraffin-embedding protocol outlined below) as a first step before manipulating other parameters. Remember, however, that ultrastructural quality will be sacrificed.

The model system on which these protocols are based is the vertebrate retina (normal, degenerating, and regenerating), but the protocols are similar to those used on a variety of tissues and cells in culture. To replicate retinal degeneration reproducibly, we use an experimental retinal detachment procedure (Anderson *et al.*, 1983; Erickson *et al.*, 1983; Lewis *et al.*, 1989). The regenerative component is accomplished by surgical reattachment of the detached retina (Anderson *et al.*, 1986; Guerin *et al.*, 1989). Because of the expense of the experiments and the desire to use fewer animals, we have continued to experiment with fixation and immunolabeling regimens for postembedding analyses. By using postembedding techniques, it is possible to go back repeatedly to earlier experimental tissue (embedded in various substrates) and probe the tissue with new antibodies or use newly developed labeling techniques. We have applied the postembedding techniques outlined below to a broad range of species, including squirrel, rabbit, cat, monkey, and human. These techniques may have to be significantly modified for use with plant tissue.

II. Protocols

All procedures in all protocols are performed at room temperature, and in glass containers, unless otherwise stated. Recipes for the solutions are found in Section III. Use a 10× volume (minimum) of solutions per volume of tissue. *Gently* swirl (do not "agitate") the tissue in the various solutions on an intermittent basis, unless otherwise stated.

A. Fixation and Embedding

1. Tissue Processing Steps: Standard Protocol for Tissue to Be Embedded in LR White Resin

1. Dissect the tissue into suitably sized pieces (e.g., 1 mm^3, but this can vary with tissue type) for traditional fixation for electron microscopy and transfer immediately to fresh primary fixative. Leave the tissue in the primary fixative for 1.0 hr. (*Note:* For retinal tissue, place the entire eye into primary fixative and excise the anterior third of the globe, near the limbus, with a razor blade. Remove any residual vitreous from the eye cup and transfer the eye cup into fresh primary fixative.)

2. Transfer the tissue into fixative wash buffer. Wash the tissue three times in fresh fixative wash buffer, 10 min per wash. (*Note:* For retinal tissue, cut the eye cup into quadrants at this time.)

3. Dehydrate the tissue:
   a. Methanol (15%): distilled H$_2$O (v/v) for 10 min
   b. Methanol (30%) for 10 min
   c. Methanol (50%) for 10 min

4. Immerse the tissue in uranyl acetate fixative for 1.0 hr.

5. Continue the dehydration:
   a. Methanol (85%) for 10 min
   b. Methanol (95%) for 10 min
   c. Methanol (100%) for 10 min

6. Infiltrate the tissue with a 1:1 ratio of 100% methanol:LR White resin (London Resin Co.; store LR White at 4°C); cap the vials containing the tissue and rotate them (at approximately 45° and about 2 rpm) overnight at 4.0°C.

7. The next day, place the tissue in 100% LR White resin, in uncapped vials, and rotate for 2.0 hr at 4.0°C.

8. Transfer the tissue to fresh LR White and rotate at room temperature (uncapped) for 2.0 hr.

9. While the tissue is rotating at room temperature, place fresh LR White resin (stored at 4.0°C) into 4- or 7-dram polyethylene vials (VWR, San Francisco, CA; use 4 or 8 ml of resin) and allow the resin to equilibrate to room temperature.

10. Place the tissue in the fresh, room-temperature LR White resin and cap the vials. There should be an air space between the top of the resin and the cap; once the resin polymerizes, the top of the resin will be clear and allow visualization of the tissue.

11. Polymerize by placing the vials at 52°C for 2 days or until hard.

2. Tissue Processing Steps: Alternative Protocol for Tissue to Be Embedded in LR White Resin

This protocol uses uranyl acetate as a fixative before dehydration and also during the entire dehydration series. The structural integrity of the tissue is improved compared to the previous protocol and preliminary evidence with numerous antibodies suggests the immunolabeling is also improved, but a rigorous quantitative study similar to that performed for the previous protocol (Erickson *et al.*, 1987) has yet to be undertaken.

1. Dissect the tissue into suitably sized pieces (e.g., 1 mm^3, but this can vary with tissue type) for traditional fixation for electron microscopy and transfer immediately to fresh primary fixative. Leave the tissue in the primary fixative for 1.0 hr. (*Note:* For retinal tissue, place the entire eye into primary fixative and excise the anterior third of the globe, near the limbus, with a razor blade. Remove any residual vitreous from the eye cup and transfer the eye cup into fresh primary fixative.)

2. Transfer the tissue into fixative wash buffer. Wash the tissue three times in fresh fixative wash buffer, 10 min per wash. (*Note:* For retinal tissue, cut the eye cup into quadrants at this time.)

3. Wash the tissue three times in maleate wash buffer (pH 5.2), 10 min per wash.

4. Immerse the tissue in 2.0% uranyl acetate in maleate buffer (pH 4.75) for 2.0 hr (secondary fixation).

5. Fix/dehydrate the tissue (*Note:* These steps are performed at 4°C):
    a. Uranyl acetate (2.0%) in 15% methanol/distilled H_2O (v/v) for 10 min
    b. Uranyl acetate (2.0%) in 30% methanol for 10 min
    c. Uranyl acetate (2.0%) in 50% methanol for 10 min
    d. Uranyl acetate (2.0%) in 70% methanol for 10 min
    e. Uranyl acetate (2.0%) in 85% methanol for 10 min

    f. Uranyl acetate (2.0%) in 95% methanol for 10 min

    g. Uranyl acetate (2.0%) in 100% methanol for 10 min

   6. Wash out the uranyl acetate in 100% methanol (three times, 10 min each, at 4°C).

   7. Infiltrate the tissue with a 1 : 1 ratio of 100% methanol:LR White resin (at 4°C); cap the vials containing the tissue and rotate them (at approximately 45° and about 2 rpm) overnight at 4°C.

   8. The next day, place the tissue in 100% LR White resin (at 4°C), in uncapped vials, and rotate for 2.0 hr at room temperature.

   9. Change to fresh, room-temperature LR White and continue to rotate at room temperature for 2.0 hr.

  10. Repeat step 9.

  11. Place the tissue in fresh LR white resin (room temperature) that is in 4- or 7-dram polyethylene vials (VWR; use 4 or 8 ml of resin).

  12. Cap the vials. There is an air space between the top of the resin and the cap; once the resin polymerizes, the top of the resin will be clear and allow visualization of the tissue.

  13. Polymerize by placing the vials at 52°C for 2 days or until hard.

3. Tissue Processing Steps: Protocol for Tissue to Be Embedded in Paraffin

   1. Dissect the tissue into suitably sized pieces (e.g., 1- to 2-mm thick slices, but this can vary with tissue type) and transfer immediately to fresh paraffin primary fixative. Leave the tissue in the paraffin primary fixative for 1.0 hr. (*Note:* For retinal tissue, place the entire eye into paraffin primary fixative and excise the anterior third of the globe, near the limbus, with a razor blade. Remove any residual vitreous from the eye cup and transfer the eye cup into fresh paraffin primary fixative.)

   2. Transfer the tissue into fixative wash buffer. Wash the tissue three times in fresh fixative wash buffer, 10 min per wash. (*Note:* For retinal tissue, cut the eye cup into quadrants at this time.)

   3. Dehydrate the tissue (perform each dehydration step at 4°C):

    a. Ethanol (15%)/distilled H_2O (v/v) for 10 min

    b. Ethanol (30%) for 10 min

    c. Ethanol (50%) for 10 min

    d. Ethanol (70%) for 10 min

    e. Ethanol (85%) for 10 min

    f. Ethanol (95%) for 10 min

    g. Ethanol (100%) for 10 min

4. Infiltrate the tissue with toluene and Paraplast Xtra (Sherwood Medical) at the indicated temperatures:

    a. Toluene (100%) for 15 min at room temperature

    b. Repeat step a

    c. Toluene (100%) for 30 min at 60°C

    d. Toluene (100%) 1 : 1 with 100% Paraplast Xtra for 30 min at 60°C

    e. Paraplast Xtra (100%) at 60°C overnight

    f. Change to fresh 100% Paraplast Xtra at 60°C for 30 min

5. Embed in fresh 100% Paraplast Xtra, placed in a plastic, aluminum, or stainless steel mold, and allow Paraplast Xtra to solidify (it turns from clear to white).

6. Once the Paraplast Xtra starts to solidify (becomes somewhat white and the tissue does not move), place the entire mold in ice water for a few minutes to speed the solidification. (*Note:* Paraplast Xtra and other paraffin blocks cut better on a microtome if they are chilled somewhat before sectioning.)

4. Other Options for Embedding

Just as there are options for fixation, there are options for embedding media. One embedding medium we have used successfully for light and electron microscopic immunocytochemistry is Lowicryl K4M (Erickson *et al.*, 1987). For light microscopic immunocytochemistry, acrylamide embedding (Johnson and Blanks, 1984) or straight immersion freezing plus cryosectioning without embedding media (Vaughan *et al.*, 1990) also work well for some applications.

B. Immunocytochemistry Controls

Because of the complexity of immunocytochemistry in general, and the limited nature of this article, we are assuming the following.

1. Characterization of the antibody and antigen: The antibodies that are being used have been well characterized, that is, they are specific for the protein/antigen of interest. These issues are covered in other articles in this volume.

2. Characterization of the antibody and tissue with which the antibody is to be used: Protein gels of whole-tissue homogenates and immunoblots of these gels (the so-called Western blots) have shown the antibodies are immunolabeling *only* the protein/antigen of interest and not some additional protein(s) within the tissue of interest (Fig. 1). (For descriptions of this technique see Smith, 1987; Winston *et al.*, 1987; see also article 6, this volume.)

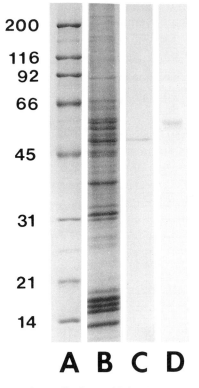

Fig. 1 A Western blot demonstrating antibody specificity. Lane A contains the molecular weight standards; lane B is a Coomassie-blue stained gel of retinal tissue homogenate; lane C is labeled with anti-GFAP (51 kDa); lane D is labeled with anti-vimentin (57 kDa). The antibodies were detected with an HRP color development reagent (Bio-Rad, Richmond, CA).

Immunolabeling Controls

It is essential that proper control experiments be performed in parallel with structural immunolabeling to ensure that tissue labeling as seen in the microscope is due to only one variable: the specific binding of the primary antibody. The following controls are appropriate (assuming the primary antibody is an IgG produced by a rabbit). Make the following substitutions for the primary antibody.

1. Preimmune serum or preimmune IgG: This serum or IgG fraction comes from the same rabbit that made the primary antibody; it is obtained *before* the rabbit makes the primary antibody.

2. Nonimmune serum or nonimmune IgG: This serum or IgG fraction comes from the same species (rabbit) that made the primary antibody, but not from the individual rabbit.

3. Phosphate-buffered saline–bovine serum albumin–sodium azide (PBS–BSA–NaN$_3$): This is the buffer into which the primary antibody is diluted, but without the antibody.

4. Nonspecific positive: This is an antibody that is specific to an antigen/protein in the tissue that is being labeled, but is *different* from the antigen/protein that is of interest. This should be an antibody known to give positive results with the tissue of interest. For example, if the antigen of interest is opsin and the primary antibody is an anti-opsin (which labels only the retinal photoreceptor cells), a reasonable nonspecific positive control would be an anti-vimentin to label the retinal Müller cells and capillary endothelial cells, but not the photoreceptor cells.

5. Nonspecific negative: This is an antibody that is not related to any antigen/protein in the tissue being labeled. This control is usually satisfied by the preimmune or nonimmune antibodies.

6. Positive tissue and negative tissue controls: Additional controls involve using a tissue preparation known to contain the antigen of interest (positive tissue control) and one known to be devoid of the antigen of interest (negative tissue control). These controls can be addressed with a single tissue if (1) it contains numerous cell types; (2) some of the cell types contain the antigen of interest and some are devoid of the antigen; and (3) the positive and negative cell types are readily identifiable and easily distinguishable from each other.

C. Light Microscopic Immunocytochemistry

This section discusses coating of glass microscope slides to enhance tissue section adhesion and three protocols for light microscopic immunocytochemistry: (1) immunofluorescence for paraffin-embedded tissue, (2) immunoperoxidase for paraffin-embedded tissue, and (3) immunogold silver enhanced for LR White-embedded tissue.

1. Coating of Glass Microscope Slides to Enhance Tissue Section Adhesion

Tissue sections do not adhere well to glass microscope slides during lengthy incubations in aqueous solutions. We have tried numerous coatings to improve the adhesion of tissue sections to slides (including gelatin–chrome alum, Haupts, low molecular weight polylysine, high molecular weight polylysine, straight gelatin, and Formvar) and have found that 3-aminopropyltriethoxysilane [APTES; Aldrich Chemical Co. (Milwaukee, WI) and Sigma Chemical Co. (St. Louis, MO); L. Angerer, personal communication] works most reliably.

1. Clean the slides: dip into distilled H$_2$O and wipe with a paper towel.
2. Add 2 ml of APTES to 98 ml of acetone (make fresh).

3. Dip the slides into the 2% APTES in acetone solution for 30 sec.

4. Dip the slides briefly in 100% acetone.

5. Dip the slides briefly in distilled H_2O (change after about 20 slides).

6. Air dry or oven dry the slides (the slides can be stored for future use).

7. The slides are now ready for paraffin or LR White sections:

    a. Place a drop of distilled H_2O on the slide.

    b. Float the tissue section on the drop of H_2O.

    c. Remove excess H_2O (e.g., with a needle/syringe).

    d. Air dry and dry on a warm plate (about 40–50°C) or in an oven (same temperature).

2. Immunofluorescence

This protocol is for immunofluorescence with paraffin-embedded tissue (Color Plate 3A and B and Fig. 2A and B).

1. Deparaffinize the tissue sections by dipping the slides in 100% xylene or a xylene:HemoD (3 : 1; HemoD is from Fisher Scientific, Pittsburgh, PA) solution (5 min, with gentle agitation about once a minute).

2. Repeat step 1 with fresh xylene or xylene:HemoD (3 : 1).

3. Repeat step 1 again, with fresh xylene or xylene:HemoD (3 : 1).

4. Rehydrate the tissue:

    a. Dip the slides into 100% ethanol for 1 min.

    b. Dip the slides into 50% ethanol for 1 min.

    c. Dip the slides into distilled H_2O for 1 min.

5. Dip the slides into PBS–BSA for 5 min.

6. Remove the excess PBS–BSA by blotting around the tissue sections with a paper towel (do not allow the tissue sections to dry out), and place sufficient blocking antibody (about 100 μl) on top of the sections so that they stay wet for 30 min. (*Note:* It is beneficial to keep the slides in a humidified chamber during the antibody incubations so that the solutions do not evaporate; an inexpensive solution is to invert a large casserole dish over the slides and have some petri dishes filled with water in the "chamber.")

7. Rinse away the blocking antibodies by gently dripping PBS–BSA onto the slide while holding the slide at a 45° angle.

8. Wipe around the tissue sections and add the primary antibody (about 150 μl) to the sections; place the sections in a humidified chamber for 2 hr.

9. Rinse dropwise with PBS–BSA, then put about 200 μl of PBS–BSA on top of the sections, letting them soak for about 5 min.

10. Repeat step 9.

11. Repeat step 9 again.

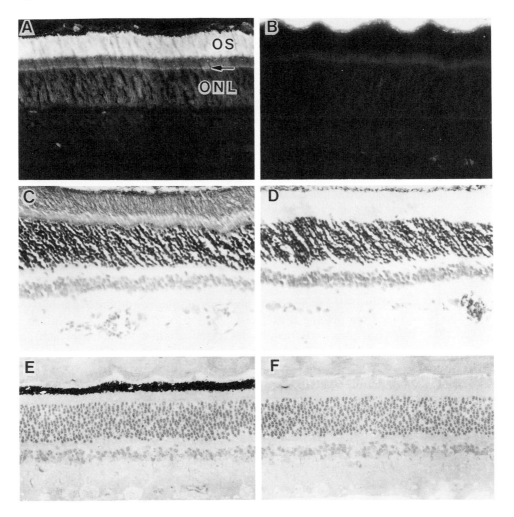

Fig. 2 Three different methods are shown for detecting an antibody against the photoreceptor-specific protein opsin (A, C, and E) along with their controls (B, D, and F). (A and B) Immunofluorescence technique on paraffin-embedded tissue. The antibody labels the photoreceptor outer segment layer (OS), photoreceptor cell bodies in the outer nuclear layer (ONL), and the Golgi apparatus found in the "inner segment" region of the photoreceptor cells (arrow). The faint labeling in the control (B) is the background level whereas the white at the top of the figure is the autofluorescence from the reflective tapetal region. (C and D) Immunoperoxidase technique on paraffin-embedded tissue. The outer segment layer labeling is observed as in (A); however, the nuclei are counterstained with hematoxylin, making them appear black (C). The control (D) shows no labeling of the outer segment layer. (E and F) Immunogold silver-enhanced technique on LR White-embedded tissue. Tissue preservation is improved but, because only the surface of the section is labeled, antigen binding is reduced; the black reaction product appears only over the outer segments and the faint labeling of the Golgi and cell bodies seen in (A) is not evident. These sections are counterstained with basic fuchsin to add tissue contrast. (Magnification: × 162.)

12. Rinse dropwise with PBS–BSA, wipe the excess from around the tissue sections, and add the secondary fluorescent antibody (about 150 μl) for 1 hr [place in humidified chamber; to reduce fading (bleaching/quenching) of the fluorescently tagged secondary antibody, cover the chamber with aluminum foil].

13. Rinse dropwise with distilled H_2O and then place about 200 μl of distilled H_2O on top of the sections and let soak for about 5 min (cover the slides with aluminum foil).

14. Repeat step 13.

15. Repeat step 13 again.

16. Mount in 5% *n*-propyl gallate (Sigma, St. Louis, MO) in glycerol (NPGG) by blotting excess H_2O from around the tissue sections, adding the NPGG to the sections, and coverslipping.

17. Seal the edges of the coverslips with clear nail polish.

18. Store the slides in the dark at 4°C.

Notes

During the antibody incubations, a smaller volume of antibody solution can be used by placing a coverslip on top of the antibody solution (for overnight incubations, seal the edges of the coverslip with rubber cement; on the next day simply peel off the rubber cement). To remove the coverslip, dip the slide vertically in PBS–BSA and agitate gently (use different containers for different antibodies to protect against antibody cross-contamination of the tissue sections) or rinse gently by applying PBS–BSA from a squirt bottle until the coverslip slides off. Rinse gently (do not "jet wash"). There are mounting media other than NPGG that may reduce fading (bleaching/quenching) more effectively, depending on the particular fluorochrome (see, e.g., Bock *et al.,* 1985), but good results can be obtained with NPGG.

3. Avidin–Biotin Conjugate Immunoperoxidase

This protocol is for paraffin-embedded tissue (Color Plate 3C and D and Fig. 2C and D). We have tried many variations of this protocol and because of cost, quality, consistency, and ease of use we have settled on using the elite ABC kit (Vector Laboratories, Burlingame, CA) for the biotinylated antibody and avidin–biotin conjugate (ABC) reagents and follow their protocols for dilutions. We also use the Vector Laboratories diaminobenzidine (DAB) substrate kit for horseradish peroxidase (HRP) and follow their protocol.

1. Deparaffinize the tissue sections by dipping the slides in 100% xylene or xylene:HemoD (3 : 1; obtainable from Fisher Scientific) solution (5 min, with gentle agitation about once a minute). The HemoD acts as a clearing agent.

2. Repeat step 1 with fresh xylene or xylene:HemoD (3 : 1).

3. Repeat step 1 again, with fresh xylene or xylene:HemoD (3 : 1).

4. Rehydrate:
   a. Dip the slides into 100% ethanol for 1 min.
   b. Dip the slides into 50% ethanol for 1 min.
   c. Dip the slides into distilled H_2O for 1 min.

5. Block endogenous peroxidase, for 15 min, by adding 0.03% H_2O_2 in methanol to the tissue sections. This can be done by adding the H_2O_2 dropwise to the sections or by dipping the slides into a container holding the H_2O_2.

6. Dip the slides into PBS–BSA for 5 min.

7. Remove the excess PBS–BSA by blotting around the tissue sections with a paper towel (do not allow the tissue sections to dry out), and place sufficient dilute blocking serum (about 100 μl) on top of the sections so that they stay wet for 30 min. (*Note:* It is beneficial to keep the slides in a humidified chamber during the antibody incubations so that the solutions do not evaporate; an inexpensive solution is to invert a large casserole dish over the slides and have some petri dishes filled with water in the "chamber.")

8. Rinse away the dilute blocking serum by gently dripping PBS–BSA onto the slide while holding the slide at a 45° angle.

9. Wipe around the tissue sections and add the primary antibody (about 150 μl) to the sections and place them in a humidified chamber for 2 hr.

10. Rinse dropwise with PBS–BSA, then put about 200 μl of PBS–BSA on top of the sections and let them soak for about 5 min.

11. Repeat step 10.

12. Repeat step 10 again.

13. Rinse dropwise with PBS–BSA, wipe the excess from around the tissue sections, and add secondary biotinylated antibody (about 150 μl; diluted 1:200 in PBS-BSA; Vector Laboratories) for 1 hr (in a humidified chamber).

14. Repeat step 10 three times with PBS (no BSA, because most BSA is contaminated with minor amounts of biotin that might interfere with the next step).

15. Rinse dropwise with PBS, wipe the excess from around the tissue sections, and add ABC reagent (about 150 μl; Vector Laboratories) for 1 hr (in a humidified chamber). *Note:* The ABC reagent must be prepared at least 30 min before using at this step.

16. Repeat step 10 three times with PBS (no BSA).

17. Rinse dropwise with PBS (no BSA), wipe the excess from around the tissue sections, and add DAB reagent (about 100 μl; Vector Laboratories) for 5 min.

18. Repeat step 10 with distilled H_2O.

19. Dip the slides briefly in 50% ethanol, then in 100% ethanol, and then soak in xylene until ready to coverslip. Do not let them air dry. Coverslip with Permount (Fisher Scientific).

Notes

During the antibody and ABC incubations, a smaller volume of solution can be used if a coverslip is placed on top of the solution (for overnight incubations, seal the edges of the coverslip with rubber cement; on the next day simply peel off the rubber cement). To remove the coverslip, dip the slide vertically in PBS–BSA and agitate gently (use different containers for different antibodies to protect against antibody cross-contamination of the tissue sections) or rinse gently with PBS–BSA from a squirt bottle until the coverslip slides off. Rinse gently (do not "jet wash"). If counterstaining is desired, do so after step 17, and use aqueous hematoxylin (2 min; Biomeda Corp., Foster City, CA) for cytoplasmic antigens (because hematoxylin is a nuclear stain) and eosin (1 min; Lerner Laboratories, Pittsburgh, PA) for nuclear antigens (because eosin is a cytoplasmic stain). After counterstaining, repeat step 10 three times, using PBS only, then move on to step 18.

4. Immunogold with Silver Enhancement

This protocol is for tissue embedded in LR White resin (Figs. 2E and F and 3E and F).

1. Dip the slides into PBS–BSA for 5 min.

2. Remove the excess PBS–BSA by blotting around the tissue sections with a paper towel (do not allow the tissue sections to dry out), and place sufficient blocking antibody (about 100 μl) on top of the sections so that they stay wet for 30 min. (*Note:* It is beneficial to keep the slides in a humidified chamber during the antibody incubations so that the solutions do not evaporate; an inexpensive solution is to invert a large casserole dish over the slides and have some petri dishes filled with water in the "chamber.")

3. Rinse off the blocking antibodies by gently dripping PBS–BSA onto the slide while holding the slide at a 45° angle.

4. Wipe around the tissue sections, add the primary antibody (about 150 μl) to the sections, and place them in a humidified chamber for 2 hr.

5. Rinse dropwise with PBS–BSA, then put about 200 μl of PBS–BSA on top of the secretions and let them soak for about 5 min.

6. Repeat step 5.

7. Repeat step 5 again.

8. Rinse dropwise with PBS–BSA, wipe the excess from around the tissue sections, and add the secondary immunogold antibody (about 150 μl) for 1 hr (in a humidified chamber).

9. Repeat step 5 three times, using PBS only (no BSA).

10. Fix the antibodies to each other and the tissue by wiping the excess PBS from around the tissue sections and adding about 150 μl of glutaraldehyde fixative for 30 min.

11. Repeat step 5 three times, using PBS only (no BSA).

12. Repeat step 5 three times, using distilled H_2O. [*Note:* At this point distilled H_2O can remain on the tissue sections while other slides are being silver enhanced a few at a time (which is the next step in this protocol).]

13. Just before use, combine equal amounts of silver enhancing solutions [e.g., IntenSE II or IntenSE M (Amersham, Arlington Heights, IL), light microscopy kit (BioCell U.S. Distributor is Goldmark Biologicals, Phillipsburg, NJ), or LI silver (Nanoprobe, New York, NY); follow the instructions of the manufacturer]. Use this silver enhancing solution immediately and time each slide so that they each have the enhancing solution on the tissue sections for identical times (it helps to stage the slides about 30 sec apart; do not attempt too many slides at the same time, and make up a fresh batch of silver enhancing solutions for each new set of slides). These two solutions (initiator and enhancer, 1 : 1) are then placed on the tissue sections after wiping the excess distilled H_2O from around the tissue sections. Incubation times with the silver enhancing solution will be determined empirically, but try to stay within the 5- to 25-min range to help control the immunolabeling results (if too fast, it is difficult to coordinate slide timing; if too long, self-nucleation of the silver occurs and the background is ''noisy''). What really determines the timing is how intense the antibody labeling is prior to silver enhancing (abundant signal requires less enhancing).

14. Repeat step 5 three times, using distilled H_2O.

15. It is possible to silver enhance again if the signal is too low (the signal can be observed with a light microscope during the enhancing); if repeating step 13 is desired, do so now. Once silver enhancing is completed, rinse the tissue sections one more time with distilled H_2O and let air dry; once dry, counterstain (e.g., with basic fuchsin) for a few seconds if desired, rinse with distilled water, and air dry again; coverslip with Permount (Fisher Scientific).

Notes

During the antibody incubations, a smaller volume of solution can be used if a coverslip is placed on top of the solution (for overnight incubations, seal the edges of the coverslip with rubber cement; on the next day simply peel off the rubber cement). To remove the coverslip, dip the slide vertically in PBS–BSA and agitate gently (use different containers for different antibodies to protect against antibody cross-contamination of the tissue sections) or rinse gently with PBS–BSA from a squirt bottle until the coverslip slides off. Rinse gently (do not ''jet wash'').

D. Electron Microscopic Immunocytochemistry

The following sections present techniques routinely used in our laboratory. There are many variations on these techniques, for example, with the proper equipment frozen sections can be used for electron microscopy, ferritin can be used as an electron-dense label, enzymes and chromagens [such as peroxidase and DAB (see Section II,C)] can be used, and so on. Immunogold (the electron-dense label described in the following section) is an excellent label for electron microscopy due to its spherical structure, its ease of detection, and the fact that different-sized gold spheres can be used for labeling more than one antigen (Fig. 3).

1. Single-Label Immunoelectron Microscopy

For the sake of convenience the following protocol is separated into three 1-day segments. All of the immunocytochemistry procedures are done on a laboratory bench at room temperature unless otherwise stated.

Day 1
1. Clean a few glass slides, dip them into 0.5% Formvar (e.g., LADD Research Industries), and let them air dry. Float the Formvar onto distilled water by scraping the surface of the glass slide around the edges, thus allowing the Formvar film on top of the glass slide to float off the top once the slide is gently submerged into the H_2O at about a 30° angle; place 200-hex nickel grids (e.g., SPI Supplies, Westchester, PA; gold grids can also be used, but do not use copper grids because of the formation of electron-dense precipitates during the protocol) onto the Formvar. Lift the grids and Formvar off the distilled H_2O by touching a piece of filter paper (that is larger than the Formvar) onto the top of the Formvar and grids, and let them dry in a covered petri dish.

2. Cut thin sections (silver to pale gold, approximately 70 nm thick) and place on the Formvar-coated nickel grids.

Note
It is also possible simply to put the thin sections directly onto the grids; if these grids are submerged at each step during the following protocol, the antibodies will label both sides of the section and the labeling density will be higher; careful rinsing of the grids between steps is essential if this variation is used because of the thin sections not being supported by a Formvar film and because of antibody binding between the grid bars on the underside of the tissue sections.

Day 2
1. Make the following solutions (see Section III):

> Phosphate-buffered saline (PBS)
> Phosphate-buffered saline plus bovine serum albumin plus sodium azide
>   (PBS–BSA–NaN_3)
> Dilute blocking serum
> Dilute primary antibody
> Dilute control antibodies

2. Float the grids onto dilute blocking serum (e.g., normal goat serum) for 15–30 min. We usually use 12-well ceramic plates; each well easily holds 200 μl of solution and numerous grids (for grids without Formvar, submerge the grids).

3. Use filter paper to blot the dilute blocking serum from the grids (hold the grids with jeweler's forceps and also blot between the forcep blades; do not rinse or completely dry the grid) and float or submerge the grids onto (or into) the primary antibody [e.g., rabbit antiglial fibrillary acidic protein (GFAP) or non-immune IgG or other control antibody]. Let the sections incubate on (or in) the primary antibody overnight; cover the grids and primary antibody to ensure that the primary antibody does not evaporate (e.g., invert a casserole dish); include petri dishes of water in the enclosed environment to increase the humidity.

Note

It is important to do a dilution series with the primary antibody (see Section III). Although it is possible to incubate the tissue with the primary antibody for a shorter period of time, this requires a more concentrated antibody solution. Overnight incubation, at a reduced antibody concentration, frequently results in a higher signal-to-noise ratio (specific labeling to nonspecific labeling).

Day 3

1. Make the following solutions (see Section III):
   Dilute secondary antibody
   Aqueous uranyl acetate (1% uranyl acetate in distilled H_2O)
   Reynold's lead citrate;
   Osmium tetroxide in sodium phosphate buffer

2. Rinse the grids, gently, with several drops of PBS–BSA–NaN$_3$, immerse in PBS–BSA–NaN$_3$ for at least 10 min and then rinse dropwise again with PBS–BSA–NaN$_3$.

3. Incubate the grids in secondary antibody, [e.g., goat–anti-rabbit IgG complexed to 5-nm gold spheres (Gar-G5; Amersham)] for 1.0 hr (because the grids are wet from rinsing, they probably will not float; simply submerge them, tissue side up, in the secondary antibody solution).

Note

A dilution series is appropriate here also; however, it is possible to obtain an excellent signal-to-noise ratio with a 1.0-hr incubation.

4. Rinse dropwise with, and then immerse in, PBS–BSA–NaN$_3$.

5. Rinse dropwise with distilled H_2O, blot the grids with filter paper, and set aside to air dry.

6. Stain the sections with aqueous uranyl acetate for 10 min.

7. Stain with Reynold's lead citrate for 10 min (do this in an enclosed petri dish; do not breathe on the lead citrate or lead precipitate may form on the tissue sections).

8. Stain with the vapors of OsO_4–$NaPO_4$ in an enclosed environment, in a fume hood, for 1.0 hr. (*Caution:* OsO_4 vapors are very toxic and *will* fix corneas, skin, lungs, etc., so be careful to keep the vapors in the fume hood. *Never* pipette OsO_4, by mouth.) If the tissue contrast is sufficient to omit this step, do so (before processing the grids through this step, look at a grid in the electron microscope and determine if extra contrast is needed; if needed, perform this step).

9. Carbon coat the sections.

10. View in the transmission electron microscope.

2. Double-Label Immunoelectron Microscopy

Several methods may be used to immunolabel two antigens simultaneously, on the same tissue section, with two different primary antibodies. The method covered here works according to the following theory: If the primary antibodies are made in different species (e.g., species "R" is rabbit and species "M" is mouse), the different primary antibodies can be distinguished with a species-specific secondary antibody (e.g., goat–anti-R will recognize R but not M, whereas goat–anti-M will recognize M but not R). If goat–anti-R has a 5-nm gold sphere attached to it (GAR-G5) and goat–anti-M has a 15-nm gold sphere attached to it (GAM-G15), they can be distinguished based on the size of the gold shperes. Thus antigen "A" will be labeled with primary antibody "A" made in rabbit (R), which will be labeled by GAR-G5, and antigen "B" will be labeled with primary antibody "B" made by a mouse monoclonal antibody (M), which will be labeled by GAM-G15. In the electron microscope two sizes of gold spheres will appear overlying the tissue: 5-nm spheres locating antigen A and 15-nm spheres locating antigen B.

This method works well (Fig. 3), but requires the following:

1. One primary antibody has been made in one species [e.g., a rabbit polyclonal antibody: (1 ° Ab-R)].

2. The other primary antibody has been made in a different species [e.g., a mouse monoclonal antibody: (1 ° AB-M)].

3. Single-label immunoelectron microscopy for *each* of the primary antibodies has been performed:

a. Antibody–antigen specificity is well characterized.

b. Antibody–tissue specificity is well characterized.

c. Control experiments give the expected results.

d. The optimum concentration of each of the primary antibodies is known (from the dilution series when the single-label immunoelectron microscopy was performed).

e. The optimum concentration for each secondary antibody is known (from the dilution series when the single-label immunoelectrom microscopy was performed).

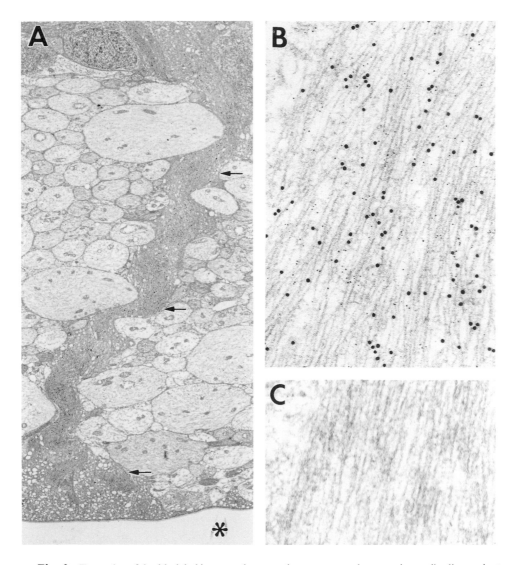

Fig. 3 Examples of double-label immunoelectron microscopy are shown, using antibodies against GFAP and vimentin on normal cat retina embedded in LR White. Both of these intermediate filament proteins are normally present in the cytoplasm of the specialized retinal astrocytes known as Müller cells. A portion of a Müller cell process (arrows) is shown, at low magnification, traversing through bundles of ganglion cell axons (A). The gold spheres are barely visible at this magnification. At higher magnification, anti-GFAP (small, 15-nm gold spheres) and anti-vimentin (large, 30-nm gold spheres) can be seen over the intermediate filaments (B). No gold spheres are present over the intermediate filaments in control sections (C). *, Vitreous cavity of the eye. [Magnification: (A) × 3300; (B) × 58,500; (C) × 78,000.]

Protocol: Day 1
Repeat the single-label immunocytochemistry protocol.

Protocol: Day 2
1. Make the mixture of primary antibodies solution (see Section III).
2. Repeat the day 2 single-label immunocytochemistry protocol, but substitute the mixture of primary antibodies for the single primary antibody.

Protocol: Day 3

1. Make the mixture of secondary antibodies solution (Section III).

2. Repeat the day 3 single-label immunocytochemistry protocol, but substitute the mixture of secondary antibodies for the single secondary antibody

III. Recipes

Stock sodium phosphate buffer (0.172 M NaPO$_4$, pH 7.2)

To make the listed volume of buffer, use the given amounts of monobasic and dibasic sodium phosphate. Add the phosphates sequentially to the desired (final) volume of distilled H$_2$O; adjust to pH 7.2. (*Note:* Save at least 100 ml of this 0.172 M NaPO$_4$ buffer for the osmium tetroxide.

		Volume
Chemicals	500 ml	1000 ml
NaH$_2$PO$_4$	3.41 g	6.82 g
Na$_2$HPO$_4$	8.66 g	17.32 g

Primary fixative (1% formaldehyde plus 1% glutaraldehyde in 0.086 M NaPO$_4$, pH 7.2)

1. To 90 ml of distilled H$_2$O at 80 ° C add 2.0 g of paraformaldehyde; stir until dissolved (in a fume hood). If needed, add a *few* drops of 1.0 N NaOH to help the paraformaldehyde go into solution. After it is completely dissolved, cool to room temperature (use this formaldehyde in solution in the next step).

2. Add 100 ml of stock sodium phosphate buffer (0.172 M NaPO$_4$, pH 7.2) to the formaldehyde in solution.

3. Add 2.86 ml of 70% glutaraldehyde (or a sufficient amount of a different percentage glutaraldehyde to bring the final volume of 200 ml up to 1% glutaraldehyde].

4. Bring the final volume up to 200 ml. This yields a 1.0% formaldehyde–1.0% glutaraldehyde solution in 0.086 M NaPO$_4$ buffer, pH 7.2.

Paraffin primary fixative (4% formaldehyde in 0.086 M NaPO$_4$, pH 7.2)

1. To 90 ml of distilled H$_2$O at 80°C add 8.0 g of paraformaldehyde; stir until dissolved (in a fume hood; if needed, add a *few* drops of 1.0 N NaOH to help the paraformaldehyde go into solution. After it is completely dissolved, cool to room temperature (use this formaldehyde in solution in the next step).

2. Add 100 ml of stock sodium phosphate buffer (0.172 M NaPO$_4$, pH 7.2) to the formaldehyde in solution.

3. Bring the final volume up to 200 ml. This yields a 4.0% formaldehyde solution in 0.086 M NaPO$_4$ buffer, pH 7.2

Glutaraldehyde fixative (approximately 2% glutaraldehyde in 0.086 M NaPO$_4$, pH 7.2)

To 25 ml of stock sodium phosphate buffer (0.172 M NaPO$_4$, pH 7.2), add 23.5 ml of distilled H$_2$O and 1.5 ml of 70% glutaraldehyde.

Note: This fixative is used to fix the antibodies to each other and to the tissue section before silver enhancing; consequently it is not critical that it be exactly 2% glutaraldehyde.

Fixative wash buffer (0.137 M NaPO$_4$, pH 7.2, 315 mOsm)

To 4 parts stock sodium phosphate buffer (0.172 M NaPO$_4$, pH 7.2) add 1 part distilled H$_2$O. This will yield a 0.137 M NaPO$_4$ buffer, pH 7.2, of approximately 315 mOsm (thus isotonic with mammalian blood) and will be used after primary fixation of the tissue.

Maleate wash buffer (0.05 M Na–H–maleate–NaOH buffer, pH 5.2)

Stock solution (0.2 M maleate buffer): 100 ml of distilled H$_2$O (final volume), 2.32 g of maleic acid (or 1.96 g of maleic anhydride), 0.8 g of NaOH

For the pH 5.2 buffer (200 ml): 142.8 ml of distilled H$_2$O, 50.0 ml of stock solution from above, 7.2 ml of 0.2 M NaOH; adjust to a final pH of 5.2

Note: Maleate buffer is used here to wash out the phosphate buffer, because the uranyl acetate will precipitate in phosphate buffers. It is also used to buffer the uranyl acetate because ". . .the maleate buffer buffers the uranyl solution better than veronal acetate buffer. . .and may form a weak complex with the uranyl ions, mitigating against precipitation" (Karnovsky, 1967).

Uranyl acetate 2% in maleate buffer (0.05 M pH 4.75)

Note: The uranyl acetate is mildly radioactive; follow the recommendations of the manufacturer (Electron Microscopy Sciences, Fort Washington, PA) for handling and dispose of the solutions in radioactive waste.

Combine the following:

1. Distilled H_2O 123.1 ml
2. Stock solution (0.2 *M* maleate buffer)   50.0 ml
3. NaOH (0.2 *M*)   26.9 ml

Adjust the pH to 6.0 and add

4. Uranyl acetate   4.0 g

(*Note:* Addition of uranyl acetate drops the pH to about 4.75; it is not necessary to adjust.)

Uranyl acetate fixative (2% uranyl acetate in 70% methanol–H_2O)
   Note: The uranyl acetate is mildly radioactive; follow the recommendations of the manufacturer (Electron Microscopy Sciences) for handling and dispose of the solutions in radioactive waste.
   To prepare, add 2.0 g of uranyl acetate to 100 ml of a 70% methanol–H_2O, stir until dissolved.
   (*Note:* For the alternative fixation, make each of the uranyl acetate–methanol fixatives by this same basic method.)

Aqueous uranyl acetate (1% uranyl acetate in distilled H_2O)
   Note: The uranyl acetate is mildly radioactive; follow the recommendations of the manufacturer (Electron Microscopy Sciences) for handling and dispose of the solutions in radioactive waste.

	Volume	
Chemical	50 ml	100 ml
$UO_2(C_2H_3O_2-2H_2O$	0.5 g	1.0 g

Reynolds' lead citrate (Reynolds, 1963):

	Volume	
Chemical	50 ml	100 ml
Distilled H_2O	30 ml	60 ml
$Pb(NO_3)_2$	1.33 g	2.66 g

1. Stir the above well until thoroughly dissolved. Then add the following:

	50 ml	100 ml
$Na_3(C_6H_5O_7)$	1.76 g	3.52 g

2. Shake for 1 min.
3. Let stand 30 min with intermittent shaking. Then add the following:

	50 ml	100 ml
1.0 N NaOH	8.0 ml	16.0 ml

4. Shake twice; the solution should clear.
5. Make up to a final volume of 50 or 100 ml with distilled H_2O.
6. Keep refrigerated and tightly capped.

Basic fuchsin (Electron Microscopy Sciences)
1. Add 10 ml of 100% ethanol; to 390 ml of distilled H_2O.
2. Add, while stirring, 2 g of basic fuchsin.
3. Filter through a paper towel before use.
4. Apply to tissue sections on glass microscope slides for a few seconds up to a few minutes depending on the type of resin and tissue.
5. Rinse with distilled H_2O and dry on a warm hot plate.

Osmium tetroxide (OsO_4)
   Caution: OsO_4 is extremely toxic and will fix corneas, lungs, and so on. *Open and use only in a fume hood.* Dissolve 2.0 g of OsO_4 crystals in 50.0 ml of distilled H_2O (in a fume hood, with a magnetic stirrer or sonicator).

Osmium tetroxide–sodium phosphate buffer (OsO_4–$NaPO_4$)
   Caution: OsO_4 is extremely toxic and will fix corneas, lungs, and so on. *Open and use only in a fume hood.* To 50 ml of stock sodium phosphate buffer (0.172 M $NaPO_4$, pH 7.2) add the 50 ml of 4% OsO_4. This yields 2% OsO_4 in 0.086 M $NaPO_4$ buffer and is used to osmicate the tissue sections following immunocytochemistry.
   Note: This solution can also be used as a secondary fixative for conventional electron microscopy, but should not be used as a fixative for immunocytochemistry. It is probably not necessary to use buffered OsO_4 for the staining of thin sections. Others have simply placed OsO_4 crystals in the container with the sections. Because this is a routine fixative in most electron microscopy laborato-

ries it is conveniently kept in buffer solution. Also, the solution is much easier and safer to handle than are the crystals.

Phosphate-buffered saline (PBS)
For a 10 × solution:

Chemicals	Volume		
	200 ml	250 ml	750 ml
NaCl	17.53 g	21.92 g	65.75 g
NaH_2PO_4	0.53 g	0.67 g	2.00 g
Na_2HPO_4	2.30 g	2.88 g	8.63 g
$(NaN_3)^a$	(1.00 g)	(1.25 g)	(3.75 g)

[a] Optional: Sodium azide, NaN_3, inhibits peroxidase, therefore do not use for diluting peroxidase-labeled antibodies as in the immunoperoxidase ABC protocol.

Note: To use the 10× PBS or 10× PBS–NaN_3, stir well before measuring and dilute with distilled H_2O to the desired concentration (1:9) and volume. *Adjust the pH to 7.4.*

Phosphate-buffered saline–bovine serum albumin–sodium azide (PBS–BSA–NaN_3)
Supplement the diluted PBS–NaN_3 with 0.5% BSA. Use this PBS–BSA–NaN_3 to dilute the antibodies and to wash the sections after the primary and secondary antibody incubations.

Dilute blocking serum:
Use serum from the species in which the secondary antibody was produced. For example, if the secondary antibody was produced in a goat, then use normal goat serum. This blocking serum serves three purposes: (1) if there are any free aldehydes on the tissue sections, they will react with the proteins in this serum and not with the primary antibody thus decreasing nonspecific primary antibody "labeling"; (2) the proteins in the serum adsorb to the sections, thus blocking nonspecific adsorption; and (3) if there happens to be any antigenic sites recognized by the goat serum, these will be blocked by this serum and will reduce background labeling with the specific secondary antibody.
Dilute the normal serum (e.g., goat) with PBS–BSA–NaN_3 (150 μl of serum diluted into 10.0 ml of buffer, 1:67, works well).

Dilute primary antibody and nonimmune IgG (primary antibody and IgG)
The primary antibody used clearly depends on the antigen of interest. For example, we were interested in the intermediate filaments of glial cells (Müller

cells and astrocytes), which are usually composed of (at least) glial fibrillary acidic protein (GFAP). We bought a commercially available anti-GFAP (Dako Corp., Carpenteria, CA); GFAP that was purified from bovine CNS was injected into rabbits and the rabbits made the polyclonal anti-GFAP. Dako Corporation has characterized this anti-GFAP and shown it to be specific for GFAP and no other intermediate filament protein. We characterized the antibody/tissue specificity with Western blots of whole-retina homogenates. This characterization showed that this antibody is not only specific for GFAP, but also does not label any other detectable retinal antigens by Western analysis, and can be used for immunocytochemistry with retina tissue.

The dilution that yields the optimum signal-to-noise ratio (specific labeling to nonspecific background) for each primary antibody used for tissue immunocytochemistry will have to be determined empirically. For example, with the polyclonal rabbit anti-GFAP, we ran a dilution series of 1:50, 1:100, 1:200, 1:300, 1:400, 1:500, and 1:750 (antibody to buffer).

For the experimental controls, run a dilution series with nonimmune IgG (if preimmune IgG is not available), nonspecific positive antibodies, and nonspecific negative antibodies. Buffer without primary antibody (PBS–BSA–NaN$_3$) should also be run as a control.

Once the optimum concentration of primary antibody (e.g., 1:400) has been determined, and the controls are clean (i.e., no background labeling), then it is possible to start cutting back on the number of control experiments required when repeating the immunolabeling. For example, just run the primary antibody at 1:400, nonimmune IgG (e.g., 1:400), and PBS–BSA–NaN$_3$. However, continue using a nonspecific positive antibody in case primary antibody labeling does not work for some unknown reason. A positive control that *does* work at least indicates the protocol is still working.

Mixture of primary antibodies

Determine the optimum concentration for *each* of the primary antibodies [e.g., rabbit–anti-GFAP (1:400) and mouse–anti-vimentin (1:300); then to a single container of PBS–BSA–NaN$_3$ add the following.

1. Enough rabbit–anti-GFAP to bring the final concentration up to 1:400
2. Enough mouse–anti-vimentin to bring the final concentration up to 1:300

Dilute secondary antibody

Which secondary antibody is used depends on the species and class of the primary antibody. For example, if the primary antibody is an IgG from rabbit, then an anti-rabbit IgG is used. To continue with the previous example for a primary antibody that is an IgG rabbit–anti-GFAP, the secondary antibody could be a goat–anti-rabbit IgG (GAR). This secondary antibody can be purchased from a number of vendors (e.g., Amersham) and is complexed to various sizes of gold spheres [e.g., 5 nm (GAR-G5)].

It is also important to do a dilution series with the secondary antibody, because gold spheres, if too concentrated, will be seen everywhere (nonspecific binding; this can occur even without a primary antibody). We usually use dilutions ranging from 1:25 to 1:100; consider a dilution series around these concentrations, but remember that using less will save money as long as the protocol works.

Secondary fluorescent antibody

Follow the guideline presented in the previous section (Dilute secondary antibody), but substitute a secondary antibody that has a fluorochrome attached rather than a gold sphere. There are numerous vendors of secondary antibodies with fluorescent tags; for single labeling, we have been using the Cappel fluorescein- or rhodamine-conjugated, affinity-purified goat–anti-mouse (or rabbit) IgG, heavy and light chain specific. Newer, brighter fluorochromes are now being offered that promise greater sensitivity (e.g., Cy3). For double labeling, see the next section (Mixture of secondary antibodies), especially the caveat regarding cross-reactivity. Also see other articles in this volume.

Mixture of secondary antibodies

Once the optimum concentration for each of the secondary antibodies has been determined [e.g., GAR-G5 (1:25), labeling rabbit–anti-GFAP; GAM-G15 (1:40) labeling mouse–anti-vimentin] and each of the secondary antibodies is a different size (i.e., 5 and 15 nm) then add the following to a single container of PBS–BSA–NAN$_3$:

1. Enough GAR-G5 to bring the final concentration up to 1:25
2. Enough GAM-G15 to bring the final concentration up to 1:40

Note: One caveat regarding double labeling is the possibility of cross-reactivity between secondary antibodies directed toward different primary antibodies. For example, goat–anti-mouse IgG can recognize epitopes on rat IgG antibodies and goat–anti-rabbit IgG can recognize epitopes on guinea pig IgG antibodies. To determine if the secondary antibodies to be used have this characteristic (undesirable for double labeling), apply the inappropriate secondary antibody in a single labeling experiment (e.g., use a primary antibody made in mouse with a secondary antibody that detects rabbit primaries but supposedly does not detect mouse primaries). If cross-reactivity is detected between the inappropriate primary and secondary antibodies, try secondary antibodies that have been further purified by absorption against the primary antibody species to be used in double labeling.

References

Anderson, D. H., Stern, W. H., Fisher, S. K., Erickson, P. A., and Borgula, G. A. (1983). Retinal detachment in the cat: The pigment epithelial–photoreceptor interface. *Invest. Ophthalmol. Visual Sci.* **24**, 906–926.

Anderson, D. H., Guerin, C. J., Erickson, P. A., Stern, W. H., and Fisher, S. K. (1986). Morphological recovery in the reattached retina. *Invest. Ophthalmol. Visual Sci.* **27**, 168–183.

Bock, G., Hilchenbach, M., Schauenstein, K., and Wick, G. (1985). Photometric analysis of anti-fading reagents for immunofluorescence with laser and conventional illumination sources. *J. Histochem. Cytochem.* **33**, 699–705.

Erickson, P. A., and Fisher, S. K. (1988). Intermediate filament composition distinguishes retinal Müller cells from astrocytes. *Invest. Ophthalmol. Visual Sci.* **29**, 205a.

Erickson, P. A., Fisher, S. K., Anderson, D. H., Stern, W. H., and Borgula, G. A. (1983). Retinal detachment in the cat: The outer nuclear and outer plexiform layers. *Invest. Ophthalmol. Visual Sci.* **24**, 927–942.

Erickson, P. A., Anderson, D. H., and Fisher, S. K. (1987). Use of uranyl acetate *en bloc* to improve tissue preservation and labeling for post-embedding immunoelectron microscopy. *J. Electron Microsc. Tech.* **5**, 303–314.

Guerin, C. J., Anderson, D. H., Fariss, R. N., and Fisher, S. K. (1989). Retinal reattachment of the primate macula: Photoreceptor recovery after short-term detachment. *Invest. Ophthalmol. Visual Sci.* **30**, 1708–1725.

Hyatt, M. A. (1981). Uranyl preparations. *In* "Principles and Techniques of Electrom Microscopy: Biological Applications," 2nd Ed. Vol. 1, pp. 327–340. University Park Press, Baltimore.

Johnson, L. V., and Blanks, J. C. (1984). Application of acrylamide as an embedding mecium in studies of lectin and antibody binding in the vertebrate retina. *Curr. Eye Res.* **3**, 969–974.

Karnovsky, M. J. (1967). The ultrastructural basis of capillary permeability studied with peroxidase as a tracer. *J. Cell Biol.* **35**, 213–236.

Lenard, J., and Singer, S. J. (1968). Alteration of the conformation of protein in red blood cell membranes and in solution by fixatives used in electron microscopy. *J. Cell Biol.* **37**, 117–121.

Lewis, G. P., Erickson, P. A., Guerin, C. J., Anderson, D. H., and Fisher, S. K. (1989). Changes in the expression of specific Müller cell proteins during long-term retinal detachment. *Exp. Eye Res.* **49**, 93–111.

Pearse, A. G. E. (1980). The chemistry and practice of fixation. *In* "Histochemistry—Theoretical and Applied," 4th Ed., Vol. 1, pp. 97–158. Churchill Livingstone, New York.

Pelletier, G., Puviani, R., Bosler, O., and Descarries, L. (1981). Immunocytochemical detection of peptides in osmicated and plastic-embedded tissue: An electrom microscopic study. *J. Histochem. Cytochem.* **29**, 759–764.

Reynolds, E. S. (1963). The use of lead citrate at high pH as an electron opaque stain in electron microscopy. *J. Microsc. (Paris)* **8**,761–766.

Schwendemann, G., Wolinsky, J. S., Hatzidimitriou, G., Merz, D. C., and Waxhan, M. N. (1982). Postembedding immunocytochemical localization of paramyxovirus antigens by light and electron microscopy. *J. Histochem. Cytochem.* **30**, 1313–1319.

Silva, M. T., Carvalho Guerra, F., and Magalhaes, M. M. (1968). the fixative action of uranyl acetate in electron microscopy. *Experientia* **24**, 1074.

Silva, M. T., Santos Mota, J. M., Melo, J. V. C., and Carvalho Guerra, F. (1971). Uranyl salts as fixatives for electron microscopy. Study of the membrane ultrastructure and phospholipid loss in bacilli. *Biochim. Biophys. Acta* **233**, 513–520.

Smith, J. A. (1987). Elecctrophoretic separation of proteins. *In* "Current Protocols in Molecular Biology" (F. M. Ausubel *et al.,* ed.), pp. 10.2.1–10.2.9. Green Publ. Assoc. and Wiley (Interscience), New York.

Smith, P. F., and Keefer, D. A. (1982). Acrolein/glutaraldehyde as a fixative for combined light and electron microscopic immunocytochemical detection of pituitary hormones in immersion-fixed tissue. *J. Histochem. Cytochem.* **30,** 1307–1310.

Terzakis, J. A. (1968). Uranyl acetate, a stain and a fixative. *J. Ultrastruct. Res.* **22,** 168–184.

Vaughan, D. K., Erickson, P. A, and Fisher, S. K. (1990). Glial fibrillary acidic protein (GFAP) immunoreactivity in rabbit retina: Effect of fixation. *Exp. Eye Res.* **50,** 385–392.

Winston, S. E., Fuller, S. A., and Hurrell, J. G. R. (1987). Western blotting. *In* "Current Protocols in Molecular Biology" (F. M. Ausubel *et al.,* ed.), pp. 10.8.1–10.8.6. Green Publ. Assoc. and Wiley (Interscience), New York.

CHAPTER 16

Electron Microscopy Immunocytochemistry Following Cryofixation and Freeze Substitution

John Z. Kiss* and Kent McDonald†,[1]

*Department of Botany
Miami University
Oxford, Ohio 45056

†Laboratory for Three-Dimensional Fine Structure,
  Department of Molecular, Cellular, and Developmental Biology
University of Colorado
Boulder, Colorado 80309

[1] Present address: Electron Microscopy Laboratory, University of California, Berkeley, California 94720.

I. Introduction

In this article we present methods of immunolabeling after cryofixation that have worked for us. These methods are by no means definitive and, in fact, are evolving constantly even in our own laboratories. There is probably no one formulation that will work for every tissue and perhaps the only generalization we can make is that every situation will be different. However, not all variables are of equal importance and, in our experience, the most crucial is the reactivity of the antibody used. It should have reasonably high titer and, ideally, react with antigens after fixation with glutaraldehyde in order to obtain the best structural preservation. When one or both of these criteria are not met, the chances of obtaining highly specific labeling with good ultrastructure diminish accordingly.

The next most important variable is the quality of cryofixation. In the ideal case, one would like to have cells with their water frozen in the vitreous (noncrystalline) state without cryoprotectants, but this is not really possible except with thin (1–5 μm) samples (Dubochet *et al.*, 1987). In practice, well-frozen tissues probably contain small ice crystals (5 nm or less) that do not perturb the ultrastructure of the cell, and with the most readily available cryofixation methods (plunge freezing and freezing against a cooled copper block), this is difficult to achieve beyond a depth of about 10 μm without cryoprotectants. Double-propane jet freezing can increase the depth to 40 μm, and propane jet freezers are becoming more readily available. To achieve good freezing in tissues up to 600 μm in size, one needs to use the technique of high-pressure freezing (Moor, 1987). We have been fortunate to have this technology available to us and in this article we will show how cryofixation and immunolabeling can be applied to relatively large tissues such as plant root tips and *Drosophila* embryos. Although this equipment is not now readily available to most researchers, we believe that it will soon become more common and a discussion of these methods is appropriate at this time.

Because of the surge in interest in both cryofixation and electron microscopy (EM) immunolabeling with colloidal gold, there are excellent review articles and books that cover some of the same ground as this article. For cryotechniques, see Gilkey and Staehelin (1986), Menco (1986), Robards and Sleytr (1985), and Steinbrecht and Zierold (1987). For immunogold methods, see the two volumes edited by Hayat (1989a), and the book edited by Verkleij and Leunissen (1989), especially the chapter by Humbel and Schwarz (1989).

Samples are commonly prepared for EM immunolabeling in three ways: (1) labeling of cryosections, (2) preembed labeling, and (3) postembed labeling.

The first of these methods, immunolabeling of cryosections, can give excellent results when it works, but it is not an easy method to master. Cells or tissues are "lightly" fixed with aldehydes, infiltrated with 2.3 M sucrose, frozen, and sectioned on a cryomicrotome. The sections are immunolabeled, then stabilized with LR White resin or methyl cellulose prior to examination in the electron microscope. The method has the advantage that one can use large tissue sizes and label antigens that are in the cell interior. The disadvantages are that this method requires expensive equipment (a cryomicrotome) that is tricky to use, and the images are often difficult to interpret because they are in negative contrast (i.e., the membranes, cytoskeleton, etc., are white against a dark background). Because this method does not require freeze substitution, we will not discuss it further here. This method is sometimes called the Tokuyasu method after its originator and more information on its application can be found in several review articles (Tokuyasu, 1986; Sitte *et al.*, 1989; Leunissen and Verkleij, 1989; van Bergen en Henegouwen, 1989).

The preembedding method of immunolabeling is the simplest method for EM preparation. In this procedure, the cells are chemically fixed, lysed, labeled with primary and secondary antibodies, and then treated as a routine (ultrathin section) EM preparation. The main disadvantage of preembed labeling is that the lysis procedures can cause considerable disruption of the ultrastructure of the cell, especially membranous components.

The focus of this article is on postembedding labeling of cells and tissues that have been ultrarapidly frozen (or cryofixed), freeze substituted, and embedded in resins. We emphasize high-pressure freezing as a cryofixation method because it is applicable to the widest range of samples. We briefly discuss two techniques that permit low-temperature embedding of material that has been chemically fixed at room temperature (Carlemalm *et al.*, 1982; Kellenberger *et al.*, 1980; van Genderen *et al.*, 1991). These techniques seem to be reasonable compromises for those who do not have access to cryofixation equipment.

II. Systems and Goals

A. *Drosophila* and Other Genetic Systems

The research on the genetic systems *Drosophila melanogaster, Caenorhabditis elegans, Saccharomyces cerevisiae,* and *Schizosaccharomyces pombe* is among the most exciting in modern biology. Efforts to understand the molecular biology of the cell cycle and of development in these organisms have resulted in the production of many antibodies against specific gene products and the localization of these gene products during the cell cycle or development by immunofluorescence light microscopy. The localization by EM immunocytochemistry has lagged behind in part because preservation of these organisms for even conventional EM by chemical fixation is difficult. The cell walls or protective embryonic layers restrict the penetration of fixatives, dehydration fluids, and resins necessary for routine EM specimen preparation. Removal by hand of the outer embryonic layers in *Drosophila* and *C. elegans* results in improved ultrastructure but this is not always an easy task in practice, and the number of embryos that can be prepared this way is relatively small. Likewise, one can improve the ultrastructure of the yeasts by digesting away the cell wall prior to fixation or by cryofixation and freeze substitution (Tanaka and Kanbe, 1986). In the latter case, relatively few cells can be frozen at one time. One of the goals of our research has been to develop methods of specimen preparation that will yield large numbers of well-fixed embryos or yeast cells that could be used for routine EM characterization or EM immunocytochemistry. Use of the Balzers HPM 010 high-pressure freezer has allowed us to achieve the first of these goals, and we are now using *Drosophila* embryos as a test system for EM immunolocalization studies.

B. Plant Cells

Another of our goals is to characterize the ultrastructure and biochemistry of the plant cell wall during gravitropic curvature. Gravitropism is a complex phenomenon that results in the directed growth of plants in response to a gravitational stimulus, and the stages of gravitropism include perception, transduction, and response (Evans *et al.*, 1986; Sack and Kiss, 1989). We are specifically interested in changes in the structure of the cell wall during the response stage of gravitropism in both lower and higher plants. One method to assay the structure of the cell wall is to use specific antibodies against various cell wall components.

As in the case of *Drosophila* embryos, adequate cryofixation of plant cells has been difficult. The main problem seems to be that plant cells have a high water content. However, in our studies we have been able to obtain excellent cryofixation of entire root tips by high-pressure freezing (Kiss *et al.*, 1990; Staehelin *et al.*, 1990). In addition to improving structural preservation, cryofixation improves preservation of the antigenicity of cell wall matrix polysaccharides in ultrathin sections (compare Fig. 1A and B) (Kiss and Staehelin, 1991). We are now at the stage at which cryofixation techniques in combination with immunocytochemistry can be used to help answer some interesting biological questions.

III. Cryofixation, Freeze Substitution, and Embedding

A. Cryofixation

Cryofixation or freeze fixation is a physical technique to preserve cellular ultrastructure that is an important alternative to conventional chemical fixation.

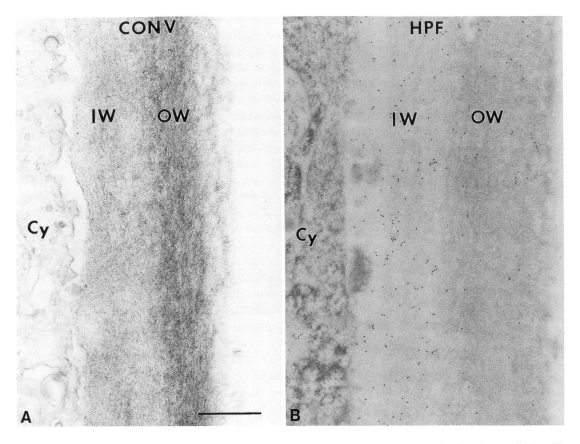

Fig. 1   Immunolabeling of *Chara* rhizoid cell wall with the JIM7 antibody (primary) followed by goat anti-rat secondary antibody conjugated to 10-nm gold. The cell wall consists of a lighter staining inner wall (IW) and a darker staining outer wall (OW). Cy, Cytoplasm. (Bar: 0.5 μm.) (A) This cell was prepared by conventional (CONV) chemical fixation in 2% glutaraldehyde and 1% osmium tetroxide. (B) This cell was prepared by high-pressure freezing (HPF)/freeze substitution in 1% osmium tetroixide in acetone. Note the higher density of immunolabeling in this sample and the higher density of labeling in the inner cell wall (IW).

The main reasons to use cryofixation rather than chemical fixation are that fixation is fast (milliseconds) and that cellular components are simultaneously stabilized. In contrast to this, chemical fixatives act slowly (seconds to minutes) and may alter cellular morphology (Mersey and McCully, 1978; Gilkey and Staehelin, 1986). Therefore it is assumed that cryofixation, compared to conventional chemical fixation, provides images that are more likely to reflect the native structure of living cells. An additional benefit of cryofixation is that it may improve immunocytochemical localization (Fig. 1A and B) (Ichikawa *et al.*, 1989; Kandasamy *et al.*, 1991; Nicolas, 1991).

1. Methods

The major cryofixation methods include plunge freezing, cold metal block freezing, propane jet freezing, and high-pressure freezing. The goal of all of these methods is to preserve cellular ultrastructure without the formation of damaging ice crystals. These methods will be briefly discussed below, but the interested reader should see several excellent works that comprehensively review cryofixation methods (Gilkey and Staehelin, 1986; Menco, 1986; Robards and Sleytr, 1985; Steinbrecht and Zierold, 1987).

The specific protocols for various cryofixation techniques are beyond the scope of this article. The following is a listing of reviews (which include detailed protocols) for the particular cryofixation method: plunge freezing (Costello, 1980; Elder *et al.*, 1982), cold metal block freezing (Heuser *et al.*, 1979; Boyne, 1979), propane jet freezing (Gilkey and Staehelin, 1986), and high-pressure freezing (Moor, 1987; Dahl and Staehelin, 1989).

a. Plunge Freezing

Plunge freezing is the simplest and least expensive of any of the cryofixation techniques that will be discussed. A sample is plunged at a rapid rate into some type of liquid cryogen. Some of these cryogens include supercooled liquid nitrogen, various halocarbons, propane, liquid helium, and Freon 22 (Robards and Sleytr, 1985). In its simplest form, a sample is plunged into the cryogen by hand. However, a variety of devices have been built to facilitate rapid immersion of the specimen into the cryogen (see, e.g., Robards and Sleytr, 1985; Lancelle *et al.*, 1986; Ridge, 1990). The major limiting factor of plunge freezing is that good freezing is generally limited to a few micrometers from the surface, so that only small specimens have been studied. However, with certain specimens, excellent ultrastructural preservation can be achieved (see, e.g., Lancelle *et al.*, 1987; Kiss *et al.*, 1988; O'Toole and McIntosh, 1991).

b. Cold Metal Block Freezing

In cold metal block freezing, the specimen is slammed (hence this method is also called slam freezing) onto a cold metal surface (usually silver or copper) that is cooled by liquid nitrogen or liquid helium. Liquid helium, despite its expense, is preferred because the sample can be cooled more rapidly than with liquid nitrogen (Robards and Sleytr, 1985). Two of the most commonly used devices were designed by Heuser *et al.* (1979) and Boyne (1979). The latter apparatus is called the "Gentleman Jim" device. Although specimens can be preserved up to a depth of 10–15 μm, the area of immediate impact with the metal block may be damaged and other subtle artifacts may result (Pinto da Silva and Kachar, 1980). Nevertheless, superb results have been achieved by using cold metal block freezing (see, e.g., Heuser *et al.*, 1979; Heuser and Steer, 1989).

c. Propane Jet Freezing

In propane jet freezing, a specimen is sprayed from two sides by jets of liquid propane that is cooled by liquid nitrogen (Gilkey and Staehelin, 1986). The specimen is in a "sandwich" formed by two copper specimen holders. In general, a greater depth of good preservation (20–40 μm) can be obtained by using propane jet freezing compared to the previous two methods. Applications of propane jet freezing that yielded exciting new information include Giddings *et al.* (1980), Fernandez and Staehelin (1985), and Staehelin and Chapman (1987). A potential problem with double jet freezing is that the jets must hit the specimen holder simultaneously, otherwise only single-sided cooling occurs and the depth of good freezing is reduced accordingly.

To extend the depth of good preservation, many laboratories have treated cells and tissues with cryoprotectants before freezing their specimens. However, pretreatment with cryoprotectants such as glycerol and dimethyl sulfoxide causes numerous artifacts (reviewed in Gilkey and Staehelin, 1986) and defeats the whole purpose of cryofixation. However, a study by Ding *et al.* (1991) has shown that sucrose pretreatment of plant cells before propane jet freezing can increase the diameter of successful freezing to 100 μm without any apparent artifacts.

d. High-Pressure Freezing

High-pressure freezing is similar to propane jet freezing in that the sample is sprayed from two sides by a cryogen (liquid nitrogen). Immediately before freezing, the sample is pressurized to 2100 atm to improve the depth of preservation. Although high-pressure freezing is the most costly technique in that it requires a relatively expensive machine, this technique allows a depth of preservation up to 600 μm (Moor, 1987; Dahl and Staehelin, 1989; Kiss *et al.*, 1990; Michel *et al.*, 1991). In practice, the depth of preservation may be on the order of a few hundred micrometers in samples with a high water content (e.g., plant cells). A variety of specimens have been successfully preserved by this technique (reviewed in Kiss *et al.*, 1990), and it offers great promise for extending the advantages of cryofixation to larger tissues that could not be preserved by the other methods.

2. Limitations

As indicated by Gilkey and Staehelin (1986), specimen handling and preparation before cryofixation could still be a limitation despite application of these techniques. Even though cryofixation takes place on the order of milliseconds, if the tissue or cell is disrupted by handling, artifactual images may result. This may be more of a problem with the first three methods discussed above because of a smaller depth of preservation, but even with high-pressure freezing the sample may be damaged during specimen loading.

B. Handling the Frozen Specimen

Following cryofixation, the specimen may be immediately used in further processing steps. However, in some cases it may be necessary to store the specimen for future use. Generally, storing the specimen in liquid nitrogen is the best approach, and a number of containers for storing specimens under liquid nitrogen have been developed (reviewed in Robards and Sleytr, 1985). Our approach is to store frozen specimens in commercially available cryovials made of polypropylene, and keep these vials in large liquid nitrogen Dewar vessels. It is important not to accidentally heat the frozen specimens in any handling steps (e.g., using room-temperature forceps), and it is best to do all transfer steps under liquid nitrogen.

C. Freeze Substitution

The major ways to process a specimen after cryofixation include (1) direct viewing of frozen specimens, (2) cryoultramicrotomy, (3) freeze fracture, and (4) freeze substitution. Although direct viewing of frozen specimens in vitrified (noncrystalline) water is an important method that has been used to great advantage in crystallographic studies of proteins (Dubochet et al., 1987), these techniques are difficult, time consuming, and require a great deal of specialized equipment. In addition, the first three of the above methods are difficult to combine with immunocytochemical localization. Therefore, we have used cryofixation extensively in conjunction with freeze substitution in our work.

Freeze substitution is the process of dissolution of ice in a frozen specimen by an organic solvent at low temperature and usually takes place in the presence of a secondary fixative (Steinbrecht and Müller, 1987). The temperature at which freeze substitution occurs should be low enough (below $-70°C$) to avoid secondary ice crystal growth. After freeze substitution is completed, the temperature can be raised without risk of ice crystallization because water is now absent from the specimen. Generally, except for low-temperature embedding, infiltration of a plastic resin is done at room temperature and is followed by heat polymerization. At the end of this process, the result is a plastic-embedded specimen that can be sectioned in the usual manner. We usually do immunocytochemical staining of the sections once they are on an electron microscope grid.

1. Substitution Media

The two most commonly used media are acetone and methanol (Steinbrecht and Müller, 1987). Methanol has the advantage that it can substitute specimens in the presence of substantial amounts of water and that substitution is faster at low temperatures in methanol than in acetone. In typical protocols (performed at $-80°C$), freeze substitution in methanol can take about 18 hr whereas freeze substitution in acetone requires 2.5 days. We have found that substitution in methanol causes heavy extraction of cytoplasm in our samples (Kiss *et al.*, 1990), but others have had good preservation with methanol (see, e.g., Hippe, 1985; Meissner and Schwarz, 1990). Other substitution media include ethanol, diethyl ether, and several additional organic compounds (for reviews see Robards and Sleytr, 1985; Steinbrecht and Müller, 1987).

2. Freeze-Substitution Equipment

A large number of devices, both commercial and homemade, can be used as freeze-substitution chambers. Some of these are reviewed in Steinbrecht and Müller (1987). Two simple ways to perform freeze substitution are to use either a dry ice–acetone bath or a −80°C freezer. We have had excellent results with the dry ice–acetone bath, which provides good thermal contact with the cryovials at a constant temperature of −78.5°C (Kiss *et al.*, 1990).

For lower substitution temperatures, we use a homemade device consisting of an aluminum block with holes drilled in it to accommodate standard 1.5-ml cryotubes. Wrapped around the block is heater wire, which is connected to a temperature regulator. This assembly is lowered into a Dewar that is filled with liquid nitrogen to a level just below the aluminum block. The temperature regulator is set to −90°C, and the sample is left at this temperature for 2–3 days. When the nitrogen runs out, the samples begin to warm slowly and we can remove them at any temperature for further processing. For low-temperature embedding we would remove them at −70 or −35°C, and for embedding in conventional resins we remove them at 0°C or room temperature. A diagram of our device is shown in Fig. 2. A list of the suppliers for the components of this particular equipment is provided below the figure.

3. Fixatives

The fixative chosen for use during freeze substitution can have a profound effect on the amount and specificity of immunogold labeling as well as on the quality of ultrastructural preservation (Newman and Hobot, 1989). The main variable that determines which fixation to use is the sensitivity of the antibody. Some antibodies are so reactive and/or the concentration of antigen is so high that good localization can be achieved on sections of material that have been freeze substituted in 5% osmium tetroxide–acetone and embedded in Epon (Nicolas, 1989). At the other extreme, some antibodies are so sensitive to fixatives or the antigen concentration is so low that postembedding labeling does not work. When it is a case of antibody sensitivity to fixation, one can sometimes achieve good labeling after freeze substitution in methanol or acetone alone and embedding in low-temperature resins (Monaghan and Robertson, 1990; Usuda *et al.*, 1990, 1991; Lancelle and Hepler, 1989; Humbel and Schwarz, 1989). When it is a matter of antigen concentration, it is possible to make calculations to

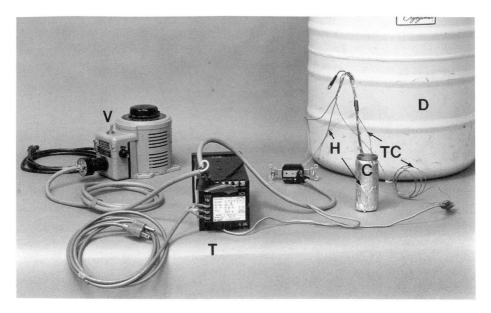

Fig. 2   The components of our temperature-controlled freeze-substitution device shown outside their housing in order to illustrate interconnections. The temperature controller (T) is connected to a heater wire (H) that is wrapped around a brass hollow cylinder (C) that will hold four standard 1.8-ml cryotubes. The cylinder is lowered into the neck of a Dewar (D) to a level just above the liquid nitrogen, which acts as the countercoolant. The temperature is set at the controller and monitored by a thermocouple (TC) that is inserted into methanol, which fills the lower part of the brass cylinder. The temperature controller is plugged into a Variac (V), which allows us to adjust the current going to the heater wire. Vendors for component parts: Temperature controller [Beckmann Industrial Company (Instrumentation Products Div., San Diego, CA) or Omega Engineering (Stamford, CD)], variable autotransformer powerstat (type 116B/120 V (in), 140 V (out), 1.4 kVA; Fisher Scientific), storage Dewar (MVE Cryogenics, New Prague, MN), thermocouples (Omega Engineering), heater wire (with a 3/16-in. flexible electric heating cord; Cole-Parmer Instrument Co., Chicago, IL).

determine the threshold below which gold labeling is unlikely (Kellenberger *et al.*, 1987).

Most antibody–antigen fixation requirements fall between the extremes discussed above. If cells are to be embedded in a low-temperature resin by ultraviolet (UV) polymerization, it is necessary to omit osmium or use low concentrations because osmium will block the polymerizing action of UV light. Standard fixatives would include paraformaldehyde and/or glutaraldehyde. The concentrations should be as low as possible for the particular antigen being studied. To test the sensitivity of antigen to different fixatives, one can do Western analysis on fixed blots.

The most common fixative used in freeze substitution is osmium tetroxide (usually 1–2%) in acetone (see, e.g., Heath *et al.*, 1984; Lancelle *et al.*, 1987;

Kiss *et al.*, 1990). Addition of uranyl acetate (see, e.g., Heath and Rethoret, 1982; Dahmen and Hobot, 1986) or tannic acid (see, e.g., Tiwari and Gunning, 1986) to osmium tetroxide in the substitution medium will often enhance contrast. Another way to enhance contrast is to substitute first with tannic acid in acetone followed by substitution with a mixture of osmium tetroxide and uranyl acetate in acetone (Ding *et al.*, 1991). Examples of the use of fixatives other than osmium include glutaraldehyde (see, e.g., Ichikawa *et al.*, 1989) and potassium permanganate (see, e.g., Heath *et al.*, 1985).

If one must use a protocol in which freeze substitution is in acetone or methanol alone (in order to maintain antigenicity), it is necessary to check the quality of freezing by a technique that gives a more "recognizable" image of the cytoplasm. Therefore, in addition to a freeze-substitution protocol that is optimal for immunocytochemistry, another protocol should be used for conventional EM imaging (i.e., freeze substitution in osmium tetroxide and embedding in an epoxy-based resin) that provides a way to check the quality of freezing.

D. Embedding Resins

Except where one can stain Epon or Spurr's resin sections, the resins of choice for immunolabeling are LR White and the Lowicryl resins. Neither group was designed for immunogold methods but have become quite useful. Together they offer a wide range of methods, ranging from simple to complex, for preparing cells for postembedding immunolabeling (Newman and Hobot, 1989).

The Lowicryl resins were originally developed to improve structural preservation in material that needed to be embedded and sectioned (Carlemalm *et al.*, 1980, 1982; Kellenberger *et al.*, 1980). The logic was to avoid use of denaturing heavy metals, to have a water-compatible embedding medium that would allow proteins to retain their shell of hydration and that would polymerize at low temperature to minimize disruptions due to thermal vibrations. These same properties were soon discovered to be highly advantageous for postembedding labeling of sections with gold probes (Roth *et al.*, 1981). The original formulations (Carlemalm *et al.*, 1982) included a polar resin (K4M) and a hydrophobic resin (HM20) that could be polymerized by UV irradiation or chemical catalysis at $-35°C$ (K4M) or $-50°C$ (HM20) or by heat up to 60°C. To take full advantage of cryofixation and substitution, two new resins were developed (Carlemalm *et al.*, 1985; Acetarin *et al.*, 1986) that could be polymerized by UV light at -60 to $-70°C$. These are the polar K11M and nonpolar HM23 formulations. These resins are manufactured by Chemische Werke Lowi (Waldkraiburg, Germany) and can be obtained from most EM suppliers. The K4M resin is the one most used for immunolabeling studies; however, HM20 may offer a number of advantages, including better sectioning properties and embedding at lower temperature ($-50°C$), although one must take more precautions against nonspecific labeling (Hobot, 1989). The use of K11M and HM23 formulations is beginning to increase in immunolabeling studies (Durrenberger *et al.*, 1988; Durrenberger, 1989; Hobot, 1989).

LR White (London Resin Company, Ltd., Basingstoke, Hampshire, England) is another acrylic resin that is commonly used for immunolocalization studies, both light and electron microscopic. Like Lowicryl, LR White was not originally designed as a resin for immunocytochemistry (it was conceived as a nontoxic alternative to epoxy resins), but has found wide application because of its particular properties. When using LR White, one is able to stop alcohol dehydration at 70% or less and apply LR White directly. The disadvantage, when compared to Lowicryl, is that the resin becomes viscous below −20°C. In addition to using LR White, some workers have obtained excellent results with the related resin LR Gold (see, e.g., Berryman and Rodewald, 1990).

Whether one chooses to use LR White, LR Gold, or one of the Lowicryls will depend on factors such as the sensitivity of the antigen to fixatives and whether the material can or cannot be rapidly frozen without cryoprotectants (Newman and Hobot, 1989; Newman, 1989). If the cells can be freeze substituted into a medium with high (>0.5%) concentrations of aldehyde fixatives, and if antigenicity would not be destroyed by heat, then LR White is convenient to use because it is nontoxic and is a one-step resin (i.e., it does not require the mixing of several components).

IV. Antibodies and Gold Markers

A. Antibodies

Whether or not one has success with EM immunolocalization of antigens depends to a great extent on the characteristics of the antibody used. If one uses the most sophisticated of specimen preparation regimes, that is, cryofixation followed by dehydration without fixatives and embedding at −80°C, and still has no labeling, then about the only course left is to make and screen more antibodies. If there is a lot of nonspecific labeling, it may be possible to purify and characterize the antibody further.

One of our scientific interests concerns the EM localization of molecules associated with cytoskeletal elements (i.e., microtubules, intermediate filaments, and actin) such as cytoplasmic dynein, myosin, and kinesin. These proteins are a challenge to localize by postembedding labeling with immunogold because they usually exist only in trace amounts and their availability for labeling is further reduced by being only at the surface of sections. For the purposes of illustration in this article, we have typically used an antibody against microtubule protein (tubulin), because (1) the extent of labeling can be correlated with an easily recognized organelle, and (2) when cross-sections of MTs are labeled (Fig. 3), the resolution of the labeling is more apparent. The tubulin antibody used here is a monoclonal antibody made in mouse.

Fig. 3 Cross-section through a mitotic spindle from a *Drosophila* embryo that was high-pressure frozen, freeze substituted in 0.2% glutaraldehyde in methanol at −90°C, and embedded in Lowicryl K4M at −35°C; 80-nm sections were labeled with anti-tubulin primary and goat anti-mouse secondary conjugated to 10-nm gold. The distance of the gold particles from the microtubule wall is an indicator of the size of the antibody–secondary complex. The number of unlabeled microtubules is an indicator of the labeling efficiency of this antibody under the conditions used (see text). (Bar: 0.1 μm.)

We are also involved in a project to localize gene products that affect the development of anterior–posterior axes in *Drosophila,* and have been using a polyclonal antibody made in rabbits against the *vasa* gene product (kindly provided by P. Lasko, Department of Biology, McGill University, Canada). This antibody stains polar granules (Fig. 4), as has been shown in previously published work by Hay *et al.* (1988).

Plant cell walls are composed of cellulose, matrix polysaccharides, and glycoproteins (Staehelin *et al.,* 1988). The antibodies utilized in our studies are against a variety of matrix polysaccharides. Two polyclonal antibodies used were anti-xyloglucan (anti-XG), against a hemicellulose, and anti-polygalacturonic acid/rhamnogalacturon-I (anti-PGA/RGI), against a pectin. These two antibodies have been described and characterized by Moore *et al.* (1986). In addition, the two monoclonal antibodies used in these studies were against unesterified pectin (JIM5) and methyl-esterified pectin (JIM7). These were characterized by Knox *et al.* (1990).

B. Gold Markers

An extensive discussion of the making and characterization of immunogold probes is beyond the scope of this article. Many excellent references on the subject are to be found in Hayat (1989a) and Verkleij and Leunnesen (1989). Although it may be desirable to make one's own probes, most investigators use commercially available products. For the *Drosophila* and mammalian tissue culture cell results reported here, we have used 5- and 10-nm gold particles conjugated to goat anti-rabbit or goat anti-mouse secondary antibodies. These were obtained from Amersham (Arlington Heights, IL).

For the polyclonal cell wall matrix antibodies, 10-nm protein A–gold particles were used. For the monoclonal antibodies against pectins, goat anti-rat 10-nm gold particles were used. Both of these gold conjugates were made by BioCell Research Laboratories and are distributed by Ted Pella, Inc. (Redding, CA).

V. Protocols

A. High-Pressure Freezing

Although high-pressure freezing allows a depth of preservation up to 600 μm (Moor, 1987), the specimen cups (designed by Craig *et al.,* 1987) must be filled with a fluid because any air bubbles would collapse under pressure, causing disruptions to the tissues (Craig and Staehelin, 1988; Kiss *et al.,* 1990). A major limiting factor of high-pressure freezing is finding an appropriate external cryoprotectant that itself would not induce artifacts (and thus defeat the whole purpose of cryofixation). We have obtained excellent results by filling the speci-

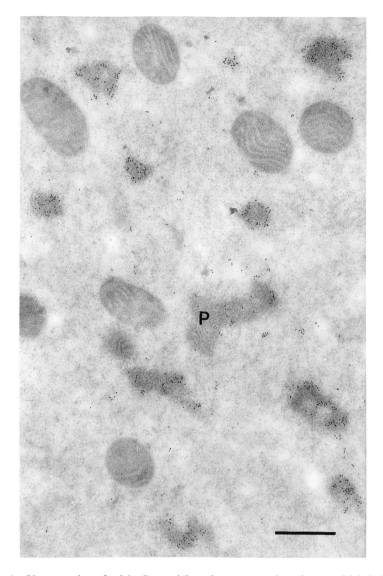

Fig. 4   An 80-nm section of a 2-hr *Drosophila* embryo prepared as above and labeled with an antibody against vasa protein, and goat anti-rabbit secondary conjugated to 10-nm gold. The *vasa* gene is required for development of the germ plasm in *Drosophila*, and the gene product maps to structures, in the posterior cytoplasm, called polar granules (P) (Lasko and Ashburner, 1990), which are identified here by 10-nm gold particles. (Bar: 0.5 μm.)

men cups with 15% dextran (M_r 38,800 or 79,100). Dextran solutions are effective freezing media because dextran has a relatively high molecular weight and low osmotic activity, and does not penetrate across the plasma membrane (Dahl and Staehelin, 1989).

With some plant cell samples we have also obtained good freezing by using sucrose solutions (100–200 mM). Sucrose at these concentrations provided excellent freezing, and the cells did not appear to suffer adversely (e.g., undergo plasmolysis) as assayed by light microscopy with differential interference-contrast optics. Ding *et al.* (1991) had similar good freezing quality with propane jet freezing when they pretreated their cells with 200 mM sucrose.

For the *Drosophila* studies, embryos were collected at timed intervals and the chorion layer removed by washing for 2 min in a 50% commercial bleach solution. After rinsing thoroughly with tap water, the embryos were placed in tap water in a small petri dish, where they collected at the air–water interface. A No. 0 camel hair brush was used to pick up the embryos and transfer them to the specimen holder. We used a mixture of yeast paste made with 10% methanol to a semifluid consistency as the material to fill air spaces between embryos in the high-pressure freezer specimen holder. Although we would not normally recommend using a 10% alcohol solution as a "filler," it seems to have had no detrimental effects in this case, perhaps because the embryos were still surrounded by a vitelline membrane. Yeast paste works better than dextran for space filling because the embryos separate from each other in the substitution and embedding media and can therefore be flat embedded or embedded individually in gelatin capsules. If one wanted to keep the samples together for some reason (e.g., cryosectioning of frozen material), dextran would be a good filler because it "cements" the contents of the holder together during freeze substitution.

B. Freeze Substitution

1. Specimen Storage

One of the advantages of cryofixation is that samples can be kept without deterioration for long periods of time (years) by storage in liquid nitrogen and retrieved as needed. One must simply take precautions not to warm the sample accidentally between the storage container and freeze substitution apparatus (e.g., by using room-temperature forceps to handle the specimen during transfer). A variety of liquid nitrogen storage containers are available, from large Dewars with canisters and "canes" to hold cryotubes, to liquid nitrogen "refrigerators" with automatic level controls and alarm systems. Which one is most appropriate depends primarily on the volume of samples stored, the cost, and to some extent on the convenience of keeping the system filled with liquid nitrogen.

2. Preparation of Fixatives

Fixatives can be conveniently prepared ahead of time and stored in liquid nitrogen. Osmium tetroxide crystals are dissolved in acetone or methanol in a vessel at room temperature, and the vessel is immediately placed in a dry ice–acetone bath. Aliquots are taken from this vessel, and stored in liquid nitrogen. Osmium dissolves in a few seconds in organic solvents (in contrast to aqueous systems), and the solution should be a light or straw-colored yellow. If the solution turns a deep yellow or darker (brown or black), then it should be discarded. If at any time a brown or black precipitate develops in the solution, it also should be discarded.

Alternatively, the osmium crystals can be dissolved in solvent cooled to lower temperatures (down to dry ice temperatures) but the dissolution will take longer. We make up stock solutions of 4% osmium tetroxide in acetone and methanol and store them in liquid nitrogen. These stocks are warmed, diluted to the desired concentration (usually 0.01–1.0%), and refrozen prior to adding the sample for freeze substitution.

Preparation of aldehyde fixatives is similar except that we use 70% concentrated glutaraldehyde and dilute it to a final concentration of 0.2% in methanol or acetone and freeze these in cryotubes for storage until needed. Anhydrous glutaraldehyde (in methanol) is available by special order from Ladd Research Laboratories (Burlington, VT); however, we have not seen any difference in cell preservation or immunoreactivity when compared to fixative made from aqueous stocks. Unlike acetone, methanol can accept relatively large amounts of water and higher concentrations of glutaraldehyde in methanol can be achieved. Paraformaldehyde or acrolein can also be used, either alone or in combination with glutaraldehyde. We dilute 8% paraformaldehyde stocks into methanol to the appropriate fixative concentration.

We use standard 1.8-ml cryotubes for freeze substitution. Because the volume of sample in the high-pressure freezer specimen holders is no more than 0.5 μl, we can put up to three such holders in one cryotube and still meet the requirement of a 1 : 1000 ratio between the volume of the sample and substitution fluid (Steinbrecht and Müller, 1987).

3. Substitution Procedure

Samples at liquid nitrogen temperature are transferred to fixative vials also at liquid nitrogen temperature, placed in the substitution chamber where they warm to either $-78.5°C$ if a dry–ice acetone bath is used, or $-90°C$ if the device described in Section IIIC2 is used. Substitution is allowed to take place over 3–4 days.

The plant samples are substituted for 2.5 days at $-78.5°C$ in a dry ice–acetone bath in a Styrofoam container (which is in a cold room to minimize dry ice evaporation). The samples are then warmed according to the following schedule: 2 hr at −20°C (standard freezer), 2 hr at 0–4°C (refrigerator), and 2 hr at 23°C (room temperature). Some workers prefer a much slower warming schedule, on the order of several days (for a review see Robards and Sleytr, 1985). The samples are rinsed several times in dried acetone, transferred to dried ethanol, infiltrated, and embedded into LR White resin. If one uses antibodies against carbohydrates, then the samples may be embedded into other epoxy-based resins, such as Spurr's, because antigenicity usually will be maintained even in these resins.

The *Drosophila* samples were allowed to warm passively to different temperatures, depending on the type of embedding to be used. For LR White embedding, they were warmed to either −20°C and UV polymerized, or to room temperature and heat polymerized at 50°C. For Lowicryl K4M they were warmed to −35°C, and for Lowicryl HM20 they were warmed to −50°C.

C. Embedding and Resin Polymerization

1. Lowicryl

a. Preparation

We follow the recommendation of the manufacturer for mixing Lowicryl K4M and HM20 for UV polymerization at −35 or −50°C.

K4M (−35°C)	HM20 (−50°C)
Cross-linker A, 2.70 g	Cross-linker D, 2.98 g
Monomer B, 17.30 g	Monomer E, 17.02 g
Initiator C, 0.10 g	Initiator C, 0.10 g

In a fume hood, weigh out the first two components of resin into an amber bottle and mix by gently bubbling a stream of dry nitrogen gas through the resin. After mixing for about 5 min, add initiator and mix for an additional 10 min or until initiator is dissolved. It is important to avoid mixing oxygen into the resin as this will impede polymerization. Cool to the desired temperature.

b. Infiltration

The resin infiltration is carried out in cryotubes, which are placed in holes in an aluminum block that is immersed in a methanol–dry ice bath cooled to the desired temperature (−35 or −50°C). The bath is made in a Styrofoam box and the assembled apparatus is placed in a fume hood, where all subsequent operations are done. The aluminum block also has holes to contain the tubes holding different rinse and resin solutions. Temperature is monitored by a thermocouple in a cryotube filled with solvent and placed in the aluminum block. Dry ice chips are added to the bath as needed to keep the temperature in the desired range. The infiltration schedule is as follows.

1. Rinse in solvent three times over 20 min.
2. Infiltrate as follows:

1 part resin:1 part solvent	2 hr
2 part resin:1 part solvent	2 hr
100% resin	Overnight
100% resin	2 hr

c. Flat Embedding

We find it helpful to have our *Drosophila* specimens flat embedded in thin (1–2 mm) layers of resin so we can screen with the light microscope for developmental stage and evaluated the quality of preservation (e.g., for excess shrinkage or possible mechanical damage incurred during handling). To achieve this goal we sandwich our resin and samples between 22 × 22 mm Thermanox coverslips (Electron Microscopy Sciences, Fort Washington, PA), which have the property (unlike glass coverslips or slides) of allowing UV light penetration for polymerization. The procedure we use is as follows.

1. Spray one side of each coverslip with Teflon-based spray release agent (MS-122; Miller-Stephenson, Los Angeles, CA). Dry for 5 min and wipe clean with cloth or tissue. Marking the coated side with a small scratch mark can be useful.

2. Cut strips of another coverslip for spacers. Add spacers to the edge of the bottom coverslip (coated side up) and place in half of large plastic petri dish in the precooled polymerization chamber (see the next section). The bottom coverslip is propped up on toothpicks in case the resin overflows. Failure to do this may mean that the sample is bonded to the bottom of the dish during polymerization. Put dry ice chips around the edges of the dish to maintain a dry atmosphere and to prevent water condensation onto the coverslip. Place the top coverslip in the dish as well, to allow it to equilibrate to the appropriate temperature.

3. Withdraw samples (in our case *Drosophila* embryos) from the cryotubes with a precooled Pasteur pipette and add small drops of resin plus embryos onto the bottom coverslip. About 0.5 ml of resin gives a good resin thickness when one spacer is used. Place the top coverslip on (release agent side down), taking care not to spill much resin.

Samples that do not need to be flat embedded can be polymerized in gelatin capsules that are suspended in wire loops in the polymerization chamber (Chiovetti, 1982).

d. Polymerization

The apparatus described below can be used for low-temperature embedding of either Lowicryl or LR White resins. A number of commercially available alternatives are also available (see listings in Robards and Sleytr, 1985).

A small (2 ft high × 1 ft wide × 1.5 ft long) insulated box (one can use a Styrofoam container) is lined with aluminum foil or other highly reflective surface. Ultraviolet lights (two 15 W, 360-nm "long" wavelength) are placed in the top and a piece of foil placed between the lights and the specimen platform so that the samples do not receive direct UV illumination. A stainless steel bowl (diameter approximately 18 cm) that will fit inside the box is wrapped with heater wire connected to a temperature controller. A wire mesh is placed in the top of the bowl as a platform for samples, for example, the dish with coverslips for flat embedding. To provide better thermal contact, the bowl is filled with methanol to the level of the dish. A thermocouple connected to the temperature regulator is placed at the level of the dish to maintain the proper temperature. The bowl sits on dry ice chips, which act as the counter coolant.

The initial polymerization at −35 or −50°C takes place over a period of 24–48 hr. Thereafter the chamber is allowed to warm to room temperature, at which point the samples are given *direct* UV illumination for an additional 2–3 days. With flat-embedded samples, the top Thermonox coverslip is then peeled off and the enmbryos in resin are screened with phase optics on a light microscope. Selected embryos can be removed from the wafer with a scalpel and remounted in known orientation for sectioning.

2. Other Resins

As mentioned in Section III,D, LR White is another resin that is commonly used in immunocytochemistry. In cases in which the antibodies are reactive, Spurr's resin and Epon can be used. In the case of antibodies against carbohydrates, good immunolocalization can frequently be achieved in Spurr's resin. Standard protocols for the use of LR White, Epon, and Spurr's resins are found in a variety of sources, including Hayat (1989b).

D. Sectioning

Immunolabeling reagents react with copper grids, therefore nickel or gold grids need to be used in order to pick up sections. We find that processing for immunolabeling also can cause the Formvar film to wrinkle unless the grids are carbon coated prior to picking up the sections. If serial sections (and thus filmed grids) are not needed, it is better to pick up the sections on unfilmed grids. This also means that both sides of the section can be stained with immunolabel.

The Lowicryl nonpolar resins and LR White are sufficiently hydrophobic that they section much like the epoxy resins. Lowicryl K4M (and K11M), however, are hydrophilic and tend to swell when floating in the water of the sectioning boat. They also tend to slip over the knife edge when the boat is filled to "normal" levels, but if the meniscus at the knife edge is lowered the sections will form ribbons normally.

E. Antibody Labeling on Sections

1. Indirect (Two Step) Labeling of Plant Specimens in LR White

The following protocol is for the polyclonal antibodies anti-XG and anti-PGA/RGI (Moore *et al.*, 1986). [Information pertaining to the monoclonal antibodies JIM5 and JIM7 (Knox *et al.,* 1990) is included parenthetically in brackets.] The procedures outlined below are modifications of methods that were presented in a variety of sources, including Craig and Goodchild (1984), Moore and Staehelin (1988), and Knox *et al.* (1989).

1. Standard silver to gold ultrathin sections are collected onto Formvar/carbon-coated *nickel* grids. Do not use copper grids or heavy precipitation on the grids will result.

2. All staining steps take place in a humid chamber (e.g., a petri dish lined with moist filter paper). Use 5–10 μl of staining solutions on Parafilm in the humid chamber.

3. Block nonspecific binding sites with 5% (w/v) nonfat dry milk in PBST [10 mM sodium phosphate at pH 7.0, 500 mM NaCl, 0.1% (v/v) Tween-20] for 20–30 min. Other blocks include bovine serum albumin (10 mg/ml) or ovalbumin (10 mg/ml). [For monoclonals, we block with 10% (v/v) fetal calf serum in PBST with 50 mM NaCl and 0.05% Tween-20.]

4. Blot excess blocking solution from the grid with filter paper. Do not wash, and do blot the grid completely dry.

5. Incubate the grids in primary antibody diluted in PBST with 0.1% Tween for 1 hr at room temperature. Try different dilutions of antibody to obtain the highest levels of specific label and the lowest background. The usual dilutions that we have used range from 1 : 5 to 1 : 20. [We use the monoclonal antibodies undiluted in this step, but some monoclonal antibodies may have to be diluted in buffer.]

6. Wash the grids with PBST with 0.5% Tween for 1 min. In this step, hold the grid with forceps, and use a continuous stream of PBST from a squirt bottle. Blot excess solution from the grid with filter paper, and do blot the grid completely dry. [For monoclonal antibodies, we wash with PBST with 50 mM NaCl and 0.05% Tween.]

7. Incubate the grids in protein A–gold (10 nm) diluted in PBST with 0.5% Tween for 20 min. Again, try to optimize staining without creating nonspecific background. In our case, we use a 1 : 25 dilution. [For monoclonal antibodies, we use a 1 : 10 dilution of goat anti-rat gold (10 nm).]

8. Repeat step 6.

9. Wash with distilled water at approximately 50°C. At this point, the grids may be blotted dry with filter paper.

10. Poststain with uranyl acetate and lead citrate. In some cases, freeze-substituted material may have low contrast, and it may be necessary to stain with 2% (w/v) uranyl acetate in methanol followed by lead staining.

2. Indirect Labeling of *Drosophila*/Lowicryl Sections

1. Make up blocking buffer (BB) in phosphate-buffered saline (PBS):

> Cold-water fish gelatin 0.1% (w/v) (Amersham)
> Bovine serum albumin, 0.8% (w/v) (Sigma, St. Louis, MO)
> Tween-80, 0.02% (v/v)

2. Incubate the grids in BB for 30 min at room temperature (all steps are at room temperature). Immerse the grids if both sides of the section can be labeled.

3. Blot quickly, taking care not to let the grids dry out between steps. Apply about 20 μl of primary antibody diluted in BB. Incubate about 1–1.5 hr in a moist chamber.

4. Blot, rinse once with BB and twice with PBS over a total of 5 min.

5. Incubate in a moist chamber for 1–1.5 hr in secondary gold conjugate diluted in BB.

6. Blot and rinse as in step 3.

7. Fix in 0.5% (v/v) glutaraldehyde in PBS for 5 min.

8. Rinse once in PBS, and twice in distilled H_2O over 2 min.

9. Air dry.

10. Poststain the grids with uranyl acetate and lead citrate. With K4M resin, normal staining times may give too much staining, so it may be necessary to experiment with shortened times for optimal contrast.

F. Controls for Immunocytochemistry

A good experiment is characterized by having a variety of controls. For an excellent example of the thorough use of controls to test for antibody specificity, see Craig and Millerd (1981). Some negative controls to validate specific labeling of the antigen in an indirect immunolabeling experiment would include the following.

1. Omitting the primary antibody and replacing it with buffer

2. Serial dilution of the primary antibody to demonstrate a decrease in labeling

3. Treating the section with preimmune serum rather than the primary antibody. If preimmune serum is not available, a nonimmune serum (from the

species of animal in which the antibody was generated) should be used as a control.

4. Treating the section with an inappropriate antibody (rather than the primary antibody) that was raised in the same animal but against an antigen known to be absent in the tissue (e.g., the primary antibody is against a protein found only in plants and the replacement antibody is against some type of protein that is found only in animals)

5. Preabsorbing the primary antibody with an excess of its respective antigen (e.g., anti-XG is preabsorbed with XG; Moore *et al.*, 1986)

In addition to the above negative controls, a known positive control should be used to demonstrate the validity of the method and of the primary antibody. For example, if it is known that a tubulin antibody reacts intensely with the microtubules on a ciliated protozoan, immunolabel sections of the protozoan and compare the results to results obtained with one's own specimen of interest.

The nature of other controls would depend on the particular antigen. For instance, in our experiments with antibodies against a variety of plant carbohydrates, one control that may be used is to treat sections with saturated sodium metaperiodate before immunostaining (see, e.g., Moore *et al.*, 1986). This control ensures that we are specifically labeling carbohydrates, because the metaperiodate alters carbohydrate structure by cleaving hydroxyl groups.

The above discussion of controls is not meant to be comprehensive, nor does one necessarily have to use all of these controls in a series of experiments. Nevertheless, the more controls that are performed, the more convincing the specificity of immunolabeling.

G. Low-Cost Alternatives to Cryofixation

Even the simplest ultrarapid freezing devices (short of hand plunging with forceps) require some investment of time and energy to develop, and the more sophisticated machines, such as high-pressure freezers, are expensive. Alternatives such as the Tokuyasu technique eliminate the need for specialized freezers but require expensive cryomicrotomes (Tokuyasu, 1986). The developers of the Lowicryl resins (Kellenberger *et al.*, 1980; Carlemalm *et al.*, 1982) worked out a method that gives good ultrastructural preservation with improved antigenicity and allows embedding in low-temperature resins. This technique is called the PLT method (for progressive lowering of temperature) and it works by lowering the temperature as the concentration of organic solvent is raised during dehydration. For example, changing into 70% ethanol occurs at $-35°C$, a temperature low enough to prevent the "collapse" phenomena associated with dehydration (Kellenberger, 1987; MacKenzie, 1972). Prior to PLT, the samples are lightly fixed with aldehydes; therefore even large tissue, which would have to be rapidly frozen by high-pressure freezing, can be prepared for Lowicryl embedding. Results from these studies illustrate that much of the loss of ultrastructural detail and antigenicity that takes place in EM specimen preparation comes during the dehydration steps (see also the results in Small, 1981). A detailed history of the PLT technique and its applications is beyond the scope of this article but there is an excellent review in Hobot (1989).

A key step in using the PLT technique for immunocytochemistry is the choice of fixative (Hobot, 1989). In general, aldehyde fixatives should be used, but at the lowest concentration and for the shortest time that will still give good ultrastructure, unless the antigen–antibody reaction under consideration is not affected by aldehydes. If good localization is achieved at low concentrations but the ultrastructure is only fair, the aldehyde concentration can be increased to the point at which loss of antibody specificity begins.

We have used the PLT technique on tissue culture cells because at one point we were having difficulty with our plunge freezing device. The protocol we used was as follows.

1. Make up solutions of PBS, PHEM buffer, and fixative:

PBS: 2.68 mM KCl, 1.50 mM KH$_2$PO$_4$, 0.49 mM MgCl$_2$·6H$_2$O, 137 mM NaCl, 8.0 mM Na$_2$HPO$_4$·7H$_2$O

PHEM buffer: 60 mM piperazine-N,N'-bis(2-ethanesulfonic acid) (PIPES), 2.5 mM N-2-hydroxyethylpiperazine-N'-2-ethanesulfonic acid (HEPES), 2 mM MgSO$_4$, 10 mM ethylene glycol-bis (β-aminoethyl ether)-N,N,N',N'-tetraacetic acid (EGTA); adjust pH to 6.9 with NaOH

Fixative: 2% paraformaldehyde, 0.1% glutaraldehyde, 0.1 M PIPES, 1 mM EGTA, 1 mM MgSO$_4$; adjust pH to 6.9 with NaOH

2. Grow PtK cells to 80% confluency on UV-sterilized Thermanox coverslips.

3. Rinse the cells briefly with PHEM buffer—about 1 min at room temperature.

4. Aspirate the rinse buffer and add fixative to dishes. Fix 15–30 min at room temperature.

5. Rinse with PBS, three times over 10 min.

6. If the previous processing has been done in plastic tissue culture dishes, transfer at this point to glass or solvent-resistant plastic dishes.

7. Dehydrate at progressively lower temperatures. We use a dry ice–methanol bath for the −35°C steps.

Ethanol concentration	Temperature (°C)	Time (min)
30%	0	10
50%	−20	10
70%	−35	20
95%	−35	20
100%	−35	20 (×3)

8. Infiltration with Lowicryl K4M.

Resin Mix	Temperature (°C)	Time (hr)
1 part resin:1 part 100% ethanol	−35	1.5
2 part resin:1 part 100% ethanol	−35	1.5
100% resin	−35	2
100% resin	−35	1

9. Flat embed and polymerize as described above (Section V,C,1,c and d) for *Drosophila* embryos.

An illustration of the results achieved with PLT embedding and immuno-labeling of sections with anti-tubulin is shown in Fig. 5.

Another method that has been developed as an alternative to high-pressure freezing for large tissue samples is a hybrid between Tokuyasu method fixation and Lowicryl embedding (van Genderen *et al.*, 1991). Samples are fixed, infiltrated with either glycerol or 2.3 *M* sucrose, then frozen in liquid propane or liquid nitrogen and freeze substituted in methanol and embedded in Lowicryl HM2O at −45°C. Postembedding labeling of sections with immunogold shows a significant improvement in cell morphology and specific labeling over cryosections of the same material prepared by the Tokuyasu method. This technique was used rather than the PLT method because the authors were immunolabeling lipids and freeze substitution at −90°C is a way of reducing the extraction of lipids by organic solvents to a minimum. For protein localization studies, the PLT method or cryofixation–freeze substitution methods are probably preferable because the potentially damaging effects of infiltration with glycerol or sucrose are avoided.

VI. Conclusions and Perspectives

In many research areas, investigators are making antibodies against specific gene products and localizing them by immunofluorescence in the light microscope. Relatively few studies carry the localization to the electron microscope level of resolution even though many of the mechanisms they are trying to understand are molecular in nature and require high-resolution data. Thus, EM immunocytochemistry is the important next step for those investigators who have the distributions of their antibodies characterized by immunofluorescence microscopy. It is our belief that the best methods for localizing antibodies at the EM level include cryofixation and postembedding labeling on sections, as we have outlined here.

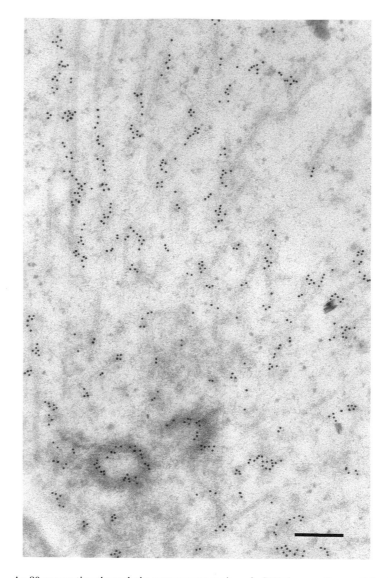

Fig. 5   An 80-nm section through the centrosome region of a PtK tissue culture cell that has been prepared by the PLT method, as discussed in text, and labeled with anti-tubulin and 10-nm gold conjugated with goat anti-mouse secondary. The microtubules are fairly well labeled but there is more extraction of the cytoplasm by this method than with cryofixation–freeze substitution. (Bar: 0.2 μm.)

Acknowledgments

We would like to thank Mary Morphew for invaluable technical assistance, Dr. L. Andrew Staehelin for the unlimited use of his Balzers HPM 010 high-pressure freezer, Dr. Robert Boswell for providing *Drosophila* embryos, Dr. Paul Lasko for the antibody to vasa, and Dr. J. R. McIntosh for the tubulin antibody. Thanks are also due to Dr. T. H. Giddings and Dr. Sam Levy for valuable discussions concerning this work. Portions of this work were supported by NIH Grant RR00592 to the Boulder Laboratory for Three-Dimensional Fine Structure (J. R. McIntosh, PI) and by a NASA Research Associate Fellowship to J. Z. Kiss.

References

Acetarin, J.-D., Carlemalm, E., and Villiger, W. (1986). Developments of new Lowicryl resins for embedding biological specimens at even lower temperatures. *J. Microsc. (Oxford)* **143**, 81–88.

Berryman, M. A., and Rodewald, R. D. (1990). An enhanced method for post-embedding immunocytochemical staining which preserves cell membranes. *J. Histochem. Cytochem.* **38**, 159–170.

Boyne, A. F. (1979). A gentle, bounce-free assembly for quick-freezing tissues for electron microscopy: Application to isolated *Torpedine* ray electrocyte stacks. *J. Neurosci. Methods* **1**, 353–364.

Carlemalm, E., Gravito, R. M., and Villiger, W. (1980). Advances in low temperature for electron microscopy. *Electron Microsc., Proc. Eur. Congr., 7th* **2**, 656–657.

Carlemalm, E., Gravito, R. M., and Villiger, W. (1982). Resin development for electron microscopy and an analysis of embedding at low temperature. *J. Microsc. (Oxford)* **126**, 123–143.

Carlemalm, E., Villiger, W., Hobot, J. A., Acetarin, J. D., and Kellenberger, E. (1985). Low temperature embedding with Lowicryl resins: Two new formulations and some applications. *J. Microsc. (Oxford)* **140**, 55–63.

Chiovetti, R. (1982). "Instructions for Use, Lowicryl K4M and Lowicryl HM 20." Chemische Werke Lowi GmbH, P. O. Box 1660, D-8264 Waldkraiburg, Germany.

Costello, M. J. (1980). Ultra-rapid freezing of thin biological samples. *Scanning Electron Microsc.* **2**, 361–370.

Craig, S., and Goodchild, D. J. (1984). Golgi-mediated vicilin accumulation in pea cotyledon cells is re-directed by monensin. *Protoplasma* **122**, 91–97.

Craig, S., and Millerd, A. (1981). Pea seed storage proteins—immunocytochemical localization with protein A–gold by electron microscopy. *Protoplasma* **105**, 333–339.

Craig, S., and Staehelin, L. A. (1988). High pressure freezing of intact plant tissues. Evaluation and characterization of novel features of the endoplasmic reticulum and associated membrane systems. *Eur. J. Cell Biol.* **46**, 80–93.

Craig, S., Gilkey, J. C., and Staehelin, L. A. (1987). Improved specimen support cups and auxiliary devices for the Balzers high pressure freezing apparatus. *J. Microsc. (Oxford)* **148**, 103–106.

Dahl, R., and Staehelin, L. A. (1989). High pressure freezing for the preservation of biological structure: Theory and practice. *J. Electron Microsc. Tech.* **13**, 165–174.

Dahmen, H., and Hobot, J. A. (1986). Ultrastructural analysis of *Erysiphe graminis* haustoria and subcuticular stroma of *Venturia inaequalis* using cryosubstitution. *Protoplasma 131,* 92–102.

Ding, B., Turgeon, R., and Parthasarathy, M. V. (1991). Routine cryofixation of plant tissue by propane jet freezing for freeze-substitution. *J. Electron. Microsc. Tech.* **19**, 107–117.

Dubochet, J., Adrian, M., Chang, J.-J., Lepault, J., and McDowall, A. W. (1987). Cryoelectron microscopy of vitrified specimens. In "Cryotechniques in Biological Electron Microscopy" (R. A. Steinbrecht and K. Zierold, eds.), pp. 114–131. Springer-Verlag, Berlin.

Durrenberger, M. (1989). Removal of background label in immunocytochemistry with the apolar Lowicryls by using washed protein A–gold precoupled antibodies in a one-step procedure. *J. Electron Microsc. Tech.* **11**, 109–116.

Durrenberger, M., Bjornsti, M. A., Uetz, T., Hobot, J. A., and Kellenberger, E. (1988). The intracellular location of the histone-like protein HU in *Escherichia coli*. *J. Bacteriol.* **170,** 4757–4768.

Elder, H. Y., Gray, C. C., Jardine, A. G., Chapman, J. N., and Biddlecombe, W. H. (1982). Optimum conditions for cryoquenching of small tissue blocks in liquid coolants. *J. Microsc. (Oxford)* **126,** 45–61.

Evans, M. L., Moore, R., and Hasenstein, K.-H. (1986). How roots respond to gravity. *Sci. Am.* **254,** 112–118.

Fernandez, D. E., and Staehelin, L. A. (1985). Structural organization of ultrarapidly frozen barley aleurone cells actively involved in protein secretion. *Planta* **165,** 455–468.

Giddings, T. H., Brower, D. L., and Staehelin, L. A. (1980). Visualization of particle complexes in the plasma membrane *Micrasterias denticulata* associated with the formation of cellulose fibrils in primary and secondary cell walls. *J. Cell Biol.* **84,** 327–339.

Gilkey, J. C., and Staehelin, L. A. (1986). Advances in ultrarapid freezing for the preservation of cellular ultrastructure. *J. Electron Microsc. Tech.* **3,** 177–210.

Hay, B., Ackerman, L., Barbel, S., Jan, L. Y., and Jan, Y. N. (1988). Identification of a component of *Drosophila* polar granules. *Development* **103,** 625–640.

Hayat, M. A., ed. (1989a). "Colloidal Gold: Principles, Methods, and Applications," 2 vols. Academic Press, San Diego.

Hayat, M. A. (1989b). "Principles and Techniques of Electron Microscopy: Biological Applications," 3rd Ed. CRC Press, Boca Raton, Florida.

Heath, I. B., and Rethoret, K. (1982). Mitosis in the fungus *Zygorhynchus molleri:* Evidence for stage specific enhancement of microtubular preservation by freeze substitution. *Eur. J. Cell Biol.* **28,** 180–189.

Heath, I. B., Rethoret, K., and Moens, P. B. (1984). The ultrastructure of mitotic spindles from conventionally fixed and freeze-substituted nuclei of the fungus *Saprolegnia*. *Eur. J. Cell Biol.* **35,** 284–295.

Heath, I. B., Rethoret, K., Arsenault, A. L., and Ottensmeyer, F. P. (1985). Improved preservation of the form and contents of wall vesicles and the Golgi apparatus in freeze substituted hyphae of *Saprolegnia*. *Protoplasma* **128,** 81–93.

Heuser, J., and Steer, C. J. (1989). Trimeric binding of the 70-kD uncoating ATPase to the vertices of clathrin triskelia: A candidate intermediate in the vesicle uncoating reaction. *J. Cel Biol.* **109,** 1457–1466.

Heuser, J. E., Reese, T. S., Dennis, M. J., Jan, Y., Jan, L., and Evans, L. (1979). Synaptic vesicle exocytosis captured by quick freezing and correlated with quantal transmitter release. *J. Cell Biol.* **81,** 275–297.

Hippe, S. (1985). Ultrastructure of haustoria of *Erysiphe graminis* f. sp. *hordei* preserved by freeze-substitution. *Protoplasma* **129,** 52–61.

Hobot, J. A. (1989). Lowicryls and Low-temperature embedding for colloidal gold methods. *In* "Colloidal Gold: Principles, Methods, and Applications" (M. A. Hayat, ed.), pp. 75–115. Academic Press, San Diego.

Humbel, B., and Schwarz, H. (1989). Freeze-substitution for immunocytochemistry. *In* "Immuno-Gold Labeling in Cell Biology" (M. A. Hayat, ed.), Vol. 2, pp. 115–134. CRC Press, Boca Raton, Florida.

Ichikawa, M., Sasaki, K., and Ichikawa, A. (1989). Optimal preparatory procedures of cryofixation for immunocytochemistry. *J. Electron Microsc. Tech.* **12,** 88–94.

Kandasamy, M. K., Parthasarathy, M. V., and Nasrallah, M. E. (1991). High pressure freezing and freeze substitution improve immunolabelling of S-locus specific glycoproteins in the stigma papollae of *Brassica*. *Protoplasma* **162,** 187–191.

Kellenberger, E. (1987). The response of biological macromolecules and supramolecular structures to the physics of specimen cryopreparation. *In* "Cryotechniques in Biological Electron Microscopy" (R. A. Steinbrecht and K. Zierold, eds.), pp. 35–63. Springer-Verlag, Berlin.

Kellenberger, E., Carlemalm, E., Villiger, W., Roth, J., and Gravito, R. M. (1980). "Low Denaturation Embedding for Electron Microscopy of Thin Sections." Chemische Werke Lowi GmbH, P. O. Box 1660, D-8264, Waldkraiburg, Germany.

Kellenberger, E., Durrenberger, M., Villiger, W., Carlemalm, E., and Wurtz, M. (1987). The efficiency of immunolabel on Lowicryl sections compared to theoretical predictions. *J. Histochem. Cytochem.* **35,** 959–969.

Kiss, J. Z., and Staehelin, L. A. (1993). Structural polarity in the *chara* rhizoid: A reevaluation. *Am. J. Bot.* **80,** 273–282.

Kiss, J. Z., Vasconcelos, A. C., and Triemer, R. E. (1988). The intramembranous particle profile of the paramylon membrane during paramylon synthesis in *Euglena* (Euglenophyceae). *J. Phycol.* **24,** 152–157.

Kiss, J. Z., Giddings, T. H., Staehelin, L. A., and Sack, F. D. (1990). Comparison of the ultrastructure of conventionally fixed and high pressure frozen/freeze substituted root tips of *Nicotiana* and *Arabidopsis*. *Protoplasma* **157,** 64–74.

Knox, J. P., Day, S., and Roberts, K. (1989). A set of cell surface glycoproteins forms an early marker of cell position, but not cell type, in the root apical meristem of *Daucus carota* L. *Development* **106,** 47–56.

Knox, J. P., Linstead, P. J., King, J., Cooper, C., and Roberts, K. (1990). Pectin esterification is spatially regulated both within cell walls and between developing tissues of root apices. *Planta* **181,** 512–521.

Lancelle, S. A., and Hepler, P. K. (1989). Immunogold labelling of actin on sections of freeze-substituted plant cells. *Protoplasma* **150,** 72–74.

Lancelle, S. A., Callaham, D. A., and Hepler, P. K. (1986). A method for rapid freeze fixation of plant cells. *Protoplasma* **131,** 153–165.

Lancelle, S. A., Cresti, M., and Hepler, P. K. (1987). Ultrastructure of the cytoskeleton in freeze-substituted plant cells. *Protoplasma* **150,** 72–74.

Lasko, P., and Ashburner, M. (1990). Posterior localization of vasa protein correlates with, but is not sufficient for, pole development. *Genes Dev.* **4,** 905–921.

Leunissen, J. L. M., and Verkleij, A. J. (1989). Cryoultramicrotomy and immuno-gold labeling. *In* "Immuno-Gold Labeling in Cell Biology" (A. J. Verkeleij and J. L. M. Leunissen, eds.), pp. 95–114. CRC Press, Boca Raton, Florida.

MacKenzie, A. P. (1972). Freezing, freeze-drying and freeze-substitution. *In* "Scanning Electron Microscopy" (O. Jahari, ed.), Vol. 2, pp. 273–280. Scanning Electron Microsc., Inc., AMF O'Hare, Chicago.

Meissner, D. H., and Schwarz, H. (1990). Improved cryoprotection and freeze substitution of embryonic qual retinal: A TEM study on ultrastructural preservation. *J. Electron Microsc. Tech.* **14,** 348–356.

Menco, B. P. (1986). A survey of ultra-rapid cryofixation methods with a particular emphasis on applications to freeze-fracturing, freeze-etching, and freeze-substitution. *J. Electron Microsc. Tech.* **4,** 177–240.

Mersey, B., and McCully, M. E. (1978). Monitoring of the course of fixation of plant cells. *J. Microsc. (Oxford)* **114,** 49–76.

Michel, M., Hillmann, T., and Müller, M. (1991). Cryosectioning of plant material frozen at high pressure. *J. Microsc. (Oxford)* **163,** 3–18.

Monaghan, P., and Robertson, D. (1990). Freeze-substitution without aldehyde or osmium fixatives: Ultrastructure and implications for immunocytochemistry. *J. Microsc. (Oxford)* **158,** 335–363.

Moor, H. (1987). Theory and practice of high pressure freezing. *In* "Cryotechniques in Biological Electron Microscopy" (R. A. Steinbrecht and K. Zierold, eds.), pp. 175–191. Springer-Verlag, Berlin.

Moore, P. J., and Staehelin, L. A. (1988). Immunogold localization of the cell-wall-matrix polysaccharides rhamnogalacturonan I and xyloglucan during cell expansion and cytokinesis in *Trifolium pratense* L.; implications for secretory pathways. *Planta* **174,** 433–445.

Moore, P. J., Darvill, A. G., Albersheim, P., and Staehelin, L. A. (1986). Immunogold localization of xyloglucan and rhamnogalacturonan I in the cell walls of suspension cultured sycamore cells. *Plant Physiol.* **82**, 787–794.

Newman, G. R. (1989). LR White embedding medium for colloidal gold methods. *In* "Colloidal Gold: Principles, Methods, and Applications" (M. A. Hayat, ed.), Vol. 2, pp. 47–73. Academic Press, San Diego.

Newman, G. R., and Hobot, J. A. (1989). Role of tissue processing in colloidal gold methods. *In* "Colloidal Gold: Principles, Methods, and Applications" (M. A. Hayat, ed.), Vol. 2, pp. 33–45. Academic Press, San Diego.

Nicolas, G. (1991). Advantages of fast-freeze fixation followed by freeze-substitution for the preservation of cell integrity. *J. Electron Microsc. Tech.* **18**, 395–405.

Nicolas, M.-T. (1989). Immunogold-labelling after rapid-freezing fixation and freeze-substitution: Application to the detection of luciferase in dinoflagellates. *In* "Immuno-Gold Labeling in Cell Biology' (A. J. Verkleij and J. L. M. Leunissen, eds.) pp. 277–289. CRC Press, Boca Raton, Florida.

O'Toole, E. T., and McIntosh, J. R. (1991). A plunge-freezing method for imaging mammalian tissue culture cells. *Proc.—Annu. Meet. Electron Microsc. Soc. Am.* **49**, 58–59.

Pinto da Silva, P., and Kachar, B. (1980). Quick freezing vs. chemical fixation: Capture and identification of membrane fusion intermediates. *Cell Biol. Int. Rep.* **4**, 625–640.

Ridge, R. W. (1990). A simple apparatus and technique of the rapid-freeze and freeze-substitution of single-cell algae. *J. Electron Microsc.* **39**, 120–124.

Robards, A. W., and Sleytr, U. B. (1985). "Low Temperature Methods in Biological Electron Microscopy." Elsevier, Amsterdam.

Roth, J., Bendayan, M., Carlemalm, E., Villiger, W., and Gravito, R. M. (1981). Enhancement of structural preservation and immunocytochemical staining in low temperature embedded pancreatic tissue. *J. Histochem. Cytochem.* **29**, 663–671.

Sack, F. D., and Kiss, J. Z. (1989). Plastids and gravity perception. *In* "Physiology, Biochemistry, and Genetics of Nongreen Plastids" (C. D. Boyer, J. C. Shannon, and R. C. Hardison, eds.), pp. 171–181. Am. Soc. Plant Physiol., Rockville, Maryland.

Sitte, H., Neumann, K., and Edelmann, L. (1989). Cryosectioning according to Tokuyasu vs. rapid-freezing, freeze-substitution and resin embedding. *In* "Immuno-Gold Labeling in Cell Biology" (A. J. Verkleij and J. L. M. Leunissen, eds.), pp. 63–93. CRC Press, Boca Raton, Florida.

Small, J. V. (1981). Organization of actin in the leading edge of cultured cells: Influence of osmium tetroxide and dehydration on the ultrastructure of actin meshworks. *J. Cell Biol.* **91**, 695–705.

Staehelin, L. A., and Chapman, R. L. (1987). Secretion and membrane recycling in plant cells: Novel structures visualized in ultrarapidly frozen sycamore and carrot suspension-culture cells. *Planta* **171**, 43–57.

Staehelin, L. A., Giddings, T. H., and Moore, P. J. (1988). Structural organization and dynamics of the secretory pathway of plant cells. *Curr. Top. Plant Biochem. Physiol.* **7**, 45–61.

Staehelin, L. A., Giddings, T. H., Kiss, J. Z., and Sack, F. D. (1990). Macromolecular differentiation of Golgi stacks in root tips of *Arabidopsis* and Nicotiana seedlings as visualized in high pressure frozen and freeze-substituted samples. *Protoplasma* **157**, 75–91.

Steinbrecht, R. A., and Müller, M. (1987). Freeze-substitution and freeze-drying. *In* "Cryotechniques in Biological Electron Microscopy" (R. A. Steinbrecht and K. Zierold, eds.), pp. 149–172. Springer-Verlag, Berlin.

Steinbrecht, R. A., and Zierold, K. (1987). "Cryotechniques in Biological Electron Microscopy." Springer-Verlag, Berlin.

Tanaka, K., and Kanbe, T. (1986). Mitosis in the fission yeast *Schizosaccharomyces pombe* as revealed by freeze-substitution electron microscopy. *J. Cell Sci.* **80**, 253–268.

Tiwari, S. C., and Gunning, B. E. S. (1986). Development and cell surface of a non-synctial invasive tapetum in *Canna:* Ultrastructural, freeze-substitution, cytochemical and immunofluorescence study. *Protoplasma* **134,** 1–16.

Tokuyasu, K. T. (1986). Application of cryo-ultramicrotomy to immunocytochemistrry. *J. Microsc. (Oxford)* **143,** 139–149.

Usuda, N., Hongjun, M. A., Hanai, T., Yokota, S., Hashimoto, T., and Nagata, T. (1990). Immunoelectron microscopy of tissues processed by rapid freezing and freeze-substitution fixation without chemical fixatives: Application to catalase in ratliver hepatocytes. *J. Histochem. Cytochem.* **38,** 617–623.

Usuda, N., Yokota, S., Ichikawa, R., Hashimoto, T., and Nagata, T. (1991). Immunoelectron microscopic study of a new *d*-amino acid oxidase-immunoreactive subcompartment in rat liver peroxisomes. *J. Histochem. Cytochem.* **39,** 95–102.

van Bergen en Henegouwen, P. M. P. (1989). Immunogold labeling of ultrathin cryosections. *In* ''Colloidal Gold: Principles, Methods, and Applications'' (M. A. Hayat, ed.), Vol. 1, pp. 191–215. Academic Press, San Diego.

van Genderen, I. L., van Meer, G., Slot, J. W., Geuze, H. J., and Voorhout, W. F. (1991). Subcellular localization of Forssman glycolipid in epithelial MDCK cells by immunoelectron microscopy after freeze-substitution. *J. Cell Biol.* **115,** 1009–1019.

Verkleij, A. J., and Leunissen, J. L. M., eds. (1989). ''Immuno-Gold Labeling in Cell Biology.'' CRC Press, Boca Raton, Florida.

CHAPTER 17

Microinjection of Antibodies in the Analysis of Cellular Architecture and Motility

Brigitte M. Jockusch and Constance J. Temm–Grove

Cell Biology Group
University of Bielefeld
DW-4800 Bielefeld, Germany

I. Introduction: The Scope of the Play

Numerous functions performed by eukaryotic cells depend on the generation and maintenance of a distinct cell type with a specific morphology. This is quite obvious for many tissue-forming cells in animals. For example, the polarized organization of epithelial cells is the basis of vectorial transport of molecules performed by renal tubules, the intestine, or endocrine organs. Even cells that at first glance seem to lack a distinct morphology, but show a striking flexibility of shape, such as the vagabonding blood cells or protozoan amebas, depend at least temporarily on a defined cellular shape, because their chemotactic migration is based on their ability to polarize. In amebas and animal cells, generation, maintenance, and dynamics of cellular architecture depend on the interaction of cytoskeletel elements with each other and with the plasma membrane. The identification of the cytoskeleton as the molecular basis for cellular morphology and structure has, of course, only shifted the problem of understanding these phenomena to another level. There is a wealth of biochemical and biophysical data available for a still-growing catalog of individual cytoskeletal proteins, but we have yet to learn how these molecules form specific time- and spatially regulated supramolecular structures that control cell polarity, cellular junctions, and the various forms of cellular motility.

One approach that allows more insight into this problem is the modulation of the level of individual proteins inside the cell and the analysis of the effects that this manipulation might have. Technically, several methods can be considered to reach this goal. With suitable DNA probes, homologous recombination ("gene targeting" experiments) may terminate transcription and translation of selected proteins (DeLozanne, 1989). Gene transfection may be used to overexpress individual proteins. At the mRNA level, experiments can be designed with sense or antisense riboprobes that might modulate the translation of cytoskeletal proteins (Knecht, 1989). Analogously, at the protein level, either the homologous protein or a specific, high-affinity antibody against it may be injected into recipient cells (Jockusch *et al.*, 1986, 1991). The power of these different methods is now recognized, and the number of studies based on these techniques is growing rapidly. Rather than giving an exhaustive review of all relevant reports published so far, we would like to emphasize here two conclusions drawn from this work: (1) experiments using these strategies have already led to new concepts on the role of individual cytoskeletal elements in different cell types. For example, the alteration of cytoskeletal protein levels in the "primitive" eukaryote *Dictyostelium* by gene targeting and antisense technology has taught us that, with respect to the actomyosin system, this ameba is anything but primitive, showing a high degree of redundancy (Gerisch *et al.*, 1989); (2) tackling a specific question with more than one of the techniques outlined above can give unexpected results that may not be easy to reconcile, but that lead to further insights into the complexity of the problem. For example, a transient overexpression in fibroblasts of gelsolin, an actin filament-severing protein, to only 125% of the control value led to the disruption of the actin filament bundles (which are essential for cellular adhesion and fibroblastic morphology in culture) and to an increase in locomotor activity (Cunningham *et al.*,

1991). In contrast, microinjection of exogenous gelsolin, yielding much higher intracellular concentrations, had no effect on these parameters (Huckriede *et al.*, 1990), indicating that there must be cellular mechanisms controlling the gelsolin activity, and that these mechanisms can be sidestepped only when the protein is synthesized inside the cell. Therefore it seems that these different approaches all have their own merits and should be pursued in parallel. Thus cell and molecular biologists can design a set of experiments as though writing a script for a play, and also act as producers and stage directors. But as the actors are living cells rather than man-made puppets, the final acts of such plays may take a few turns that are surprising and unexpected—for both spectators and playwrights.

In this article, we concentrate on the microinjection of antibodies for the analysis of the role(s) of individual cytoskeletal proteins in cellular architecture. Many aspects of the methodology and technical details that are important for the injection of tissue culture cells with glass capillaries in general have been described (Kreis and Birchmeier, 1982; Wang, 1989; McNeil, 1989). Here, the script is specially designed to make use of the specificity and selectivity of antibodies for studying the role of individual cytoskeletal proteins in living cells.

II. Actors and Requisites in the Play

A. Cells

To be eligible actors, cells must fulfill a number of criteria. Those of primary importance are size and adhesiveness. The technique described here works best with relatively large cells (50–100 μm in diameter) attached to a solid, transparent support. Primary cells or permanent lines of fibroblastic, epithelial, and endothelial cells of avian or mammalian origin are usually good candidates for microinjection, as they adhere strongly to the glass coverslip on which they are seeded. Highly motile cells such as amebas, macrophages, or granulocytes can also be injected, but this requires more practice and patience: because of their less firm attachment, they may be pushed by the needle rather than being penetrated. For injecting into the cytoplasm, well-spread cells are advantageous over round, more compact cell types: in the latter case (e.g., with nerve or glia cells) the needle must be lowered at a rather steep angle to avoid the nucleus, and the depth of penetration within the cytoplasm is difficult to control microscopically.

Large cells, like oocytes, again need special care: they usually require a second microcapillary for holding them in place, and injection needles with a wider tip opening. They will not be considered further in this article.

There is no general answer to the question as to what cell species are suitable for microinjection. Physically, microinjection puts some strain on the cell: it will be exposed for some time period to microscopic light and probably some changes in temperature. Its dorsal plasma membrane will be penetrated by glass, and its cytoplasmic volume will be increased by about 10% by the micro-injection of fluid. All microinjected cells, therefore, will undergo a few initial reactions as a consequence of this treatment. Mammalian epithelial and fibro-blastic cells will develop perinuclear vesicles and withdraw their marginal veils and ruffles immediately after injection. These visible "wounding reactions" will disappear within 5–10 min after removal of the capillary. Highly motile cells of lower vertebrates, such as fish keratocytes or amphibian epithelial cells, may be less tolerant: frequently, they stop moving altogether and die when the needle penetrates the membrane (Jockusch *et al.*, 1991). Such observations suggest differences in the capacity of various cells to reseal their plasma membranes after damage.

B. Antibodies

The selection of antibodies used for the microinjection "script" should follow the general principles guiding the use of antibodies in cell biology: homogeneous and well-characterized antibodies of high affinity are the requisites of choice. Polyclonal, affinity-purified antibodies or their $F(ab')_2$ fragments as well as monoclonal antibodies of the IgG class have been used successfully to interfere with cytoskeletal components in living tissue culture cells and protozoa (see section V). The injection of complete serum or whole IgG fraction is less advisable, as the titer of antibodies against the poorly immunogenic structural proteins of the cytoskeleton is usually not very high. We do not use antibodies of the IgM type, because of their large molecular volume and poor stability. Low-affinity monoclonal antibodies may not be able to form stable complexes with the antigen inside the cell. The *in vitro* characterization of the antibodies should not be restricted to methods in which the antigen is denatured [solid-phase enzyme-linked immunosorbent assay (ELISA) or immunoblotting with sodium dodecyl sulfate (SDS)-denatured proteins], but include assays in which the native conformation of the antigen is presumably better preserved (immuno-precipitation, immunofluorescence) and therefore more likely to mimic the situation inside the cell. The concentration of the antibody to be injected is, of course, important and should ideally be adjusted to yield at least an equi-molar intracellular concentration with respect to the antigen. The purified anti-body should be dialyzed against a buffer that has been shown not to cause adverse effects in recipient cells (e.g., most cells tolerate phosphate-buffered saline). Then, using a collodion bag (Sartorius, Göttingen, Germany) suspended in the final buffer and either a water-stream vacuum or a membrane pump (KNF Neuberger, Freiburg, Germany), the solution can be concentrated by pressure dialysis to between 2 and 30 mg/ml without severe clumping or precip-itation of the antibody. Alternatively, using a Centricon filter (Amicon, Wit-ter, Germany), the antibody may be dialyzed and concentrated in the same system.

C. Controls

The best controls are of course mock injections. Buffer injections should demonstrate the tolerance of the recipient cells for the particular buffer used. Fluorescently labeled bovine serum albumin (BSA) or dextran can be used to visualize successful injection and distribution of macromolecules in the injected cells. Antibodies of the same immunoglobin type that do not react with endogenous proteins (as determined by immunofluorescence and immunoblotting) can serve as controls for the reactive antibodies.

III. The Set

A. Stage Construction

The most prominent object required on stage for the performance is the microinjection system. The set-up we use consists of three items: (1) an inverted microscope with a large distance between the stage and the condensor, equipped with a microscope stage temperature regulator (Zeiss, Thornwood, NY), (2) a pressure injection monitor (Eppendorf, Hamburg, Germany) that can be operated either with compressed air or nitrogen, and (3) a micromanipulator. This is the most important item, as it will determine the number of successful injections per needle, and the velocity (injected cells per minute). The one we use (Eppendorf micromanipulator 5170) can perform automatic injections triggered by a foot pedal or can be manually operated.

B. Production of Needles

Our needles are produced with an electrically operated horizontal microcapillary pulling system. Such an instrument, described in detail by Graessman *et al.* (1980), may be purchased or a simple model may be built in a basic machine shop. The instrument used here consists of two horizontal needle holders attached to springs with a heating coil between them, which may be turned on by pushing a button and turned off by the release of a small lever. In other versions, the needle holders are vertical and are pulled by a weight or a magnet (e.g., from Bilaney Consultants GmbH, Düsseldorf). Glass microcapillaries of 0.58 mm inner diameter are used that contain a thin filament that runs the length of the rod (Science Products Trading GmbH, Frankfurt, Germany). The glass capillary is slid into the grooves on the needle holders, which have been pushed together and secured. The coil is turned on, and as it heats the glass melts, the needle holders return to the resting position, and the capillary breaks, thus creating two needles with very fine tips. The coil is automatically turned off by the release of the small lever, which had been depressed onto an electrical contact by one of the needle holders. The length and shape of the capillary tip portion can be altered by adjusting the tension applied to the spring or the amount of current used to heat the coil. The two tips created from one capillary with this system will never be identical because one of them has the additional friction of running over the contact lever; however, usually both can be used. The shape of the tip is important during the loading of antibody. When it is too long, there is often a problem of an air bubble becoming caught at the tip of needle. The best length is between 2 and 3 mm from the beginning of the narrowing to the tip. After the melting of the microcapillaries the tips are usually closed and must be carefully broken to obtain needles with a suitable opening. This is done under microscopic control. Using glass capillaries with a greater diameter opening or a thinner glass wall may lead to the tips being open more often. However, we control all our tips microscopically before using them. Any inverted microscope with a X32–X40 objective can be used for this purpose. The needle is held in place on the microscope stage with a piece of modeling clay and the focus is set on the tip. The fixed tip is simply touched with the shaft of another microcapillary; it will then break. The object is to remove the area appearing black in dispersed light (Color Plate 4, arrow). When the breaking is properly accomplished, there is no rough, slanted, or jagged edge and the color yellow should be seen at the tip. In scanning electron micrographs, the inner opening of microcapillaries with such an appearance was measured to be between 0.8 and 1.2 μm in diameter (Füchtbauer, 1984), which is suitable for microinjection into animal tissue culture cells. With practice, at least 80% of the pulled needle tips will be broken successfully.

The needles may then be stored dust free by pushing the shafts sideways into some modeling clay (point down) that has been attached to the inner rim of a glass jar, and covering them with an inverted beaker. In this condition, they may be easily transported as well.

C. Cell Preparation

The cells must be seeded onto coverslips at least 24 hr prior to microinjection to ensure optimal adhesion. The shape of the coverslip may need to be considered, depending on the future analysis planned. In general, 20 to 22 mm^2 coverslips are appropriate. However, if they are to be critical point dried (CPD) later, then the size of the CPD chamber is a limiting factor and 15-mm diameter round or halved coverslips may be better. For convenience, the coverslips may be etched with a diamond point in order to be able to relocate the injected cells. There are also etched coverslips on the market; however, the self-etched ones are considerably more economical. The marked area should be large enough to accommodate approximately 100 well-dispersed cells and still leave room for them to multiply (if they must be allowed to grow for a day or two after injection). The cell culture medium may be exchanged for a HEPES-buffered medium to prevent serious pH problems while exposed to room atmosphere.

D. Antibody Solution

When the needles and cells have been prepared, about 10–20 μl of properly diluted antibody is filtered by centrifugation through a Millipore (Bedford, MA) filter of 0.45-μm pore size. Alternatively, the solution may be centrifuged [e.g., in an airfuge (Beckman, Fullerton, CA)]. Both of these procedures serve primarily the purpose of removing dust, bacteria, or clumps of antibodies from the preparation. There is always a loss of antibody in these steps, which must be taken into account when calculating antibody-to-antigen ratios in the recipient cells.

IV. The Performance

A. Loading the Needle

At first, the loading process should be practiced a few times before exposing the cells to the ravages of the environment outside an incubator. There are two methods for loading the microinjection needle. The easiest method is to set the blunt end of the needle (tip up) in the solution. Because of the inner filament, the fluid will be drawn to the tip. This requires a few minutes, but this may be used in getting the cells, focusing the microscope, and finding the etched area. However, this method is not sterile and allows the introduction of dust or other particles to the solution. The other method may be done under sterile conditions if desired, but requires a steady hand. Long, thin micropipette tips suitable for pipetting 1–10 μl can be used with an automatic pipette to fill the microcapillary from the blunt end by pressure. The most serious drawback of this method is that, on ejecting the solution, a small air bubble may become trapped in the tip of the needle. The air bubble is virtually impossible to remove, and precious antibody as well as a good needle may be wasted. To avoid this problem, the solution should never be pipetted into the very tip of the capillary, but deposited in the wider part. It will be drawn to the tip by capillary force.

B. The Injection

The microscope stage is prewarmed to 37°C, and the culture dish containing the cells is placed in a fitted metal ring set into the stage. First, the cells to be injected are relocated on the coverslip, then the focus plane is raised as far above the cells as possible. Then the needle is loaded, screwed into its holder, and the holder is in turn screwed into the manipulator. (There are of course some variations in needle holders.) Without moving the focus plane, the needle is brought into focus with the micromanipulator. Gradually, the focus is lowered to the cells and the needle is continually brought to the level of the focus plane. Finally, the cells are in focus and the needle tip is out of focus a short distance above the cells.

At this point, a word should be said in regard to the pressure injection system. There are three different pressures to be considered: (1) total pressure (from our experience, 3000–5000 hPa is optimal), (2) injection pressure (100–300 hPa), and (3) constant pressure in the needle (at least 80 hPa). The constant pressure is to prevent back flow of cell growth medium into the needle. It should not be high enough to be ejecting fluid continually. The constant pressure may also be used as the injecting pressure: because the pressure of fluid inside a cell is slightly lower than that in the surrounding medium, the constant pressure of the solution in the needle can be set so that it does not release fluid into the medium, but on entering the cell fluid is injected. Then one must manually remove the needle after enough solution has entered. This can be tricky and, therefore, the automatic injection (which can be set for depth and time length) is advantageous. The depth of injection should be set at a point near the nucleus (where the cell is thickest) on the side of the cell, from which the needle is coming (Fig. 1a). The amount of solution entering the cell can be seen as a clear "wave" as for a split second the granules in the cell flow out of the way. The "wave" should be about the size of the nucleus and should not move the nucleus (Fig. 1b). The time required for this volume to enter the cell, with an appropriately broken needle tip, is 0.3–0.5 sec. This volume has been measured to be approximately 10^{-7} μl by injecting fluid into aqueous droplets under a silicone oil layer, measuring the increase in diameter of the droplet, and calculating the volume (Füchtbauer, 1984). With the cells we use, this amounts to 5–20% of the cellular volume. Another method that has been used successfully to determine the injected volume includes quantitative fluorescence microscopy of fluorescent marker molecules (Lee, 1989; Fishkind *et al.*, 1991). The total pressure may be applied to check the opening of the tip, or even to clear a plugged needle.

C. Troubleshooting

The most critical aspect of microinjection is the evaluation of possible damage to the recipient cells. In this context, the amount of injected solution is crucial. Table I summarizes some hints for checking the volume injected. In some instances, it may be difficult to see clearly the amount of fluid entering the cell or if it has entered the cell. This can be controlled by injecting a fluorescent marker [such as dextran–fluorescein isothiocyanate (FITC)] at the same time as the antibody solution.

V. The Lessons to Be Learned

The microinjection of antibodies against cytoskeletal proteins into tissue culture cells can be combined with several strategies to screen for effects on cellular architecture or motility. The first is simply to look for possible effects on the behavior and morphology of the injected, living cell. Such single-cell analy-

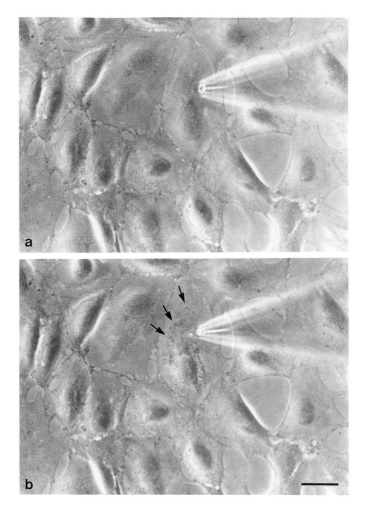

Fig. 1 An epithelial (PtK₂) cell layer as used for microinjection. One cell is seen immediately before (a) and immediately after (b) a buffer injection. Note the cytoplasmic vesicles delineating the contours of the spreading fluid droplet (arrows). The point of membrane puncture is also still visible as a white spot directly at the needle tip. The appearance of this spot, together with the vesicles, is indicative of a successful injection. (Bar: 30μm.)

ses can be carried out with cells left on the stage for many hours, provided that they can be kept there under adequate growth conditions (temperature and carbon dioxide buffering control). Alternatively, the injected cells can be returned to the incubator and examined periodically, if their identification is ensured. Low-intensity light cameras connected to a video system, and if needed, image processing can be used to follow motility phenomena such as locomotion, division, intracellular organelle transport, and shape changes in the

Table I
Criteria for Evaluation of Injected Volume

Too little	Just right	Too much
No fluorescence when label injected	Cells return to healthy state	Nucleus visibly moves during injection
Cannot see solution enter cell	Approximate volume is 1/5 to 1/10 of cell plasma volume	Cell balloons out during injection
No fluorescence after fixation and labeling	Detectable antibody or fluorescent label	Cells round up or are gone after 0.5 hr

antibody-injected cells, and to compare these parameters with adequate controls.

Another way to monitor antibody-induced changes is to inject many cells as quickly as technically feasible, and examine these cell populations at discrete time points after injection. With the Eppendorf system, one can routinely inject 100–200 cells in approximately 5 min. Skilled investigators can do even better with manual injections, without a pressure generator. The injected cells can subsequently be screened (e.g., in migration assays) or, after fixation and staining, inspected for changes in cytoskeletal organization. With such large numbers, any extreme values in the data received, which might be due to the variability in the amount of injected antibody or in the physiological state of the cells, can be balanced by statistics. With these techniques, the dynamics of the effects (i.e., onset, duration, and reversibility) can be reliably monitored, although in the case of fixed cells only with discrete populations. Figures 2 and 3 give examples of the effects of a monoclonal antibody against pig nonmuscle myosin on the organization of microfilaments in rat fibroblasts, as seen in cells fixed and stained for actin distribution as well as for the injected antibody 30 min after injection. Color Plate 5 shows an analogous experiment in which a polyclonal antibody against another microfilament protein, vinculin, was injected into a sheet of porcine epithelial cells. In both cell types and with both antibodies, the radial microfilament bundles (stress fibers) have partially or completely disintegrated. In addition, the peripheral belt of microfilaments, which is typical for epithelial cells, is disorganized and at several points disrupted (Color Plate 5).

The injection of several hundred cells within a relatively short period also allows a combination of microinjection with biochemical analyses. Thus injected populations can subsequently be subjected to radioactive tracer molecules, and their biosynthetic capacities can be analyzed with any technique suitable for small amounts of DNA, RNA, or proteins. Autoradiograms have been performed with two-dimensional gels of labeled proteins derived from as few as 50 injected cells (Schulze et al., 1989). Alternatively, labeled, injected cells can be mixed with cold controls for more convenient handling (Lamb et al., 1988).

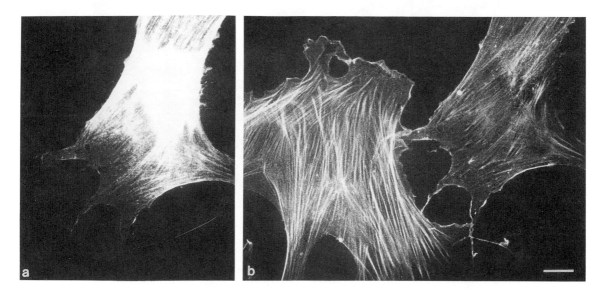

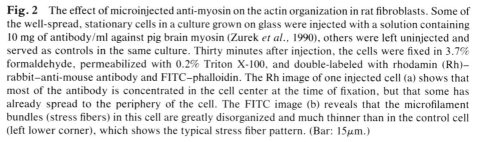

Fig. 2 The effect of microinjected anti-myosin on the actin organization in rat fibroblasts. Some of the well-spread, stationary cells in a culture grown on glass were injected with a solution containing 10 mg of antibody/ml against pig brain myosin (Zurek *et al.*, 1990), others were left uninjected and served as controls in the same culture. Thirty minutes after injection, the cells were fixed in 3.7% formaldehyde, permeabilized with 0.2% Triton X-100, and double-labeled with rhodamin (Rh)–rabbit–anti-mouse antibody and FITC–phalloidin. The Rh image of one injected cell (a) shows that most of the antibody is concentrated in the cell center at the time of fixation, but that some has already spread to the periphery of the cell. The FITC image (b) reveals that the microfilament bundles (stress fibers) in this cell are greatly disorganized and much thinner than in the control cell (left lower corner), which shows the typical stress fiber pattern. (Bar: 15μm.)

The injection of antibodies against cytoskeletal proteins into vertebrate tissue culture cells or amebas, in combination with the methods of analysis outlined above, has contributed in numerous cases to our present knowledge of the role of individual proteins and of complex cytoskeletal elements in cellular architecture and motility. A complete record of all reports in this field would clearly go far beyond the scope of this article, and earlier reviews have covered previous literature on this topic (Jockusch and Füchtbauer, 1985; Wehland and Weber, 1985; Wehland *et al.*, 1986; Jockusch *et al.*, 1986, 1991). However, a few examples are needed to illustrate the type of conclusions that can be drawn. Antibodies to the tail portion of myosin II were found to modulate cortical tension and motile activity at the plasma membrane of fibroblasts and epithelial cells, respectively (Höner and Jockusch, 1988; Höner *et al.*, 1988; Zurek *et al.*, 1989; Buss *et al.*, 1992; Temm-Grove *et al.*, 1992), and to delay cytokinesis (Zurek *et al.*, 1990). These studies emphasize the role of the actomyosin system in membrane dynamics. Interestingly, the same antibodies increased the locomotor activity in vertebrate cells drastically (Zurek *et al.*, 1990; Jockusch *et al.*,

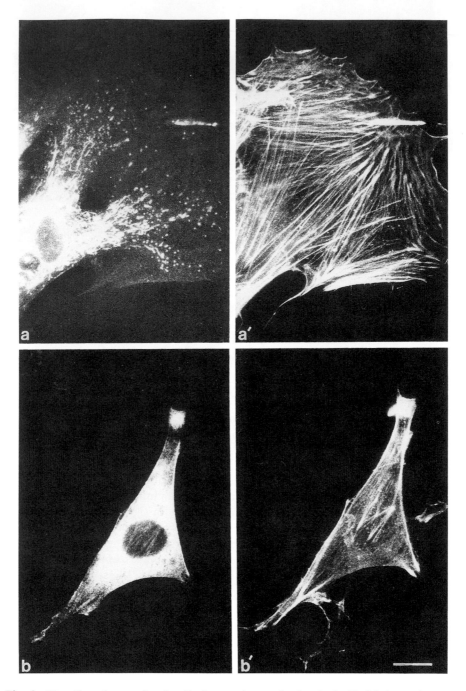

Fig. 3 The effect of an unrelated antibody on actin organization. In double-label experiments, rat fibroblasts were injected with a monoclonal antibody against *Dictyostelium* contact site A, which binds to some cellular structures (probably mitochondria) as seen after decoration with the Rh-labeled second antibody (a). The stress fiber system in this cell revealed by FITC–phalloidin

1991), whereas antibodies against *Acanthamoeba* myosin II decreased the velocity of recipient amebas (Sinard and Pollard, 1989). These apparent discrepancies possibly reflect long suspected differences between the mechanisms of locomotion in tissue-forming cells and in amebas. Antibodies against the microfilament protein vinculin were found to disrupt microfilament bundles in both fibroblasts (Westmeyer *et al.*, 1990) and epithelial cells (this study), and to release the injected fibroblasts from their junctions with the substratum (Westmeyer *et al.*, 1990). This latter finding supports the hypothesis postulated from *in vitro* experiments that vinculin is involved in organizing actin filaments at and linking them to the plasma membrane at the sites of cell–matrix adhesion.

Interesting results in regard to the structural and functional interrelationship of the three fibrillar cytoskeletal systems (microtubules, microfilaments, and intermediate filaments) were obtained by antibody injection. Antibodies against the actin-binding protein spectrin, as well as antibodies against tubulin, were found to include collapse and aggregation of the intermediate filament system (Mangeat and Burridge, 1984; Blose *et al.*, 1984; Wehland and Willingham, 1983). Antibodies against the actin-binding proteins caldesmon and tropomyosin inhibited the transport of cytoplasmic granules in fibroblasts (Hegman *et al.*, 1989, 1991), thus indicating that in addition to microtubule-based organelle transport (which has also been investigated by microinjection of antibodies against α-tubulin; Wehland and Willingham, 1983), there is also a microfilament-associated transport.

Antibodies were also used to change the posttranslational modification of cytoskeletal proteins. Such studies demonstrated that the organization of the highly dynamic microfilaments and also of the intermediate filaments depends critically on the correct state of phosphorylation. For example, it was found that antibodies against myosin light chain kinase reduced the phosphorylation of the regulatory myosin light chain and, concomitantly, caused a severe disruption of microfilament bundles (Lamb *et al.*, 1988). The inhibition of phosphorylation of myoblast intermediate filaments by injected antibodies was found incompatible with myotube differentiation (Tao and Ip, 1991). The observed differential effect of an antibody directed against an epitope common to many intermediate filament (IF) proteins on the IF organization in different cell types may also reflect differences in phosphorylation of these proteins (Meyer *et al.*, 1992). Injection of antibodies against the tyrosine ligase of α-tubulin led to a complete loss of tyrosinated microtubules in fibroblasts (Wehland *et al.*, 1986).

When injected into a recipient cell, antibody molecules directed against cytoskeletal proteins may bind to targets that are either immobilized in or at a

---

staining, is not affected (a'). In contrast, cells injected with the same antibody as in Fig. 2, fixed and processed at the same time as the cell in (a and a'), show again the disruption of their stress fibers [FITC–phalloidin staining (b)], and a concomitant change in morphology [cf. the size of the cell depicted in (a and a') with (b and b')]. Thus both these effects are due to the injected anti-myosin that is present throughout the cytoplasm and revealed by Rh–rabbit–anti-mouse antibody staining (b). (Bar: 15 μm.)

polymeric structure, or are present in a soluble pool. For IFs these soluble pools are small, and this may be the basis for the finding that many different antibodies against IF proteins were found to bind effectively to the IFs and cause lateral aggregation and coiling of the filaments (see reviews in Jockusch *et al.*, 1986; Meyer *et al.*, 1992). However, for the components of the microtubule and microfilament system, the situation is different: approximately 50% of the main structural components tubulin and actin are in the soluble fraction and, for many of the actin-associated proteins, this proportion is even higher. Thus the antibodies may bind to both the protein bound in the polymer and the soluble subunits. The observed effects on cellular architecture or motility might then be caused either by the direct attack of the antibody on polymer integrity, or by interference with the equilibrium between the cytoskeletal and the soluble state. A decision between the preferential site of antibody attack can sometimes be reached by comparing the effects of microinjection with those seen after adding the antibody to cytoskeletal preparations (cell models), or to *in vitro*-assembled polymers. By this approach, we found that a polyclonal antibody against β-tubulin caused disruption of microtubules in the living cell by directly interfering with the polymer. Because of its high affinity for β-tubulin, the antibody caused multiple fractures of microtubules and generation of unstable pieces, even when injected in substoichiometric molar ratios (antibody:tubulin, 1:50–1:200; Füchtbauer *et al.*, 1985). As such antibodies clearly act as molecular scissors, one can conclude that in this case the affinity between the polymer subunits must be lower than the affinity between the antibody and its target protein. Direct interference between an antibody and a functionally important domain on its antigen is also suggested in the case of antibody-induced inhibition of myosin light chain kinase activity (Lamb *et al.*, 1988) and of vinculin-membrane association (Westmeyer *et al.*, 1990). However, there may also be another factor involved: the disruption of the tertiary structure of the target protein by a high-affinity antibody.

As with the use of antibodies in other cell biological analyses, the results obtained with different antibodies in microinjection experiments can in most cases be compared to only a limited extent. For example, a polyclonal anti-actin raised against the N-terminal synthetic decapeptide of α-smooth muscle actin disrupted the microfilament organization in cultured smooth muscle cells effectively (Skalli *et al.*, 1990), whereas several polyclonal, affinity-purified antibodies against $\alpha/\beta/\gamma$ smooth muscle actin did not show this effect, even when injected at 20 times higher concentrations (Jockusch *et al.*, 1986). One of these latter antibodies, however, had a drastic inhibitory effect on the transcriptional activity of RNA polymerase II in amphibian oocytes, by causing the collapse of lampbrush chromosomes (Scheer *et al.*, 1984) in substoichiometric concentrations (molar ratio actin:antibody, 400:1). In this case, a low antibody concentration was apparently sufficient to interfere successfully with nuclear actin, which possibly serves as a cofactor in transcription (Smith *et al.*, 1979; Egly *et al.*, 1984). Thus antibody affinity, epitope location, as well as antigen concentration, location, and availability for interaction with the antibody are crucial parameters determining the result of antibody injections.

VI. Perspectives

Antibodies have clearly proved to be valuable requisites in the attempts to determine the functions of individual cytoskeletal proteins. With respect to the different strategies for modulating the endogenous level of such proteins (see Section I), the microinjection of antibodies is comparable to gene-targeting experiments and those using antisense mRNA probes: in all three cases, the elimination of a functional polypeptide is the aim of the experiments. So far, gene targeting is not easily applicable to diploid somatic cells, and antisense approaches have also in many cases proved exceedingly difficult (Graessmann *et al.*, 1991). Microinjection of antibodies also has some disadvantages, and the most serious ones should be emphasized once more: only certain cell types are suitable, the antibodies must be well characterized and of high affinity, and the effects are transient (as the injected cells continue with protein synthesis). However, we believe that this technique will retain and expand its position in cell biological investigations, mainly for three reasons. First, once the script has been outlined, the actors are chosen, and the stage is set, the performance itself is simple and fast. Second, the methods for biochemical analysis are becoming increasingly more sensitive, so that the limitation in cell number originally seen as a drawback of that technique is now negligible. Third, and most important, the number of antibodies against functionally important domains of target antigens is rapidly increasing, because molecular biology provides us with the protein sequences necessary to design antibodies against the relevant peptides. Thus there is good reason to believe that the show will go on.

Acknowledgments

We thank Drs. M. Schleicher and G. Gerisch (Martinsried, Germany) for the antibody against *Dictyostelium* contact site A, Dr. H. Faulstich (Heidelberg, Germany) for FITC–phalloidin, Mrs. R. Klocke for typing the manuscript, and the Deutsche Forschungsgemeinschaft (SP *Functional Domains of Structural Proteins*) for financial support of our own studies reported in this article.

References

Blose, S. H., Meltzer, D. I., and Feramisco, J. R. (1984). 10 nm filaments are induce to collapse in living cells microinjected with monoclonal and polyclonal antibodies against tubulin. *J. Cell Biol.* **98,** 847–858.

Buss, F., Temm-Grove, C., and Jockusch, B. M. (1992). Cortical microfilament proteins and the dynamics of the plasma membrane. *Acta Histochem., Supl.* **41,** 291–301.

Cunningham, C. C., Stossel, T. P., and Kwiatkowski, D. J. (1991). Enhanced motility in NIH 3T3 fibroblasts that overexpress gelsolin. *Science* **251,** 1233–1236.

DeLozanne, A. (1989). Gene targeting and cell motility. *Cell Motil. Cytoskeleton* **14**, 62–68.

Egly, J. M., Miyamoto, N., Moncollin, V., and Chambon, P. (1984). Is actin a transcription factor for RNA polymerase B? *EMBO J.* **3**, 2363–2371.

Fishkind, D. J., Cao, L., and Wang, Y.-L. (1991). Microinjection of the catalytic fragment of myosin light chain kinase into dividing cells: Effects on mitosis and cytokinesis. *J. Cell Biol.* **114**, 967–975.

Füchtbauer, A. (1984). Analyse des Mikrofilamentsystems tierischer Zellen durch Mikroinjektion. Ph.D. Thesis, Univ. of Bielefeld, Bielefeld, Germany.

Füchtbauer, A., Herrmann, M., Mandelkow, E.-M., and Jockusch, B. M. (1985). Disruption of microtubules in living cells and cell models by high affinity antibodies to beta-tubulin. *EMBO J.* **4**, 2807–2814.

Gerisch, G., Segall, J. E., and Wallraff, E. (1989). Isolation and behavioral analysis of mutants defective in cytoskeletal proteins. *Cell Motil. Cytoskeleton* **14**, 75–79.

Graessmann, A., Graessmann, M., and Müller, C. (1980). Microinjection of early SV40 DNA fragments and T antigen. *In* "Nucleic Acids," Part I (L. Grossman and K. Moldave, eds.), Methods in Enzymology, Vol. 65, pp. 816–825. Academic Press, New York.

Graessmann, M., Michaels, G., Berg, B., and Graessmann, A. (1991). Inhibition of SV40 gene expression by microinjected small antisense RNA and DNA molecules. *Nucleic Acids Res.* **19**, 53–59.

Hegman, T. E., Lin, J. L., and Lin, J. J. (1989). Probing the role of nonmuscle tropomyosin isoforms in intracellular granule movement by microinjection of monoclonal antibodies. *J. Cell Biol.* **109**, 1141–1152.

Hegman, T. E., Schulte, D. L., Lin, J. L.-C., and Lin, J. J.-C. (1991). Inhibition of intracellular granule movement by microinjection of monoclonal antibodies against caldesmon. *Cell Motil. Cytoskeleton* **20**, 109–120.

Höner, B., and Jockusch, B. M. (1988). Stress fiber dynamics as probed by antibodies against myosin. *Eur. J. Cell Biol.* **47**, 14–21.

Höner, B., Citi, S., Kendrick-Jones, J., and Jockusch, B. M. (1988). Modulation of cellular morphology and locomotory activity by antibodies against myosin. *J. Cell Biol.* **107**, 2181–2189.

Huckriede, A., Füchtbauer, A., Hinssen, H., Chaponnier, C., Weeds, A. G., and Jockusch, B. M. (1990). Differential effects of gelsolins on tissue culture cells. *Cell Motil. Cytoskeleton* **16**, 229–238.

Jockusch, B. M., and Füchtbauer, A. (1985). Funktionsanalyse zellulärer Strukturproteine durch Mikroinjektion. *In* "Aktuelle Methoden der Molekular- und Zellbiologie" (N. Blin, M. F. Trendelenburg, and E. R. Schmidt, eds.), pp. 1–12. Springer-Verlag, Berlin.

Jockusch, B. M., Füchtbauer, A., Wiegand, C., and Höner, B. (1986). Probing the cytoskeleton by microinjection. *In* "Cell and Molecular Biology of the Cytoskeleton" (J. W. Shay, ed.), Vol. 7, pp. 1–40. Plenum, New York.

Jockusch, B. M., Zurek, B., Zahn, R., Westmeyer, A., and Füchtbauer, A. (1991). Antibodies against vertebrate microfilament protein in the analysis of cellular motility and adhesion. *J. Cell Sci., Suppl.* **14**, 41–47.

Knecht, D. (1989). Application of antisense RNA to the study of the cytoskeleton: Background, principles and a summary of results obtained with myosin heavy chain. *Cell Motil. Cytoskeleton* **14**, 92–102.

Kreis, T. E., and Birchmeier, W. (1982). Microinjection of fluorescently labeled proteins into living cells with emphasis on cytoskeletal proteins. *Int. Rev. Cytol.* **75**, 209–227.

Lamb, N. J. C., Fernandez, A., Conti, M. A., Adelstein, R., Glass, D. B., Welch, W. J., and Feramisco, J. R. (1988). Regulation of actin microfilament integrity in living nonmuscle cells by the cAMP-dependent protein kinase and the myosin light chain kinase. *J. Cell Biol.* **106**, 1955–1971.

Lee, G. M. (1989). Measurement of volume injected into individual cells by quantitative fluorescence microscopy. *J. Cell Sci.* **94**, 443–447.

Mangeat, P. H., and Burridge, K. (1984). Immunoprecipitation of nonerythrocyte spectrin within live cells following microinjection of specific antibodies: Relation to cytoskeletal structures. *J. Cell Biol.* **98**, 1363–1377.

McNeil, P. L. (1989). Incorporation of macromolecules into living cells. *Methods Cell Biol.* **29**, 153–173.

Meyer, T., Weber, K., and Osborn, M. (1992). Microinjection of IFA antibody induces intermediate filament aggregates in epithelial cell lines but perinuclear coils in fibroblast-like lines. *Eur. J. Cell Biol.* **57**, 75–87.

Scheer, U., Hinssen, H., Franke, W. W., and Jockusch, B. M. (1984). Microinjection of actin-binding proteins and actin antibodies demonstrates involvement of nuclear actin in transcription of lampbrush chromosomes. *Cell* **39**, 111–122.

Schulze, H., Huckriede, A., Noegel, A. A., Schleicher, M., and Jockusch, B. M. (1989). Alpha-actinin synthesis can be modulated by antisense probes and is autoregulated in nonmuscle cells. *EMBO J.* **8**, 3587–3593.

Sinard, J. H., and Pollard, T. D. (1989). Microinjection into *Acanthamoeba castellanii* of monoclonal antibodies to myosin-II slows but does not stop cell locomotion. *Cell Motil. Cytoskeleton* **12**, 42–52.

Skalli, O., Gabbiani, F., and Gabbiani, G. (1990). Action of general and α-smooth muscle-specific actin antibody microinjection on stress fibers of cultured smooth muscle cells. *Exp. Cell Res.* **187**, 119–125.

Smith, S. S., Kelley, K. H., and Jockusch, B. M. (1979). Actin co-purifies with RNA-polymerase II. *Biochem. Biophys. Res. Commun.* **86**, 161–166.

Tao, J. X., and Ip, W. (1991). Site-specific antibodies block kinase A phosphorylation of desmin *in vitro* and inhibit incorporation of myoblasts. *Cell Motil. Cytoskeleton* **19**, 109–120.

Temm-Grove, C., Helbing, D., Wiegand, C., Höner, B., and Jockusch, B. M. (1992). The upright position of brush border-type microvilli depends on myosin filaments. *J. Cell Sci.* **101**, 599–610.

Wang, Y. L. (1989). Fluorescent analog cytochemistry: Tracing functional protein components in living cells. *Methods in Cell Biol.* **29**, 1–12.

Wehland, J., and Weber, K. (1985). Microinjection of monoclonal tubulin antibodies into living cells as an approach to evaluate the contribution of microtubules to cellular physiology. *Biochem. Soc. Trans.* **13**, 16–18.

Wehland, J., and Willingham, M. C. (1983). A rat monoclonal antibody reacting specifically with the tyrosylated form of α-tubulin. II. Effects on cell movement, organization of microtubules, and intermediate filaments, and arrangement of Golgi elements. *J. Cell Biol.* **97**, 1476–1490.

Wehland, J., Schröder, H. C., and Weber, K. (1986). Contribution of microtubules to cellular physiology: Microinjection of well-characterized monoclonal antibodies into cultured cells. *Ann. N.Y. Acad. Sci.* **466**, 609–621.

Westmeyer, A., Ruhnau, K., Wegner, A., and Jockusch, B. M. (1990). Antibody mapping of functional domains in vinculin. *EMBO J.* **9**, 2071–2078.

Zurek, B., Höner, B., and Jockusch, B. M. (1989). The role of myosin filaments in nonmuscle cells. *Verh. Dtsch. Zool. Ges.* **82**, 5–16.

Zurek, B., Sanger, J. M., Sanger, J. W., and Jockusch, B. M. (1990). Differential effects of myosin:antibody complexes on contractile rings and circumferential belts in epitheloid sheets. *J. Cell Sci.* **97**, 297–306.

CHAPTER 18

Immunoselection and Characterization of cDNA Clones

Robert A. Obar* and Erika L. F. Holzbaur†

*Alkermes, Inc.
Cambridge, Massachusetts 02139

†University of Pennsylvania
 School of Veterinary Medicine
 Philadelphia, Pennsylvania 19104

I. Introduction

Immunoselection is the process of screening complementary DNA (cDNA) expression libraries, using antibodies as probes. This article is intended as a step-by-step introduction to the practical and theoretical considerations involved in obtaining cDNA clones by immunoselection.

The concept of construction and screening of DNA expression libraries was developed by Broome and Gilbert (1978), but the construction of λgt11 (Young and Davis, 1983) simplified and generalized the process. The technique has rapidly gained acceptance, even in laboratories with little or no previous molecular cloning expertise. λgt11 is a double-stranded λ replacement vector of 43.7 kilobase pairs (kbp), with a single *Eco*RI restriction site, into which cDNA "inserts" are cloned during the construction of a library. A newer generation of vectors [such as λZAP (Stratagene, La Jolla, CA), λEXlox (Novagen, Madison, WI), and λGEM (Promega, Madison, WI)] has made the subcloning and characterization of cDNAs more rapid, and it is to be expected that further improvements in the method will also be forthcoming.

In our view, the success of an immunoselection project is based on thoughtful choice and characterization of both antibody and library, and an understanding of the statistical aspects of the screening process. As the first-time cloner will soon discover, the major difficulty lies not in the selection of clones from a library, but in distinguishing the valid clones from among those selected. We will focus on strategy and methodology of antibody and fusion protein usage, rather than the techniques of antibody production, library construction, genomic cloning, plasmid expression libraries, and routine nucleic acid and protein manipulations, which can be found elsewhere. We recommend the remainder of this volume, and Davis *et al.* (1986), Sambrook *et al.* (1989), Ausubel *et al.* (1989 and supplements), and Berger and Kimmel (1987) for treatments of these topics.

II. Overview of a Typical Immunoselection Project

The decision to embark on an immunoselection project may lead to 6 months or more of cloning, subcloning, and sequencing chores, which have little direct relevance to the biological questions motivating the project. Unforeseen experiments often arise as sidebars during immunoselection, but it is crucial that the strategy of the project ensures that the information gained by molecular cloning of the target protein is rapidly integrated with other ongoing work. The following is a generalized outline.

1. *Establishment of goals:* Schedule for selection of clones, confirmation of clones, rescreening of library, clone characterization, data analysis, and so on.

2. *Development of tools*

    a. Obtain antibody; prepare control strips and standard assay.

    b. Perform preliminary molecular biological characterization of the target protein, for example, partial amino acid sequence, developmental or tissue dependence of expression; isoform diversity.

    c. Obtain cDNA expression library.

    d. Characterize the library (if not done by supplier), using guidelines in Section IV.

    e. Perform trial run. Revise goals if necessary.

3. *Screening:* Select and plaque purify 10–20 immunopositive clones.

4. *Characterization:* Perform Western blots and polymerase chain reaction (PCR) amplifications on each clone. "Reverse purify" antibody from fusion proteins, and rule out negative clones.

5. *Collation of data:* Construct maps of fusion protein structure (from Western blots), and restriction maps of cDNA structure (from analysis of PCR products and phage DNAs).

6. *Confirmation of clones:* Choose clones for further characterization: subcloning, sequencing, and so on, to establish their identities unambiguously.

7. *Reassessment of strategy:* If one or more clones cannot be validated, revise strategy. If necessary, rescreen the library to complete the cDNA sequence.

8. *Resolution:* Return to the biological questions of interest (formulated in step 1).

III. Preparation of Antibodies for Library Screening

A. General Considerations

The choice of a successful strategy depends on the availability of appropriate, well-characterized antibodies as tools for isolation, purification, and characterization of cDNA clones. We will, therefore, go into some detail on qualities desired in antibodies to be used for these purposes.

B. Characterization of Antibodies

Whether the antibody to be used in immunoselection is monoclonal or polyclonal, purified or crude, some efforts must be made to quantify its affinity for other proteins. Antibody characterization traditionally consists of analysis of several interrelated parameters that describe the interaction between antibody and its target antigen: titer (the dilution at which the antibody can be effectively used in a particular protocol), specificity (the degree to which it can distinguish a component of interest in a complex milieu), determination of the epitope(s) or site(s) of the interaction of the antibody with its target, and stability (i.e., to variations in temperature, pH, ionic strength). In the case of monoclonal antibodies, it is also important to determine heavy chain class or subtype, so that a suitable secondary antibody system is used for detection.

Because most of these topics are treated in some detail elsewhere in this volume, we will restrict discussion to the case of antibody characteristics required for interpreting interaction with recombinant fusion proteins produced by cloned cDNAs. It is important to realize that the more time one puts into characterizing each antibody to be used, the sparser the problems likely to follow at an advanced stage of the project. Although ambitious cloning projects have occasionally seen success with poorly characterized antibodies (see, e.g., Lewis et al., 1986a,b; Garner and Matus, 1988; Holzbaur et al., 1991), this should be attempted only by those already experienced in avoiding the common pitfalls of such schemes.

Monoclonal antibodies are generally less useful for initial selection of clones from expression libraries, except in two cases: (1) if several monoclonal antibodies are available they can be used as a mixture or "synthetic polyclonal"; or (2) a monoclonal antibody can be used in conjunction with a polyclonal antibody either for "double screening" of an expression library (see Section V), or for further characterization of clones picked by screening with the polyclonal antibody.

The most important piece of information about the epitope for a given antibody is whether it is a linear sequence or a more complex structure that includes noncontiguous sequence elements. If the latter, there is a strong chance that the epitope will not be presented by clones in a bacterial expression system such as a cDNA library, or on an immunoblot. If the epitope is a single linear sequence, it has a high probability of being found by immunoselection. It should be determined whether the sequence is exposed when the antigen is in a native conformation, in denatured conformations, or both. In expression library screening, the conformation of the antigen is unknown. Therefore the most useful antibodies are those that recognize short (i.e., less than ~20 amino acids) linear sequences in both unfolded and folded structures.

Epitopes nearer the carboxyl- than the amino-terminal end of a polypeptide sequence are also preferable for three reasons: (1) many cDNA libraries are constructed wholly or partially with oligo-dT as a primer; this leads to the overrepresentation in these libraries of sequences derived from the 3′ ends of mRNAs [those closer to the poly(A) tail]; (2) when a cDNA is inserted in the sense orientation of an expression vector, it still must be in the correct reading frame to be expressed, and the likelihood of this depends in part on specific sequences within the cDNA, such as restriction sites. The "expressibility" of any portion of an open reading frame, therefore, depends on the presence of these potential fusion sites upstream. Sequences near the carboxyl terminus have more natural coding sequence upstream than do amino-terminal sequences, and a higher natural probability of being contained within an open reading frame in a fusion protein construction; (3) all of the common expression libraries manifest foreign polypeptide sequences as the carboxyl-terminal ends of fusion proteins; it is this fusion that an antibody sees during the screening process. In this context, an antibody is more likely to recognize a sequence that *evolved* as the carboxyl-terminal domain of a protein, than one that comprises some other region.

The usefulness of an antibody in immunoselection is decreased if *in vivo* or *in vitro* chemical modification of amino acid residues within the epitope affects the ability of the antibody to bind, or if the epitope occurs more than once in the target protein.

1. Procedure for Preparation of Control Strips

The most useful form of antigen control for the protocols described in this article is a set of nitrocellulose strips prepared from a Western blot. A heterogeneous protein sample (preferably a homogenate from a tissue containing the antigen of interest as well as many other unrelated proteins) is run on a preparative sodium dodecyl sulfate-polyacrylamide gel electrophoresis (SDS-PAGE) gel (Laemmli, 1970; Weber *et al.,* 1972), then electroblotted onto a nitrocellulose membrane (Towbin *et al.,* 1979). A guide strip (from a side edge of the membrane) is excised and stained with Ponceau S solution (see Section XI for this and other materials and reagents mentioned in this article), while the remainder of the membrane is incubated for 30 min at room temperature with an excess of blocking solution, and air dried on a sheet of Whatman (Clifton, NJ) 3MM filter paper. The membrane is then placed between two sheets of Parafilm and cut into vertical strips of ~2-mm thickness with a razor blade, or processed with a commercial strip cutter (Hoefer Scientific, San Francisco, CA). It may also be convenient to number the top or bottom edge of the membrane before it is cut into strips. Strips can be stored at room temperature indefinitely.

2. Procedures for Determination of Antibody Titer, Stability, Kinetics, and Thermodynamics

a. Titer

For practical purposes, the important parameter in antibody use is the dilution of an antibody stock that gives an acceptable signal-to-noise ratio in the assay of interest. For example, many commercially available antibodies are stock solutions of 0.5 mg/ml with a titer of 1 : 1000, and are used at dilutions of 1 : 1000 (i.e., at 0.5 μg/ml), at which they give acceptable results. The optimum working dilution for an antibody should be worked out with a set of "control strips" (see above). This is similar to the way one would determine antibody titer for other applications, but with several special considerations.

1. It is more important that the antibody produces a strong signal than a blank background. In fact, some background staining can be quite helpful (see Sec. V).

2. The antibody may be used several times, so a higher initial antibody concentration is preferable to a low one, other things being equal.

3. A large volume of antibody solution (>60 ml) may be needed, therefore the antibody preparation must be adjustable to this scale.

4. If two antibodies are to be used in immunoselection (i.e., with duplicate plaque lifts), the signal-to-noise ratios of the two antibodies should be comparable.

b. Stability

As mentioned, the identification of fused or unfused polypeptides from an expression library requires the reproducible reaction of a given antibody with uncharacterized targets. This makes consistency of the antibody formulation a necessity, and the highest consistency is obtained by reusing the same solution of a stable antibody. The stability of the antibody solution will be affected by the number of filters, density of plaques on these filters, the number of reactive and cross-reactive plaques, the incubation temperature, and the length of time it is in contact with target filters. It is important to monitor the stability of the antibody solution by including a control strip with the plaque or colony lifts in each round of screening.

c. Kinetics and Thermodynamics

Once an antibody concentration that works well with an overnight incubation at 4°C has been determined, it is helpful to discover whether a shorter incubation period (to a minimum of ~30 min) or a higher temperature (to a maximum of 37°C) can be used. Expression library screening can go significantly faster when the lengths of the incubations are minimized, and because the incubations can be extended without deleterious effect a clear understanding of the kinetic and thermodynamic limits of the antibody–antigen interaction, before the start of the project, is advised.

First, determine whether the specificity of the antibody–antigen interaction is retained at higher temperatures, by incubating control strips with antibody overnight at room temperature and at 37°C, and compare the pattern of background staining with that obtained with a 4°C incubation.

Next, ask what is the minimum time of incubation that gives acceptable results at 4°C, or at a higher temperature if this was found to give favorable results. Incubations of different lengths should be compared by introducing the strips into antibody solution at different times (rather than removing them at different times), and developing all the strips together.

After several incubation times and temperatures have been compared, you may find that a 3-hr incubation at 37°C is comparable to an overnight incubation at 4°C, and a decision must be made as to whether the saving in time (using the 37°C incubation) outweighs the cost in antibody, which will survive fewer treatments at 37°C than at 4°C.

C. Purification of Antibodies

Monoclonal antibodies can be used as diluted ascites fluid, culture supernatant, or purified IgG (or other subclass). Further purification prior to library screening is unnecessary, and may be counterproductive; antibodies tend to become less stable with excessive handling or reduced protein concentrations.

Polyclonal antibodies can frequently be used for library screening without purification. However, the two most common experiments used to distinguish among immunoselected clones are the challenging of fusion proteins with polyclonal antibody affinity purified against target protein, and the challenging of the target protein with antibody affinity purified against fusion proteins. If these experiments are likely to be tried, the time to work out the affinity purification method is before the first round of screening. Many methods exist for purification of polyclonal antibodies (e.g., see chapter 5 in this volume). The following general method has consistently worked well in our hands.

Procedure for Affinity Purification of Antibody from Immunoblot

a. Electroblotting

A sample containing the antigen of interest (e.g., ~50 μg of a 50-kDa protein to purify up to ~200 μg of antibody) is electrophoresed and electroblotted by standard methods onto a nitrocellulose (Cat. No. BA85; Schleicher & Schuell, Keene, NH) or polyvinylidene difluoride (PVDF; Millipore, Bedford, MA) membrane. The separated polypeptides are visualized by staining the membrane for ~2 min with Ponceau S solution, followed by a brief rinse with water. The stained membrane is placed between two sheets of Parafilm, and the band of interest is excised with a razor blade, to yield an "antigen strip."

b. Blocking and Binding

The antigen strip is washed for ~15 min with water, then incubated in blocking solution for 30 min or more at room temperature. The strip is then incubated in diluted antibody. This incubation is frequently performed as an overnight incubation at 4°C, with agitation, in a 1 : 5 to 1 : 500 dilution of antiserum, but the optimum dilution, volume, temperature, and time should be determined empirically for each antibody–antigen combination.

c. Washing

The strip is then washed four times for 5 min with TBST, then once with 1 M NaCl–TBST, followed by a brief rinse with TBST.

d. Elution

To release the antibody from the strip, add 1.0 ml of antibody elution buffer for 30 to 60 sec at room temperature.

e. Neutralization

This 1.0-ml antibody solution is then quickly applied to a NAP-10 column (Cat. No. 17-0853; Pharmacia, Piscataway, NJ) that has been previously equilibrated with four washes (3 ml each) of blocking solution, at 4°C. Once the antibody solution has entered the column, it is eluted by applying 1.5 ml of blocking solution to the top of the column. This step should be concluded within 3 min. The resulting (1.5 ml) neutralized antibody–blocking solution is then titered prior to use in library screening (see Sec. V). *Note:* The used antigen strip can now be equilibrated in several changes of TBST, reblocked, and used in subsequent antibody purifications as needed.

f. Scale-Up

The quantity of antibody required in a library screening project is most conveniently thought of as the volume of (diluted) working solution to be used within a period of ~2 weeks. For a typical project (i.e., a first round of six to ten 150–mm plates, followed by a second and third round of ten 100-mm plates each) ~60 ml of working solution is sufficient. Thus, if the purified antibody solution has a working titer of 1 : 50, only 1.2 ml is required (and the 1.5 ml produced in a single round of purification, described above, would suffice). With less potent antibody, several antigen-containing strips can be treated simultaneously.

An alternative, more gentle method of purifying antibody was developed by Hammarback and Vallee (1990) and this can be particularly useful for antibodies that are not stable to acid treatment. Briefly, an antigen strip with specific antibody bound to it is prepared as in steps a–c above. Then, rather than eluting the antibody in a separate step, the strip is incubated with a target (strips or plaque lifts); at equilibrium, a significant proportion of antibody molecules is present on the target filter(s). Other alternative methods may also be effective.

g. Preabsorption

Most animal sera contain antibodies that interact with bacterial proteins, and it is advisable that these be saturated prior to first-round screening. This can be conveniently accomplished by adding a bacterial lysate (from the bacterial strain to be used in plating) (Huynh et al., 1985) to ~500 mg/ml, and incubating at 4°C for 60 min. This step need not be repeated prior to subsequent uses of the antibody solution, but the treated solution should be stored at 4°C in the presence of 10 mM NaN$_3$ to inhibit bacterial growth.

IV. Choice and Preparation of Expression Libraries

A. General Considerations

cDNA expression libraries can be manufactured with the use of kits from a range of suppliers (e.g., Promega, Invitrogen, Clontech), obtained as gifts from other researchers, or purchased as commercial products [Stratagene, Clontech, Promega, and the American Type Culture Collection (ATCC) all offer a variety]. We will deal only with those constructed in the vectors λgt11 and λZAP; the same methods may be used with other phage vectors, but could require the use of specialized bacterial strains and/or growth media.

The biggest question in assessing the utility of a library is what is the source of the mRNA from which the library was synthesized. For example, if antibody work has established that a particular protein has an early developmental peak of expression, a library from newborn or fetal tissue would be the best choice. In some cases the expression of a protein can be induced by special conditions such as deflagellation, viral infection, heat shock, or chemical or hormonal treatment. A library prepared from cells or tissues treated in such a fashion can greatly simplify the screening procedure, and possibly aid in the identification of new genes of interest as well.

The parameters that are generally used to characterize cDNA libraries are complexity (the number of independent clones), titer [expressed as plaque-forming units (pfu) per unit volume], percentage of recombinants, the type(s) of primer used, and average cDNA insert size. These are generally most useful in comparing specific libraries from the same or similar sources, and predicting the likelihood that a clone with a particular level of representation will be present in each library. In general, a large average insert size is more important than a high level of complexity (unless the protein of interest is expected to have a small or rare message). In practice, for proteins of >1000 amino acids (i.e., encoded by >1000 codons or 3 kbp), clones corresponding to the 5' end of the message will be underrepresented; a random-primed or unique oligonucleotide-primed library may be needed to finish the cloning. One advantage of constructing a new library for a particular project is that the input cDNA may be size-selected to increase the proportion of clones containing inserts within a desired range of sizes. This increases the likelihood of finding long clones, but does so at the expense of most of the potential complexity. However, if a laboratory is engaged in the study of proteins of $M_r > 100,000$ we strongly advise the use of a size-selected library [see Sambrook *et al.* (1989) or Elliott and Green (1989) for a variation on the usual technique].

B. Procedure for Titration of Library

Before a library can be used for immunoscreening, the number of plaque-forming units per unit volume in a library stock is determined by plating a dilution series such that plates with ~30 to ~500 plaques can be counted.

Assuming a stock titer of no more than 10^{10} pfu/ml ($10^7/\mu l$) and no less than 10^7 pfu/ml (or $10^4/\mu l$), make dilutions of $1:10^2$, $1:10^3$, $1:10^4$, $1:10^5$, and $1:10^6$ in SMG. Handle the phage gently, and *do not vortex*. Mix 10 μl of each dilution, in duplicate, with 200 μl of plating bacteria (*Escherichia coli* strain Y1090 for λgt11, or strain BB4 for λZAP), and incubate for 15 min at 37°C. Mix each aliquot with 2.5–3.5 ml of molten top agar (heated to boiling, then cooled to ~48°C), and pour onto a 100-mm LB-Amp (for Y1090) or LB (for BB4) plate. When solidified, incubate the plates, inverted, at 42°C for 3.5 hr, then at 37°C overnight.

The next day, count the plaques [by spotting their locations with a marking pen or with an automatic counter such as a Manostat (New York, NY, Cat. No. 81-520-000)] on all plates where counting is possible. Average the number of plaques for each set of duplicates, and determine the titer for the dilution used on those plates. For example, if duplicate plates containing 10 μl each of a $1:10^4$ dilution of a library contain 144 and 160 plaques, the average number of plaques is 152, and the titer of this dilution is 152 pfu/10 μl, or 1.52×10^1 pfu/μl. Multiplied by the dilution factor (10^4), this means that the stock has a titer of 1.52×10^5 pfu/μl. If all the plates have more than 1000 plaques, additional, higher dilutions must be made; if none has more than 30, the library probably has deteriorated to the point of being useless, and should be replaced.

For plating and screening of the library with antibody, $1–3 \times 10^4$ plaques/150-mm plate is optimal; for the example given above, 3–10 μl of a $1:10$ dilution would be suitable (for each plate used). The complexity of a good library is $10^5–10^6$; because the practical limit for a single researcher with a single incubator is ~12 plates per round, it is generally considered adequate to screen a number of plaques representing ~50% of the library (meaning a number of plaques equal to one-half the complexity).

At a density of 2×10^4 plaques/plate, 12 plates would represent 2.4×10^5 pfu, or 50% of a library with a complexity of 4.8×10^5. With libraries of greater complexity, multiple rounds of screening or the substitution of large glass baking pans (also known as "lasagna pans") for petri plates is advised; for libraries of lower complexity, a lower plaque density (rather than a smaller number of plates) is advised, because the lower plaque density will allow for larger and more widely spaced plaques, thus reducing the number of rounds of screening required for plaque purification.

V. Immunoselection of Clones from cDNA Libraries

A. Trial Round of Screening

A trial round of screening involves probing a small fraction of the library with an antibody in order to assess its specificity. Plaque lifts from plates seeded at low plaque densities are incubated with the antibody probe along with a control strip. If planning to screen with an affinity-purified antibody, it is worthwhile to perform a trial round with both the purified and the unpurified antibodies in parallel.

1. Procedure for Plating a Trial Run

Aim for a plaque density of about 100 plaques/100-mm plate, so that the plaques are clearly separated. It is not necessary to achieve a representative sample of the library at this stage; one or two plates should be sufficient to assess the specificity of the antibody reaction (remember, a true positive signal from a plaque is not expected at this stage).

1. Dilute an aliquot of the working library stock into SMG to achieve a titer of 20 to 30 pfu/μl.

2. Mix 5 μl of this dilution with 200 μl of plating bacteria in each of two tubes. Incubate the tubes at 37°C for 15 min. While the tubes are incubating, melt LB-top agar, then cool to 50°C.

3. Plate out each tube by adding 4 ml of warm top agar, mix gently by inversion, and then carefully pour the molten agar onto a warm LB-Amp plate. Quickly swirl the plate to spread the top agar over the surface without introducing any bubbles. Allow the agar to harden, then transfer the plates to an incubator.

4. Incubate the plates at 42°C for about 4 hr. Plaques should just begin to be visible on the surface of the top agar. Then overlay the plates with filters soaked in IPTG (described in the next section).

5. Incubate the plates at 37°C for about 4 to 5 hr in order to express the fusion proteins, then proceed with the antibody-binding and visualization steps described below.

2. Procedure for Preparation of IPTG Filters

The traditional filters used for plaque lifts and antibody screening are nitrocel-lulose (0.45-μm pore size). If handled with care, these filters are satisfactory and yield consistent results. Investigators have begun using nylon membranes for screening clones. These filters are sturdier and, if used according to the recommendations of the manufacturer, should prove reliable. Alternatively, tear-resistant nitrocellulose filters are available from several suppliers.

1. First label or number each filter with a non-water-soluble ink. Filters should be handled carefully throughout, with gloves and forceps.

2. Incubate each filter for about 1 min in a 10 mM sterile solution of isopro-pylthio-β-D-galactoside (IPTG), obtained by dissolving the IPTG in water and sterilizing with a syringe filter. Soaking the filters in IPTG can be conveniently performed in a sterile, disposable petri dish. First layer the filter on the surface of the solution, and then submerge it after about 30 sec. Wait another 30 sec and then lift out the filter and blot it dry on a layer of blotting paper. After soaking all the filters needed, cover them with another sheet of blotting paper and allow them to dry at room temperature for about 30 min.

After developing the plaque lift filters from the trial run, compare the strength of the positive signal on the control strip to any background reaction that occurs with the plaques representing a sampling of the library. It is not uncommon for an antibody to show some, or even significant, levels of cross-reactivity with *E. coli* proteins. If background reactivity is observed, the easiest method to improve the signal-to-noise ratio is to preincubate the antibody solution with *E. coli* extract (see Section III).

If preabsorption of the antibody probe is not sufficient to obtain a desirable signal-to-noise ratio, then it may be necessary to perform the library screening with antibody that has been affinity purified against the antigen of interest, as described above. If the affinity-purified antibody solution is handled and stored with care then the same solution may be reused in the screening of the first and subsequent rounds.

B. First Round of Screening

The theoretical basis for library screening is straightforward—a sufficient number of plaques should be probed with the antibody to ensure a representative screening of all independent clones in the library. The complexity, or number of independent clones, of an expression library will depend on both the species and tissue of origin. The representation of the clone of interest within the library will also depend on the level of its mRNA expression at the point in development at which the mRNA was isolated to construct the library. However, even when cloning a protein one has reason to believe is highly expressed, it is desirable to screen a significant number of independent clones.

Current research in cell biology is pointing to a strong role for isotype diversity in the organization and regulation of cellular function. If multiple protein isoforms are recognized by a single antibody it may be of interest to isolate the variant cDNA clones. Thus it is worthwhile both to screen a sufficiently large number of plaques in the first round and to carry as many potential positives as possible into subsequent rounds.

For most mammalian expression libraries, it is sufficient to screen six to twelve 150-mm plates with plaque densities in the range of 5×10^4 to 1×10^5 plaques per plate. Use plates that are as dry as possible in order to avoid cross-contamination of adjacent plaques. This can be achieved by pouring the plates a day in advance and/or incubating the plates at 42°C for several hours prior to plating out the phage. Another factor that may affect the results of the screening is the purity of the plating bacteria. Even strains with antibiotic resistance tend to become contaminated on storage. This may lead to the formation of colonies or other artifacts on the plates to be screened, causing increased or uneven background when the filters are developed. If the plates do become contaminated but a clear positive signal has been obtained, then pick the plaques and proceed to the next round. The routine chloroform treatment of the phage plugs should prevent the further spread of contamination.

In general, performing duplicate lifts to screen with the same antibody preparation is not necessary for the immunoselection of clones. In some circumstances it may be desirable to perform duplicate lifts in order to simultaneously screen with two different probes. For example, one might screen with two different polyclonal antibodies to the same polypeptide, or with a polyclonal and a monoclonal antibody. Alternatively, one might screen simultaneously with an antibody and a nucleotide probe, in those cases in which sequence information is available. Specific recommendations for performing duplicate lifts are included below.

1. Procedure for Plating the Library

1. Calculate the required dilution of phage, based on the library titer, that is required to yield 5×10^4 to 1×10^5 plaques/plate. For each plate, mix the required aliquot of phage with 0.6 ml of plating bacteria. Incubate the tubes for 15 min at 37°C. Then, one tube at a time, add 8 ml of warm top agar, mix gently, and pour onto the surface of a warm 150-mm LB plate.

2. Incubate the plates at 42°C for 4–5 hr, until the plaques just become visible.

3. Overlay the plates with nitrocellulose filters that have been saturated with IPTG to induce the synthesis of fusion proteins from the *lac* promoter regulating the expression of the cloned inserts. Layer each filter carefully onto the top agar of the plate, starting in the center, and gradually allow the edges to roll down to meet the agar. Try to avoid trapping any air bubbles between the filter and the

surface of the plate. Do not shift the filter once it has come in contact with the top agar.

4. Transfer the plates to 37°C and incubate for another 4–5 hr. The optimum length of time for growth and induction of fusion protein synthesis may vary greatly, but 4–5 hr is generally sufficient. If the resulting positive reactions are particularly weak then it may be worthwhile to extend the period of induction to an overnight incubation with the IPTG-saturated filters at 37°C.

When performing duplicate lifts, incubate the first set of filters for about 3.5 hr at 37°C, carefully remove them, and then layer on a second set of filters for an overnight incubation at 37°C.

5. Briefly chill the plates to 4°C in order to facilitate the peeling of the filter from the top agar without disrupting it. Before removing the filters from the plates the orientation must be marked to assist in aligning positive signals with plaques after antibody binding and visualization. Marking the alignment can be performed either with a large-bore (18-gauge) needle or with a smaller bore needle attached to a syringe filled with India ink. The needle should be used to make a unique and asymmetric series of markings around the periphery of the plate, stabbing straight down through the filter and plate agar. Then, using forceps, carefully pull the filter away from the plate, rinse briefly in TBST, and then submerge in blocking solution (TBST with 5% nonfat dry milk). Wrap the plates and store at 4°C until it is time to select the positive plaques.

2. Binding and Visualization of Antibody

Incubation of the filters in blocking solution can be done in almost any convenient container, for example, a plastic tray or a glass baking dish. It is desirable to keep all the filters face up during this step; that is, with the face that was in direct contact with the plaques up, and the numbered side down. The blocking step, and all wash steps, should be performed with moderate agitation, either on a rotary or lateral shaking platform. Keep the volume of serum solution required to a minimum by performing the antibody incubation step with the filters in sealed plastic bags [Seal-a-Meal plastic freezer bags (Kapak Corp., Bloomington, MN) or the equivalent]. The most convenient method of incubation is to put two filters in each bag, with their top sides facing away, toward the outside of the bag. However, if the available supply of antiserum is limited, then it may be necessary to stack up to eight filters in a bag. Prevent the top sides of each filter from coming in contact by keeping the tops of the filters oriented in the same direction. Heat seal the bag closely around the filters in order to further minimize the total volume of antibody solution required. Pipette the antibody solution directly into the bag through a small opening, squeeze out any trapped air bubbles, and seal the bag. As described above, the optimum time and temperature for antibody binding may vary, but the usual conditions are overnight incubation at 4°C.

Following incubation with the antibody, the solution should be carefully removed from the bag through a small slit and saved for use in additional rounds of screening. Then cut the bag open, remove the filters with forceps, and transfer them directly into wash solution. The filters should be washed three or four times in a relatively large volume of wash solution (TBST), with care taken to ensure that the filters are not sticking to one another.

The method of visualization of the antibody–antigen complex on the filter will depend on what may work best with the particular primary antibody. In our experience reaction of the filters with the appropriate alkaline phosphatase-conjugated secondary antibody, followed by visualization on reaction with nitroblue tetrazolium (NBT) and 5-bromo-4-chloro-3-indolylphosphate (BCIP), is the quickest and most reproducible method with satisfactory sensitivity. Use of a horseradish peroxidase-conjugated secondary antibody may not provide sufficient sensitivity for detection of the desired clones, unless coupled with an enhancement system. Alternative methods of detection, such as the use of an iodinated secondary antibody, may also be employed, and are described elsewhere (Sambrook et al., 1989; Harlow and Lane, 1988; Snyder et al., 1987).

Commercially available preparations of alkaline phosphatase-conjugated secondary antibody are of sufficiently high affinity that the filters need be incubated with the appropriate dilution for only 30 min (use the dilution for Western blots suggested by the manufacturer; we recommend the Promega alkaline phosphatase-conjugated secondary antibodies, used at a dilution of 1 : 7500, for consistent and reproducible results). The incubation with secondary antibody can be performed at room temperature, either in bags as above, or more simply in shallow trays. The filters should then be washed three to four times in wash solution, and then briefly rinsed in alkaline phosphatase reaction buffer.

Incubation of the filters in the visualization solution containing the dyes should be done carefully, in batches, so that the extent of development can be monitored. Development should proceed until clear signals are visible against a stained background. In this case, unlike usual Western blots, it is desirable to have background staining on the blotted filters. This background or nonspecific staining can often be used to correlate the alignment of the filter with the plate, and in later rounds of screening may help to identify with greater certainty the specific positive plaque. Also, allowing the colorimetric reaction to proceed to this extent may assist in distinguishing between true positive reactions and apparent false reactions, which may result from irregularities in the surface of the top agar or in the plated bacterial lawn. The reactions can be stopped by rinsing the filters several times in water.

3. Selection of Positive Clones

A "classic" positive plaque would appear as a small purple ring, or doughnut (e.g., see Fig. 1). Usually the center of the plaque, where the cells have been

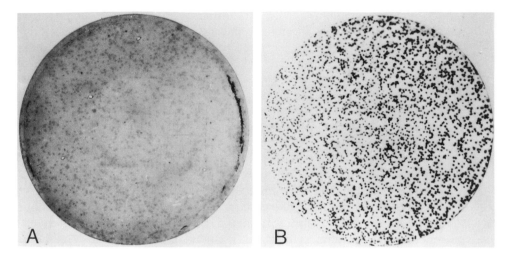

Fig. 1 Immunoselection of MAP 2 clones. (A) First round of screening of a rat brain cDNA library with a polyclonal antibody to MAP2. Note that apparent MAP2 clones are abundant in this library (i.e., more than 150 of ~40,000 plaques on this filter). (B) Filter from a later round of plaque purification of one of the clones shown in (a).

lysed for the longest time, show little or no reactivity; instead the antigen appears to be concentrated around the edge of the plaque to yield a deep purple ring. However, often in the first round of screening the plaques are so dense that they do not grow large enough to produce this type of effect. Often a true positive clone in the first round will have the appearance of a small purple dot. Further, different plaques may show a range of apparent positive reactions—some more intense and sharper, some lighter or more blurry. Such a range of reactions might be particularly evident when screening with a polyclonal antibody, which is likely to recognize several different epitopes with varying binding affinities. The number of true positive clones seen at this point may vary widely. When immunoscreening for proteins with abundant messages, it is possible that 1 of 500 plaques will be positive (Fig. 1). Screening for relatively rare cDNAs may yield only one or two potential positives from all the plates. Either way, it is worthwhile at this stage to select as many positive plaques as possible, so that the range of types of positive signals is represented. The extra work in carrying even >50 clones to the second round is relatively minimal. Isolating multiple clones may be especially important when immunoselecting for larger proteins with polyclonal sera because the various potential positive clones may represent different fragments of a single cDNA that encode distinct antibody epitopes.

If a single plaque on the plate could be aligned with the positive signals at this point, then further purification of the clone would be simplified. However, because of the usual high density at which the initial screening of the library is plated, this is not possible. Instead, after aligning the holes and/or ink on the filter with the corresponding marks on the plate, one takes a broader sampling. The wide end of a Pasteur pipette is punched straight down through the top and plate agar to yield a plug centered at the position of the observed signal on the filter lift. Such a plug may contain phage from 5 to perhaps 25 different plaques. However, this provides the greatest assurance that the plaque of interest will be selected. The plugs should be stored in individual sterile polypropylene tubes, either Eppendorfs or 15-ml conical centrifuge tubes. The phage are eluted from the plug by addition of 1 ml of SMG and incubation for several hours or overnight, preferably with gentle agitation. A drop (10 to 100 μl) of chloroform should then be added to each tube and mixed by inversion in order to kill the plating bacteria and prevent the carryover of contamination to the next round. The phage stocks can then be stored indefinitely at 4°C.

C. Subsequent Rounds of Screening

The goals of the second round of screening are to verify true positive reactions and to further purify the plaques of interest. Therefore the phage should be plated at a much lower density in order to achieve widely separated plaques. However, it is important to ensure that all clones in the plug are represented on the plate. An optimum plating density at this point would be about 50–100 plaques/100-mm plate. Unfortunately, the titer of the phage stocks isolated in the first round will vary depending on such factors as the length of time the plugs were eluted in SMG. A ballpark figure for a suggested dilution would be to plate 2–5 μl of a 10^4 dilution of the phage stock made from the plug, using SMG as a diluent. However, this dilution may have to be adjusted based on the observed titer of the actual stocks. It is worthwhile to achieve a reasonable density at this point, for if a well-spaced plaque can be identified as positive during this round it should be possible to avoid additional rounds of further purification.

The second and possibly third rounds of screening are performed as described above. If the same preparation of antibody is being used as a probe for these additional rounds, then include a control strip to ensure a continued satisfactory level of signal. Any clones that continue positive in this round should be selected more specifically from the plate; the narrow end of a sterile Pasteur pipette can be used to obtain an agar plug centered on the plaque of interest. Again, the phage from these plugs should be eluted into SMG and then treated with chloroform. Screening should be continued until in the final round every plaque on the plate yields a positive signal, thus ensuring that the phage stock is pure. It is worthwhile to ensure that the resulting phage stock is pure, because of the time and effort involved in subcloning the insert and in the further characterization of the cloned cDNA. However, as described below, the application of minigel electrophoresis and PCR methodology at this point may speed these steps significantly.

VI. Characterization of Cloned cDNAs

Subcloning and Restriction Mapping

The λ vectors usually used for library construction (Fig. 2) are not the optimal tools for the further analysis of clones of interest. Therefore the next step in most cloning protocols is to transfer, or subclone, the cDNA inserts of interest into more easily manipulated vectors. To do this, it is usually necessary to purify a sufficient quantity of the λ DNA, from which the insert can be excised.

1. Plate Lysate Method to Generate High-Titer Phage Stocks from λgt11 Clones

The first step in the preparation of λ DNA is to generate a high-titer phage stock, which can be used as an inoculum for cultures for a large-scale isolation of phage DNA.

1. Plate out the purified phage as described, but using top agarose, onto several 100- or 150-mm dishes at a high plaque density, in order to produce a confluent plate. Incubate the plates at 42°C for 6 hr to overnight, to allow phage growth and bacterial lysis.

2. Pipette several milliliters of SMG onto the surface of the top agar, using just enough volume to cover the plate. Gently swirl the SMG over the top of the plate for several hours or overnight on a rotating platform shaker, preferably at 4°C, to minimize contamination or bacterial growth.

3. Gently remove the SMG to a chloroform-resistant 15-ml conical tube, using care to avoid transferring any agar pieces. Rinse the plate with 2–3 ml of SMG, and add this to the centrifuge tube. If necessary, spin the tube briefly and decant the supernatant in order to remove any agar fragments.

4. Add 0.1 ml of chloroform and vortex, in order to kill any remaining bacteria. Recentrifuge at low speed, and transfer the supernatant to another tube. Add a drop or two (10–20 μl) of chloroform, and store the plate lysate at 4°C. The resulting phage stock should have a titer of 10^{10} to 10^{11} pfu/ml (Sambrook et al., 1989), but this should quickly be assayed by titering as described above in order to achieve the most effective multiplicity of infection for the following λ DNA prep.

2. Procedure for the Isolation of Phage DNA

The following procedure is an efficient and dependable version of the standard λ preparation (Sambrook et al., 1989; Vande Woude et al., 1979; Yamamoto et al., 1970).

1. Prepare an overnight culture of the host strain of bacteria in medium with 10 mM MgSO$_4$.

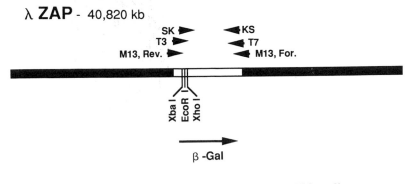

λ **ZAP** - 40,820 kb

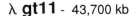

Reading frame at the EcoR I site: 5' . . . AGG **AAT TCG** . . . **3'**

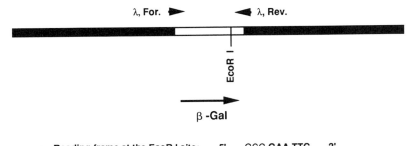

λ **gt11** - 43,700 kb

Reading frame at the EcoR I site: 5' . . . GCG **GAA TTC** . . . **3'**

Fig. 2 Restriction maps for the λ expression vectors λgt11 and λZAP, indicating the site of cDNA insertion, convenient restriction mapping sites, the relative size of the β-galactosidase contribution to fusion proteins, and the location of useful priming sites for PCR amplification of inserts.

2. In a sterile 4-liter flask (an oversized flask is recommended to allow vigorous shaking of the culture), mix 20 ml of the overnight culture with approximately 5×10^8 phage, and incubate for 5 min at room temperature.

3. Add 1 liter of prewarmed (37°C) medium with 10 mM MgSO$_4$ to the flask, and shake vigorously for 5 to 8 hr. Monitor the culture for both the initial growth of the bacteria (the culture will increase in turbidity) and then the lysis of the bacteria by the phage (the culture will appear to clear, with lysed bacteria settling to the bottom of the flask).

4. When lysis is apparent, add about 10 ml of chloroform to the culture and shake at 37°C for 5–10 min more.

5. Add sufficient solid NaCl to make the culture 0.5 M in salt, and stir to dissolve. Pellet out the debris by centrifugation at 4°C for 10 min at 6000 g, and decant the supernatant into a clean centrifuge bottle.

6. Add powdered polyethylene glycol (PEG 6000) to 10% (w/v), and stir the solution until the PEG dissolves. Incubate the centrifuge tube in an ice–water bath for 1 hr in order to precipitate the phage.

7. Pellet the phage by centrifugation at 11,000 g for 10 min at 4°C. Discard the supernatant, and remove any remaining liquid with a pipette. Then resuspend the phage pellet in 5 to 10 ml of TM buffer. Extract the phage with an equal volume of chloroform by vortexing and then centrifuge to clarify the phases.

8. Recover the aqueous supernatant, and layer onto a glycerol step gradient. A step gradient can easily be prepared in an ultracentrifuge tube [for a Beckman (Fullerton, CA) SW41 rotor, or the equivalent] by first adding 3 ml of 40% glycerol to the bottom of the tube. Then carefully overlayer this solution with 3 ml of 5% glycerol. Then layer the phage solution over the top. Gently fill the remainder of the tube with TM buffer. Centrifuge the gradient for 1 hr at 35,000 rpm and 4°C.

9. Remove the supernatant and resuspend the phage pellet in 1 ml of TM buffer with 10 μg of DNase I/ml and 10 μg of RNase A/ml, then digest at 37°C for 30 min.

10. Extract by adding an equal volume of phenol, vortex, and centrifuge. Reextract with phenol–chloroform, and then again with chloroform.

11. Precipitate the DNA by adding 2 vol of ethanol. If possible, remove the threadlike DNA from the tube by adhering it to the side of a glass Pasteur pipette (Sambrook et al., 1989). Then rinse the DNA in 70% ethanol, dry it, resuspend in sterile water or TE, and measure the OD_{260} to determine the concentration.

Alternatively, there are methods of λ DNA purification that are based on immunoprecipitation of the bacteriophage particles, for example the Lambda-Sorb method (Promega). These methods are rapid, but in our hands have resulted in unreliably low yields.

3. *In Vivo* Excision of cDNA Inserts

Many new cDNA libraries are constructed in vectors such as λZAP (Uni-ZAP from Stratagene; see Fig. 2) that are specifically designed to allow excision of the cDNA insert followed by recircularization into a more convenient plasmid vector in an *in vivo* reaction. In the case of λZAP, a straightforward protocol allows the "subcloning" of the cDNA insert into the plasmid pBluescript within a few hours (Short et al., 1988; or see Stratagene technical literature).

4. Procedure for the Isolation of cDNA Inserts by Polymerase Chain Reaction

Alternatively, it is also possible to isolate the cDNA insert of interest for subcloning by using the polymerase chain reaction (PCR) to amplify DNA from a single plaque. This technique requires the use of primers specific for common λ phage vector sequences on either side of the insert in order to specifically amplify the variable sequences (Fig. 2). The advantage of this method is that the PCR reaction can be used to subclone and restriction map λ clones before they have been purified to homogeneity. All that is required is the selection of a single, unambiguously positive plaque for each putative cDNA clone, by careful alignment of the plaque on the plate with its corresponding immunoreactive signal on filter lift (Fig. 3). The source of the DNA can be either a single plaque core or a spot excised from the plaque filter lift (the spots generally give cleaner results).

1. Add the core or spot to 0.2 ml of sterile water. Boil for 2 min, cool on ice, and spin briefly in a microfuge.

2. Mix 25 μl of the resulting supernatant with 2 μl of forward primer (λgt11 forward or M13 forward, at 1 pmol/μl; available from Promega or Stratagene), 2 μl of reverse primer (λgt11 reverse or M13 reverse, at 1 pmol/μl; available from Promega or Stratagene), 10 μl of 10× PCR buffer, 1 μl of a 10 mM stock of all four dNTPs, 9.8 μl of sterile water, and 0.2 μl (1 unit) of *Taq* polymerase.

3. The recommended cycle profile is as follows: one cycle at 95°C for 1 min; then 40 cycles at 94°C for 1 min, 53°C for 30 sec, and 72°C for 3 min; then a final cooling to 4°C.

4. The resulting PCR products can be resolved on a 1% agarose gel (Fig. 3) (see Sambrook *et al.*, 1989, for general methods).

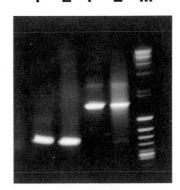

Fig. 3 PCR amplification of cDNA inserts from each of two different single plaques. DNA was amplified from either (P) a phage plug, or (L) a single immunoreactive spot on a plaque lift filter, with λgt11 forward and reverse primers. The reaction products were analyzed on a 1% agarose gel stained with ethidium bromide. (M), Molecular weight markers (Pharmacia, Piscataway, NJ).

5. Subcloning and Restriction Mapping

Once a sufficient quantity of λ DNA has been prepared, the insert can then be subcloned into a more convenient plasmid vector. Most cDNA libraries in λgt11 or λZAP are usually created by insertion of the cDNAs at the *Eco*RI site of the phage polylinker; however, this should be verified for the library being screened. Therefore, the cDNA inserts are usually removed by digestion with *Eco*RI, followed by purification of the fragment by agarose gel electrophoresis (Sambrook *et al.,* 1989).

It is possible, however, that the clone of interest may contain internal *Eco*RI sites. In this case it will be necessary to subclone each of the *Eco*RI-generated fragments, and then determine their relative orientations by restriction mapping of the original phage constructs. A second possibility is that the λ clone will have lost one of its *Eco*RI sites. Comparisons of the restriction map with the PCR products should be particularly helpful in these cases. In both these situations, pairs of other enzymes in the polylinker can be tested to determine if they release the entire insert in a single fragment from the phage DNA.

The choice of a plasmid vector for subcloning should be based on its utility in the further characterization of the insert by restriction mapping, sequencing, and expression. We have found the phagemid vector pBluescript (Stratagene) to be particularly well suited for this role, as it allows color selection of insert-containing clones (blue/white screening), double-strand sequencing from an array of commercially available primers (M13 forward and reverse primers, T3 and T7 primers, SK and KS primers; available from Stratagene), expression of β-galactosidase–fusion protein from the IPTG-inducible *lac* promoter, and *in vitro* transcription of RNA from either the T3 or T7 promoters.

Alternatively, some researchers prefer to perform single-stranded DNA sequencing. In this case the cDNA inserts of interest and subsequent subclones for sequencing are constructed in the M13 series of vectors. However, we have found that if the double-stranded DNA preparations are sufficiently pure then there is no decrease in the quality and quantity of information generated from the direct sequencing of double-stranded plasmid DNA, thus making plasmid vectors a more useful choice.

The first step in characterizing potentially interesting cDNA clones is to quickly generate a restriction map. Determine the overall size of the insert, as well as the size of fragments obtained on digestion with several restriction enzymes. Usually it is most convenient to start with the enzymes in the polylinker of the vector in order to simplify the analyses of the restriction fragments generated. The resulting restriction map should be compared with any initial data on the characterization of the clone by PCR or Western blotting.

It is becoming increasingly valuable to obtain some sequence information from potentially interesting clones quickly. The size of the database for sequence information is expanding rapidly, so the probability is increasing that some relationship between the sequence obtained and a previously characterized cDNA clone will be found. Database searches are described in more detail below (Section X,B); however, it is mentioned here in order to emphasize that at this point a scan of the database with even limited sequence information may prevent the further loss of time on an artifactual clone or on an already characterized nucleotide sequence.

Once the insert is subcloned it is easy to do a plasmid DNA minipreparation by standard methods (Sambrook *et al.*, 1989). For double-stranded DNA sequencing (Chen and Seeburg, 1985), it is important to purify the plasmid DNA further, as described by Kraft *et al.* (1988), or by using GeneClean (Bio 101, La Jolla, CA) or the equivalent. By using a purified preparation of the subclone and standard primers, it is possible to quickly sequence 200–300 bases from each end. Alternatively, PCR-generated products can be sequenced directly (Gibbs *et al.*, 1989; or see Gibbons *et al.*, 1991a). The reading frame at the 5′ end of the clone can then be determined from the frame of the β-galactosidase portion of the clone (Fig. 2).

VII. Cloned cDNAs as Hybridization Probes

Once the size of the cDNA insert in the λ clone has been determined, it is possible to estimate whether the clone is sufficiently large to encode the polypeptide of interest. If the clone was isolated from an oligo(dT)-primed cDNA library, then it should be assumed that at least some of the 3′ end of the insert encodes the 3′ untranslated region along with a poly(A) tail. If the clone was isolated from a random primed cDNA library, the average insert size of the clones may be somewhat low, and it may be more likely that only a partial clone, which may be missing either or both of the 5′ and 3′ ends of the full-length cDNA, has been isolated. In addition, selecting for in-frame fusion proteins by the immunoselection method increases the likelihood that the cDNA clones isolated will lack the 5′ untranslated region as well as the initiator methionine found in the full-length mRNA transcript of the species of interest. Therefore it is often necessary to rescreen the library in order to isolate a full-length clone. This screening can be performed in two ways, either by hybridization screening or by PCR.

A. Library Screening by Hybridization

The traditional method of rescreening a library is to label the cloned cDNA insert, usually with ^{32}P, and then to probe the library by hybridization. This allows the selection of clones by sequence, rather than by immunoreactivity, so that overlapping clones that extend the sequence can be identified. The method is straightforward, and has a high probability of success. However, if a large polypeptide (>200 kDa) is being cloned, it may be necessary to perform the second screen on a more specialized library, either a size-selected cDNA library, in which only messages larger than a certain rough cut-off size are used to construct the library, or a random-primed cDNA library, in which it is more likely that overlapping clones encoding the entire polypeptide of interest will be represented.

1. Procedure for Preparation of Probe

The cDNA insert should be isolated by restriction enzyme digestion or by PCR so that no vector sequences are included. The insert can be purified by agarose gel electrophoresis according to standard methods (Sambrook *et al.,* 1989). Following excision of the band of interest from the agarose gel, one purification method that we have found to be quick and reliable involves the use of the GeneClean kit (Bio 101), in which the agarose gel slice is dissolved in NaI and the DNA is then isolated by binding to a silica matrix. The resulting DNA is sufficiently pure for probe preparation, sequencing, or subcloning. However, several other, perhaps more economical, methods have been developed for probe purification from agarose gels, including freeze-thaw, electroelution, and electrophoresis onto DEAE membranes (Sambrook *et al.,* 1989; Ausubel *et al.,* 1989). The isolated DNA can then be labeled with ^{32}P, again by several different methods. Nick translation [see Sambrook *et al.,* 1989; or use a commercial kit such as that offered by Bethesda Research Laboratories (Gaithersburg, MD)] is reliable, and produces a probe of sufficient specific activity for most library screening and Northern blotting applications. However, for detecting messages of lower abundance, it is sometimes necessary to label probes to a higher specific activity by random-primed labeling (kits are available from Stratagene, Promega, etc.; follow manufacturer's protocols).

2. Procedure for Hybridization Screening

When screening with a radioactive probe it is recommended that duplicate lifts be performed, for artifacts such as irregularities in the top agar, or bubbles, can often give rise to apparent false positives.

1. Plate out the library as described (Section V) with 50,000 pfu/150-mm plate. Incubate the plates at 42°C for 6 to 8 hr or at 37°C overnight. Then chill the plates briefly before beginning the filter lifts.

2. Perform the first filter lift as described above, but take care to ensure that the key marks made on the plate are clearly visible. Then remove the first filter carefully, so as not to disturb the top agar. Layer on the second filter, and mark it to match the marks on the plate. Because the plate will be drier for the second lift, leave the second filter on the plate for about 10 min.

3. After peeling each set of filters from the plates, the DNA must be denatured and then neutralized. First set up two Pyrex baking dishes with two layers

of blotting paper covering the bottoms. Saturate the blotting paper in the first dish with a solution of 0.2 N NaOH and 1.5 M NaCl and in the second dish with a solution of 1 M Tris-HCl (pH 7.4) and 1.5 M NaCl, then pour off the excess. The blotting paper in each tray should be resaturated with the appropriate solution after every 5 to 10 filters. The plaque lifts should be layered carefully onto the surface of the NaOH–NaCl-saturated blotting paper. Avoid trapping any bubbles beneath the filter, which will impede fluid exchange, and submerging the filters. The face of the filter that was in contact with the plate should remain face up throughout this procedure.

4. After denaturing the DNA for 5 min, use forceps to transfer the filters to the neutralization tray. First blot off any excess fluid from the plaque lift onto the side of the baking dish, then layer the filter, DNA side up, onto the blotting paper that is saturated with 1 M Tris-HCl (pH 7.4)–1.5 M NaCl. After 1 min, transfer the filter to dry blotting paper.

5. After the plaque lifts have air dried, layer them between paper towels and bake for 2 hr in a vacuum oven at 70°C.

6. The filters can now be prehybridized, in order to prevent nonspecific binding, and then incubated overnight with the labeled probe. If using nylon membranes, follow the suggestions of the manufacturer for prehybridization/hybridization solutions and conditions to prevent high levels of background. We have found the following conditions to work well with nitrocellulose filters.

Place the filter lifts in sealed plastic bags as described above for antibody incubations. Add prehybridization solution [50% formamide, 5× SSC, 5× Denhardt's reagent, 0.1% SDS, and salmon sperm or calf thymus DNA (20 μg/ml)] to about 5 ml/filter, seal the bags, and incubate in trays in a water bath at 42°C for two or more hours. The optimum temperature for the incubation is a function of the affinity of the probe for its target. Specifically, it will depend on the length of the probe, the extent of overlap with the target clones, the GC content of the probe, and the ionic strength of the hybridization solution. In practical terms, most hybridizations can be preformed at 42°C in the presence of 50% formamide in the hybridization solution described below.

7. Label the insert from the incomplete cDNA clone by nick translation or random priming. Double-stranded DNA probes must be denatured by heating to 100°C for 5 min and then cooling rapidly in ice water immediately prior to incubation with the filters. Before adding the probe, replace the prehybridization solution in the bag, using the minimum volume required. The probe should be added to the hybridization solution in the bags to a final specific activity of about 10^6 cpm/ml.

8. After overnight incubation of the filters with the probe, the filters should be quickly rinsed in 2× SSC with 0.1% SDS and 0.1% sodium pyrophosphate, then washed twice more for 20 min each time with this solution. Then rinse twice for 20 min in 1× SSC with 0.1% SDS and 0.1% sodium pyrophosphate, and then

wash once for 20 min in 0.3× SSC with 0.1% SDS and 0.1% sodium pyrophosphate.

9. Blot excess moisture from the filters and seal them between two sheets of plastic wrap to prevent the filters from drying out. The sealed filters should then be arranged in an autoradiography cassette adjacent to a piece of X-ray film. To speed the exposure time required to visualize the signal, intensifying screens can be included in the cassette, which should then be kept at − 80°C.

10. Develop the film and align it with the plaque lifts and the key marks on the plates. True positive clones should show duplicate signals on the film from each of the two plaque lifts from the plate. In the first round of screening, isolate any potential positives by sampling the agar with the wide end of a Pasteur pipette. In later rounds it should be possible to accurately align a positive signal on the film with a single plaque on the plate.

Once the positive plaques have been purified to homogeneity, a Southern blot (Southern, 1975; Sambrook *et al.*, 1989) can be performed on the purified λ DNA in order to verify the hybridization with the probe. Comparisons of the restriction maps of the new clones with the initial immunoselected clone should indicate if any additional cDNA sequence has been obtained. These observations can then be verified by sequencing the new clones at both ends, and determining if any overlapping sequence or new sequence is observed.

B. Library Screening by the Polymerase Chain Reaction

The second method of screening is based on the development of PCR technology. Specific primers can be designed from the clone of interest so that hybridization of these probes to the library in conjunction with a common λ DNA primer will amplify sequences that overlap and extend the clone of interest. The resulting amplified DNA fragments can be sequenced directly from the products of the PCR reaction, as described above. If the sequences are determined to overlap with the clone of interest, then they can be subcloned for further characterization. The application of such methods to the cloning of a large cDNA (14.9 kb; Ogawa, 1991; Gibbons *et al.*, 1991b) encoding the β heavy chain of sea urchin dynein has been described in detail (Gibbons *et al.*, 1991a).

C. Northern Blot Analyses

If an isolated clone appears to be promising, then it is worthwhile to perform a Northern blot with the labeled cDNA insert as a probe. This experiment will reveal the size of the message of interest, which will provide further information on whether the full-length cDNA is in hand. Further, by probing mRNA isolated from specific tissues or stages of development one can determine if the clone shows the expected distribution and expression patterns for the polypeptide of interest. More complete blotting studies should perhaps be postponed until the clone is confirmed in order to avoid wasting too much time on potential artifacts.

VIII. Fusion Protein Analysis

One aspect of immunoselection methodology that is often overlooked is the extremely informative nature of fusion proteins. There are several characteristics of fusion proteins that can be of great value in ascertaining the identities of immunopositive clones and in determining the directions and interpretation of other parts of a project. Some uses are as follows: comparison of different fusion proteins, confirmation of clones by comparison of fusion proteins with the natural protein, use of fusion protein to generate antibodies, and use of the fusion protein as a surrogate for the natural protein in mapping of structural and functional sites along the primary sequence.

A. Procedure for Affinity Purification of Antibody from Plaque Lifts

This technique is less informative than affinity purification of antibody from fusion protein bands, but it is simpler, especially when a large number of clones is involved. The method is adapted from Snyder *et al.* (1987).

Each clone is plated at a density of 10^5 pfu/100-nm plate and induced with IPTG, as described in Section V, overnight. Filters are blocked and cut into pieces of ~5 × 10 mm (one piece of each should be saved as a positive control for purified anti-plaque lift antibody); the pieces are then processed as antigen strips as described in Section III above. Following elution of the "anti-plaque lift" antibody, it is used without dilution to challenge a control strip and the piece that was saved. If a positive reaction is seen on the control strip, the result must nevertheless be verified with anti-fusion protein antibody (see Sec. VIII,D).

B. SDS-PAGE and Western Blot Analysis

At the stage in plaque purification when several well-separated immunoreactive plaques are seen on each plate (possibly as early as the second round), plaque proteins should be subjected to SDS-PAGE and Western blot analysis. This will reveal the molecular weights of the immunoreactive polypeptides, their state of degradation, relative levels of expression, and in some cases orientation of the cDNA insert. A Western blot is also a necessary precursor for the "reverse affinity purification" of antibody from the fusion protein.

Procedure

From each clone of interest, pick a plaque for purification, as described in Section III above, from a plate with several well-spaced immunoreactive plaques. Then pick three plaques (a total of ~10 μl) as cores, and add them to a 1.5-ml microfuge tube containing 10 μl of 2× plaque sample buffer. The samples are boiled for 3 min, with agitation after 1.5 min, to mix the contents thoroughly. After the boiling step, the samples are quickly centrifuged and cooled to 50°C, at which temperature they will remain molten. The warm samples are then loaded rapidly, one at a time, with disposable pipette tips, into dry (i.e., empty) wells of a 7% SDS-PAGE minigel. A "positive control" sample (such as β-galactosidase) and a "negative control" from any clone that does not produce immunoreactive protein should be included. After the samples have solidified (~3 min at room temperature) the apparatus is filled with running buffer, molecular weight standards are loaded, and the gel is run as usual.

The gel is then electroblotted onto a PVDF membrane under standard blotting conditions, and the membrane is stained with Coomassie Brilliant Blue, soaked in Coomassie Brilliant Blue destain for ~10 min, and photographed, then destained to completion in methanol, soaked in H$_2$O for 10 min, and probed with antibody. A duplicate blot should be probed with anti-β-galactosidase (or antibody to other fused peptide), in order to facilitate the interpretation of the blot (Fig. 4).

If the fusion protein can be identified, it should be used for "reverse affinity purification" of antibody as soon as 50 to 100 immunopositive plaques are available. These are run on a preparative gel, blotted, and the fusion protein is treated as described in Section III above.

The antibody purified from fusion protein is then used, undiluted, to challenge Western blots of the same fusion protein, the target protein, and any other fusion proteins selected with the original antibody.

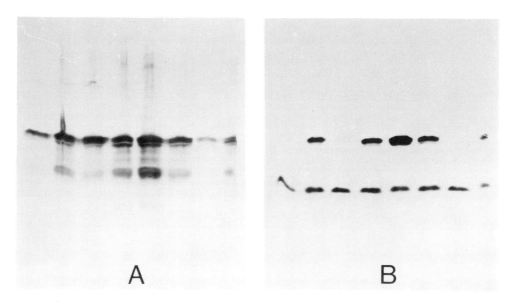

Fig. 4 Western blots of fusion proteins expressed by phage-infected *Escherichia coli*. Agar plugs of immunoreactive plaques were isolated as described, and the plaque proteins were resolved by 7% SDS-PAGE. (A) Blot probed with anti-β-galactosidase (Promega). (B) Blot challenged with the screening antibody (anti-MAP2; Theurkauf and Vallee, 1982).

C. Interpretation

Five major questions can be answered by Western blot analysis of fusion proteins: (1) What is the polypeptide that led to the selection of each clone, and how large is it? (2) Is the polypeptide expressed as a fusion construct (e.g., with β-galactosidase) or as an independent open reading frame (ORF)? (3) How abundant is the polypeptide (as a proportion of total plaque protein)? (4) Is the polypeptide degraded *in situ*, and if so, to what extent? (5) Does the polypeptide show the antigenic characteristics of the target protein it is supposed to represent?

We strongly advise the careful analysis of the sizes and immunoreactivity patterns of fusion proteins and their fragments, and the construction of maps based on these data (Mehra *et al.*, 1986; Kosik *et al.*, 1988). Ideally, all fusion proteins produced by a single vector will align at one end, but this must be empirically determined in each case. It is important to note that cryptic, bacterially active promoters may initiate transcription within a eukaryotic ORF (Snyder *et al.*, 1987), and in either the "sense" or "antisense" orientation within an engineered phage transcription unit. The ability to detect such atypical gene products is limited only by the quality of the antibody used, and even a poorly expressed polypeptide can be easily cloned by immunoselection.

In most cases the fusion protein produced by λ gt11 clone consists of two segments: an amino-terminal segment of 1008 codons of the *lacZ* gene (encoding 113 kDa of β-galactosidase protein) joined to a caraboxyl-terminal segment of cDNA. These segments are joined at the single *Eco*RI site (GAATTC) of *lacZ*, which is in the reading frame /GAA/TTC— (encoding /Glu/Phe/).

The first polypeptide to probe for on an immunoblot of a fusion protein expressed by a λ gt11 clone is a polypeptide of >114 kDa that is recognized by anti-β-galactosidase antibody. The β-galactosidase portion of the fusion protein is generally quite stable, so if more than one band is seen, the carboxyl-terminal end of the fusion protein is most likely becoming degraded *in situ*. The highest immunoreactive band should represent the full-length fusion protein, and its molecular weight should be measured accurately, with appropriate markers.

The next step is to challenge an identical blot with antibody to the target protein. The highest band recognized by both antibodies should be the same one, but the patterns of other (fragment) bands may differ. Keep in mind that all β-galactosidase epitopes lie upstream of all target protein epitopes; this will allow the construction of a linear map of each clone, based on the two antibody patterns. Frequently the anti-target protein antibody will recognize fewer bands. Each 10 kDa of fusion protein size above 114 kDa corresponds to ~90 codons or 270 bp of cDNA insert. Therefore, if the highest molecular weight polypeptide recognized by both antibodies is 140 kDa, the cDNA must contain at least 700 bp of open reading frame sequence. If monoclonal antibodies to the target protein exist, this is a stage at which they should be used to aid in verifying the identity of a clone, as well as for mapping of their epitopes on the fusion proteins (Mehra *et al.*, 1986; Dingus *et al.*, 1991).

D. Production of Anti-fusion Protein Antibodies

Fusion proteins encoded by cloned cDNAs are regarded as good immunogens for three reasons: (1) it is relatively easy to produce them in large quantities; (2) they can be produced in the absence of other eukaryotic proteins; (3) they are precisely defined polypeptides (their amino acid sequences are known, with posttranslational modifications being rare). Over 90% of *E. coli* proteins are <120,000 in molecular weight; therefore, because the fusion proteins expressed by λ gt11 clones usually include 114 kDa of β-galactosidase, they can be relatively easy to purify by sizing techniques such as SDS-PAGE. If the level of expression of fusion protein is high (i.e., at least 0.5 μg/plaque), and the fusion protein is stable, this should be used as immunogen, following electrophoresis.

The cDNA insert of the clone can also be excised (from purified phage DNA, subcloned plasmid DNA, or PCR product) and subcloned into a suitable, inducible expression vector, such as the pET, pT7, pEX, or pGEX vectors (reviewed in Ausubel *et al.,* 1989).

We have found that pET series to be useful for expression of cDNAs: a tightly regulated T7 promoter drives synthesis of an 11- to 14-amino acid portion of the T7 gene *10* protein, fused to the amino-terminal end of a cDNA-encoded polypeptide. Expression levels are usually quite good, and the series contains vectors with *Bam*HI and *Eco*RI sites in each of the three reading frames. Although the fusion proteins do not have the large sizes of β-galactosidase fusions, they can be purified with antibodies to the T7 gene *10* portion of the chimera, which are commercially available. The use of the T7 promoter makes this system amenable to the expression of sequences that were unstable or lethal when expressed as λ phage clones.

IX. Confirmation of cDNA Clones

A. General Considerations

The first step in establishing the validity of a clone selected with a polyclonal antibody is to challenge the target protein with antibody obtained by "reverse affinity purification" against the fusion protein. This can be carried out on crude plaque lifts (as described above), but for obvious reasons can be carried out earlier and better with single (purified) bands from an electroblot of plaques, isolated as described above. Plaques (up to 150/minigel) are run on a preparative gel and electroblotted, and the appropriate band is identified by staining of a guide strip from the edge of the membrane. The fusion protein band is excised and processed for affinity purification as described in Section III above. If only a few plaques are available, but the antibody is effective, the Olmsted (1986), Earnshaw and Rothfield (1985), or Hammarback and Vallee (1990) methods may also be used.

Often, more than half of the clones picked in the first round are spurious or artifactual clones, which encode fusion proteins that merely share an epitope with the target protein. These clones usually (but not always) have a short ORF and a high level of expression. They can derive from true cDNAs in a natural or unnatural reading frame or orientation, from coding or noncoding sequence, and (particularly in *Drosophila* cDNA libraries) from introns. The DNA sequences of clones with short ORFs can be informative for epitope mapping (see, e.g., Dingus *et al.*, 1991), but they are rarely worth the effort of extended characterization and confirmation.

The most unambiguous means of confirmation of the identity of a clone is to demonstrate close correspondence of the amino acid sequence of the protein with the sequence predicted by a cDNA. This is sometimes quite difficult, but acquisition of a substantial amount of sequence data is highly recommended, regardless of the size or abundance of the target protein. Even when a clone is shown to be valid, the determination of the correct reading frame over the entire sequence, or the subunit identity, often becomes controversial without corroborating amino acid sequence data (see, e.g., Mayer, 1991; Kirsch *et al.*, 1990).

B. Direct Microsequencing

This technique has achieved widespread use since the advent of miniature gels and electroblot apparatus, and efficient methods for solid-phase microsequencing from membranes. These are reviewed in LeGendre and Matsudaira (1988).

The role of microsequencing in an immunoselection project is to provide extended amino acid sequences from distinct regions of the project polypeptide. Even gaps (of known length) in an amino acid sequence have information content, and a single sequence, from the right part of the molecule, may be all that is needed. If possible, try to obtain at least three nonoverlapping amino acid sequences; the matching of primary sequences is the most important criterion for verification of cDNAs.

If the target protein is to be excised from an immunoblotted replica of an SDS-PAGE gel, it need only be purified to the extent that it is of >95% purity with regard to other polypeptides of similar molecular weight. This means, for example, that an actin-binding protein of an unusual molecular weight (e.g., 250,000) may be considered pure enough if there are no other major polypeptides between 210,000 and 280,000 in a convenient preparation. It is imperative to have at least 2 nmol of target protein at this stage (only 20–100 pmol of a fragment will be needed for sequencing), so that trial and preparative experiments can be carried out on the same batch of material.

If the target protein is in abundant supply and the cost of sequencing is not a factor, the intact protein should be electrophoresed, blotted, and subjected to sequencing directly. Approximately 80% of all cytoplasmic eukaryotic proteins are modified *in vivo* at their amino termini, which "blocks" them to Edman degradation. If there is some clue that the protein is a member of a group (such as the tubulins) that is not amino-terminally modified, then direct sequencing may be feasible and cost effective.

If the target protein is not available in purified form in solution, the best strategy is to purify the protein by preparative SDS-PAGE, then subject it to cleavage. Generally this is done by excising the appropriate bands from two identical minigels, then using one for optimization of cleavage conditions and the other for fragment production for sequencing. Several different cleavage methods should be tried to determine the best means of preparing useful levels of fragments for a given protein.

1. Preparative SDS-PAGE of Target Protein

The processing of polypeptides for sequencing requires thoughtful handling; the procedures for routine electrophoresis (Laemmli, 1970; Weber *et al.*, 1972) may be used with the following modifications.

Only fresh, (0.4-μm pore size) filtered reagents should be used. The sample(s) should be heated for 5 min at 75°C rather than boiled. Stabilization of cysteine residues by carboxymethyl or vinylpyridyl moieties prior to microsequencing is essential for complete sequence analysis of a polypeptide, but it can result in significant loss of material and can be omitted if the goal is simply to obtain useable sequence data. The lowest percentage polyacrylamide that will effect resolution of the band of interest should be used, with the shortest minigel system available. If possible, preparative gels should be run at 4°C, and the running buffer should be made 10 mM in sodium thioglycolate (a free-radical scavenger) prior to filtration.

Following electrophoresis, the gel should be stained as briefly as possible with Coomassie Brilliant Blue stain (this solution contains acetic acid, and its exposure to the gel must be minimized). The gel is then destained in methanol (40% in H$_2$O), and as soon as the band of interest is discernible it should be excised with a fresh, sterilized razor blade. The band should then be cut into slices that will fit conveniently into the wells of a subsequent stacking gel, then stored frozen or equilibrated for chemical or enzymatic cleavage.

a. Chemical Cleavage

The following methods are adapted from Goding (1986). Cyanogen bromide cleaves peptides on the carboxyl-terminal side of methionine residues, and acetic acid cleaves asparagine–proline bonds; both methods usually produce fragments large enough to resolve by SDS-PAGE with a gel of 10 to 15% acrylamide.

CNBr: A gel slice is equilibrated in 5 ml of 70% formic acid, then 340 μl of cyanogen bromide solution (i.e., a molar excess of ~40-fold over methionines) is added, and cleavage proceeds at room temperature overnight, in the dark. The gel slice is then equilibrated with several changes of water, then with SDS-PAGE sample buffer, until the pH is >6.

Acid cleavage: A gel slice is equilibrated in a 1.5-ml polypropylene tube with a minimal volume of 10% acetic acid, then boiled for 10 min. The slice is cooled, and then equilibrated with SDS-PAGE sample buffer as above.

Note: The liquid phases following chemical cleavage may contain small cleavage products that have eluted from the gel. These can be recovered by lyophilization or similar methods.

b. Enzymatic Cleavage

The following method is adapted from Gooderham (1984) and Cleveland *et al.,* (1977), and works well with chymotrypsin, V8 protease, "Lys-C" and papain.

Gel slices are equilibrated in a small volume of 1× gel slice equilibration buffer, then each slice is placed in a well of a 15-well minigel, and overlaid with 15 μl of overlay buffer. A 10-μl aliquot of protease in protease buffer (in approximately a 1:10 to 1:500 molar ratio to the substrate polypeptide) is applied gently. The gel is run slowly at room temperature, and to facilitate cleavage the power can be turned off for 15–30 min, at the time when the protease layer has penetrated the sample. This point is easy to observe visually, because the pyronin Y dye in the protease solution migrates faster than the bromphenol blue dye in the sample, and the two dyes usually converge near the bottom of the stacking gel when the enzyme catches up to the substrate. The gel is stained (for analytical digests) or electroblotted (for preparative digests).

2. Procedure for Electroblotting of Fragments for Microsequencing

Following electrophoresis, the gel is electroblotted onto a preequilibrated PVDF membrane, in either standard electroblotting buffer, or basic electroblotting buffer. In a Bio-Rad (Richmond, CA) Transblot apparatus, 70 min at 100 V works well for either buffer, but this step can be optimized with the use of prestained molecular weight markers (Diversified Biotech).

Following electroblotting, the PVDF membrane is stained briefly on Coomassie Brilliant Blue stain, and then in destain solution. When bands are visible, the membrane is photographed and dried. Bands to be subjected to sequencing are excised and washed 7–10 times with water in a microfilterfuge device (Millipore), then stored frozen.

C. Evaluating Peptides

It is frequently the case that several fragments of quality and quantity amenable to microsequencing are generated in a single experiment, and they must then be placed in order of priority. The ideal fragment would be of ~10 kDa (a small size facilitates sequencing), derived from the carboxyl terminus of the protein (which is most easily cloned, as discussed above), and reactive with one or more antibodies (for correspondence to clones selected with these antibodies). If the amino-terminal sequence contains methionines, cysteines, or tryptophans it can be back-translated into minimally degenerate oligonucleotides, should the antibody approach fail. A long stretch of sequence from a single peptide is preferable to short sequences from several peptides, and may be useful for PCR-based cloning. Any sequence information is useful, however, and validation of clones is possible with gapped or short peptide sequences.

As each sequence is obtained, it should be used. First, probe a database for related sequences. The result may indicate, for example, whether the target protein is a member of a conserved superfamily, or if the 35-kDa polypeptide one has been fervently attempting to clone is in fact a well-known component of the Krebs cycle.

Amino-terminal sequences from different peptides are extremely useful to have in hand when a clone is validated, as they can be used to define the reading frame at multiple locations within the cDNA and to prevent errors; they are also an important assay for the identity of a single gene product within a family. In addition, confirmation of sequences by analysis of different-sized peptides with identical or overlapping amino termini will add to the value of the (rather costly) sequence data.

D. Alternative Methods for Verifying Clone Identity

Unfortunately, antibody reactivity is not sufficient proof that a cDNA encodes the target protein, and after candidate cDNA clones have been isolated it is always necessary to "verify" them. It is important to note that verification is achieved only by attempting to falsify the validity of a clone, and in failing to do so. Many investigators make the easy mistake of trying to "prove" a clone is what they hope it is, and fail to do the experiments that would disprove their hypothesis (or presumption).

If a fusion protein was obtained by immunoselection, results obtained with an anti-fusion protein antibody can at best confirm the presence of common epitopes. Independent confirmation can never be obtained in this way, and the generation of antibodies against an unvalidated fusion protein should be considered a last resort: the time involved (i.e., up to several months) would be much better spent developing other assays for determining the identity of the project protein. However, if it is deemed necessary to raise such an antibody, the mouse is the system of choice. The limited quantities of serum available from mice should suffice for testing of cross-reactivity, and if the identity of the fusion protein is validated, monoclonal antibodies can be prepared.

After primary sequences, the best criteria for establishing the identity of a clone are those biochemical characteristics used in establishing the identity of the target protein itself. In some assays, the recombinant protein will show both similar characteristics to and competition with the natural protein. Blot overlay assays with radiolabeled ligands have proved to be useful and simple in this regard. *In vitro* transcription and translation, and in some cases modification of the translation products, require somewhat more effort, but are more generally applicable and more sensitive to variations in structure. In cases in which a distinctive genetic or developmental pattern of protein expression has been documented, Northern and/or Southern blots should be performed at an early stage, using plaque-derived PCR products as probes.

If sequence correspondence is lacking, many other methods may be used to attempt to validate the identification of a clone. Is the predicted amino acid sequence inconsistent in codon usage, length, amino acid composition, or predicted secondary structure, with what is known about the target protein? Does the target protein have a distinctive tissue-specific, phylum-specific, or developmentally regulated pattern of occurrence that can be addressed by RNA or DNA blotting?

Once the size of the fusion protein(s) is established, a plaque purification is continued; larger quantities of fusion proteins can be purified on preparative gels and used in comparative one- or two-dimensional peptide mapping experiments, which, when coupled to Western blotting, can be a rapid, powerful means of investigating the identity of the clone(s) (see, e.g., Yang *et al.,* 1988).

In many cases, recombinant fusion proteins contain functional binding sites for molecules other than antibodies, and this is the basis for many alternative methods for screening expression libraries when suitable antibodies are not available. When immunoselection is used as the primary screening tool, the expressed polypeptides can be rescreened for a particular functional characteristic, as a method of investigating clone identity as well as a way of establishing a molecular basis for that function. Functional assays that are known to be tolerant of a denaturing step (such as SDS-PAGE) will often work with fusion proteins. For example, recombinant pieces of rat MAP2 will bind the R_{II} regulatory subunits of protein kinase A (Obar *et al.,* 1989), and recombinant Ras binds GTP (reviewed in Barbacid, 1987); these assays could be carried out with immunoselected plaques before the clones are even plaque pure. Other simple techniques that assay unique characteristics of a protein, such as gel shift in the presence of calcium (e.g., calmodulin), or heat stability (Tau, MAP2, and MAP4) can be useful and informative; enzymatic characteristics are especially informative (see, e.g., Eberwine *et al.,* 1987; Hanks, 1987; Singh *et al.,* 1988). Last, if the chromosomal location of the gene is known through genetic analysis, *in situ* hybridization with a cDNA probe is in order (see, e.g., Itoh *et al.,* 1986).

X. Sequence Analysis of Cloned cDNAs

The decision concerning when to sequence cloned cDNAs fully must be made thoughtfully. Sequence analysis of unconfirmed clones may be unproductive, and may therefore waste valuable time. However, usually the unambiguous verification of clones is dependent on the match between protein microsequencing data and predicted protein sequence encoded by the cDNA. In general, if

Table I
Confidence Levels for Common Clone Confirmation Techniques

Technique	Confidence level (%)[a]
Reactivity of plaque lift or fusion protein with a polyclonal antibody	15–25
Reactivity of plaque lift or fusion protein with an affinity-purified polyclonal antibody	20–30
Reactivity of plaque lift or fusion protein with a monoclonal antibody	15–25
Reactivity of fusion protein with a second, independent, monoclonal, or polyclonal antibody	50
Reactivity of target protein with "reverse-purified" polyclonal antibody	40–50
Reactivity of target protein with antibody raised against fusion protein	20–30
Similar peptide maps/Western blots	20–60
Northern or Southern blot patterns that match protein size or expression	15–30
Sequence match of ≥8 amino acids between a peptide and translated cDNA	70–90

[a] These numbers are roughly additive; for example, reactivity of a fusion protein with both a monoclonal and a polyclonal antibody should put the confidence level about 50%. When a given clone reaches a confidence level of greater than ~50%, it should be fully characterized.

tests such as the complementary affinity purification of antibodies from the fusion protein and from the polypeptide of interest support the identity of the clone, and especially if these tests can be confirmed with a second independent antibody, then begin sequencing the clone (see Table I). Sequencing of the complementary strand, although necessary for the accurate determination of the nucleotide sequence, can be postponed until the clone is unambiguously verified.

A. Strategy

The overall strategy for sequencing the clone of interest depends on the size of the cDNA insert. For clones of 1 kb or less, sequencing from either end will yield almost half of the total sequence. If there are any convenient internal restriction sites for subcloning the insert or for constructing deletion subclones, then this may represent the most direct route. In theory, for a clone of 1 kb, the sequencing of four subclones yielding 250 bp of sequencing data each would be sufficient, although in practice it may be rare that the restriction sites will be convenient enough to allow such efficiency. The increasing availability and decreasing cost of custom oligonucleotide synthesis often make it worthwhile to design specific sequencing primers to quickly and efficiently fill any sequence gaps between subclones.

For larger cDNA inserts (>1 kb), it is worthwhile to take the time to construct a series of nested deletion subclones, using exonuclease III [see Ausubel *et al.*, 1989; or use a commerical kit, available from New England BioLabs (Beverly,

MA) Pharmacia, Promega, or Stratagene]. Once a series of nested, or overlapping, deletions are created, the entire sequence of the insert can be obtained quickly. Again, any small gaps in the sequence are most quickly bridged by using specific custom oligonucleotides as primers.

Dideoxy sequencing (Sanger *et al.,* 1977, 1980) is currently the method of choice, with [α-^{35}S]dATP used to label the DNA bands, followed by resolution of the DNA sequence by electrophoresis on urea–acrylamide gels with wedge-shaped spacers (Sambrook *et al.,* 1989; Ausubel *et al.,* 1989). Commercial kits of sequencing reagents are now used widely; we recommend the Sequenase kit (U.S. Biochemical, Cleveland, OH).

B. Data Analysis

Many software packages, such as GCG, IntelliGenetics, DNA Inspector, and so on, have been developed to allow the handling and analysis of sequencing data. The choice of a particular package will most likely be dictated by what is easily available. However, we can recommend the GCG package (Devereux *et al.,* 1984) as one tool in particular that allows the editing and ordering of sequence data, the analysis of motifs at both the nucleotide and protein level, and that provides convenient programs to compare sequences with the general databases, including GenBank and EMBL.

Once sequence data from the clone is available, the first priority becomes the unambiguous identification of the clone as encoding the polypeptide of interest. Methods of verification, such as affinity purification of antibody, immunoreactivity of a second independent antibody, generation of antibody to the fusion protein or a predicted peptide, and so on, are all useful, but in the end the most satisfactory identification depends on the location of peptide sequence obtained by protein microsequencing within the predicted amino acid sequence. Amino acid microsequencing data will also be useful in editing the nucleotide sequence and determining the reading frame. Multiple stretches of confirmed sequence will ensure that any frameshift errors in the sequencing data will be caught and corrected.

When the clone has been unambiguously identified, the next most powerful tool available is the comparison of the sequence to those that have been previously cloned and characterized. The identification of homologs or related sequences by this method is quickly revolutionizing the field of cell biology (see Fig. 5 for an example of a homology plot). For example, the identification of sequence similarity between kinesin heavy chain sequence and the *KAR3* locus in yeast allowed the determination of a function for a kinesin-like microtubule-based motor involved in karyogamy (Meluh and Rose, 1990). A comparison of the sequence data for MAP2 and Tau allowed the identification of conserved repeat sequences that are likely to be involved in microtubule binding by both these proteins (Lewis *et al.,* 1988). Identification of related sequences in *Drosophila* for both dynamin (Fig. 5; Obar, *et al.,* 1990) and p150[Glued] has led us to

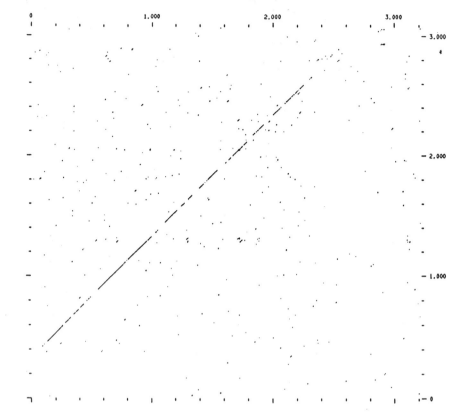

Fig. 5 Dot matrix plot describing regions of similarity between the amino acid sequences of fly and rat dynamin. Points on the diagonals represent areas of sequence identity above the threshold value.

the identification of specific mutations that should be valuable in the further characterization of the *in vivo* function of these polypeptides (Van der Bliek and Meyerowitz, 1991; Chen *et al.*, 1991; Holzbaur *et al.*, 1991).

When scanning the database with a particular sequence there are several possible outcomes. An exact match for the sequence may be found in the database. If this should happen, it still might provide information on different observed activities or tissues of origin for a known protein. The second possibility is that a similar sequence may be identified that may assist in the identification or characterization of the clone, or that might lead to the definition of a new gene family or superfamily. It is also possible that any observed similarities of the sequence to previously identified sequence may be limited to specific, and perhaps functionally significant, sequences, such as an ATP- or GTP-binding domain. There are programs available in GCG, as well as other analysis programs, that search for various motifs. Finally, no significant relationships may be identified between the sequence of interest and previously characterized polypeptides. This may at first be frustrating, but at least one now has in hand an invaluable research tool in the further elucidation of the cellular role of the target protein.

Once the complete cDNAs and sequences are in hand many types of further analysis become possible, and the choice of which to pursue will depend on the significant questions for the clone of interest. For example, detailed Southern and Northern blotting can be performed to determine (1) whether the clone is a member of a multigene family, (2) the tissue distribution of the mRNA transcript, (3) the developmental pattern of the polypeptide expression, and (4) possible alternative splicing products to produce variant polypeptides, and so on. Further immunological methods may be applied, such as the raising of antibodies to bacterially expressed fusion protein, or to synthetic peptides, as described above.

One of the major rewards from a successful cloning project is the ability to express the polypeptide of interest, either at high levels for antibody production or under controlled conditions, such as transfection into cultured cells. In the end, the most satisfying proof of a clone is the determination of functional activity in a cloned cDNA expressed either *in vitro* or in a heterologous expression system. Once a functional assay has been developed for the clone of interest, structure–function studies can be initiated in which specific sequences in the cDNA are altered and any resulting changes in activity are assayed. For example, the determination that the bacterially expressed cDNA for the kinesin heavy chain could function in an *in vitro* microtubule gliding assay (Yang *et al.*, 1990) allowed the precise mapping of the motor domain, as well as the discovery that a homologous form of the enzyme catalyzes motility along microtubules, but in the opposite direction (Walker *et al.*, 1990; McDonald *et al.*, 1990). However, the generation of an active polypeptide or enzyme in a heterologous expression system may not be trivial in all cases, as the folding of the polypeptide may easily be disrupted in nonnative conditions, or when the protein is expressed at a high level. What is clear, however, is that the successful cloning of a cDNA encoding the target protein opens new doors in the further study of its biological function.

XI. Materials and Reagents

A. Solutions for Electrophoresis and Western Blotting

SDS-PAGE sample buffer (2×): 250 mM Tris-HCl (pH 6.8), 2% SDS, 20% glycerol, 0.01% bromphenol blue, 10% 2-mercaptoethanol

Plaque sample buffer for SDS-PAGE (2×): 250 mM Tris-HCl (pH 6.8), 10%
   SDS, 0.01% bromphenol blue, 10% 2-mercaptoethanol
Standard electroblotting buffer: 12.5 mM Tris-HCl, 96 mM glycine, 10 mM
   2-mercaptoethanol, 0.01% SDS; pH ~8.3

Note: For electroblotting of peptide fragments for microsequencing, addition
of methanol to the transfer buffer (10–30% by volume) increases recovery onto
PVDF membranes.

Ponceau S solution: 1:200 dilution (of Sigma Cat. No. P7767) in water
Coomassie Brilliant Blue stain: 0.1% Coomassie Brilliant Blue R-250 (Cat.
   No. B0149; Sigma) in 40% methanol, 50% water, 10% acetic acid
Coomassie Brilliant Blue destain: 40% methanol, 50% water, 10% acetic acid
TBS: 100 mM Tris-HCl (pH 7.4), 150 mM NaCl
TBST: 100 mM Tris-HCl (pH 7.4), 150 mM NaCl, 0.05% Tween-20
Blocking solution: TBST with 5% nonfat dry milk

B. Solution for Antibody Purification

Antibody elution buffer: 0.25 M glycine, pH 3.0

C. Solutions for Protein Microsequencing

Gel slice equilibration buffer (1×): 125 mM Tris-HCl (pH 6.8), 1% SDS,
   10 mM EDTA
Gel slice overlay buffer (1×): 125 mM Tris-HCl (pH 6.8), 1% SDS, 0.01%
   bromphenol blue, 10 mM EDTA, 20% glycerol
Protease buffer (1×): 125 mM Tris-HCl (pH 6.8), 1% SDS, 0.01% pyronin Y,
   10 mM EDTA, 10% glycerol
Sodium thioglycolate (100×) stock: 1 M sodium thioglycolate
Basic electroblotting buffer: 10 mM CAPS (pH 11.0), 10% methanol
Cyanogen bromide solution: In a fume hood, place ~500 mg of cyanogen
   bromide (CNBr) crystals in a polypropylene or glass tube with a tight-fitting
   cap. Weigh the capped tube, then return to the fume hood and add a volume
   of dimethylformamide (in microliters) equal to the number of milligrams of
   CNBr.

Notes: Cyanogen bromide must be handled as an extremely hazardous
chemical, and should be disposed of only after treatment with an excess
of sodium hypochlorite. Fresh cyanogen bromide is most effective in this protocol.

D. Reagents for Growth of Phage

Plating bacteria: Use a single bacterial colony of the appropriate *E. coli* strain to inoculate a sterile flask with 100 ml of LB (with ampicillin if appropriate) with 0.2% maltose. Grow overnight at 37°C with shaking, then pellet the cells by centrifugation for 10 min at 4000 rpm. Resuspend the pelleted cells in 50 ml of sterile mM MgSO$_4$, and store at 4°C for up to 1 week (Davis *et al.*, 1986)

SMG: 50 mM Tris-HCl (pH 7.5), 100 mM NaCl, 10 mM MgSO$_4$, 0.01% gelatin

LB (for 1 liter): 10 g of Bacto tryptone (Difco, Detroit, MI) 5g of Bacto yeast extract (Difco), and 10 g of NaCl; adjust pH to 7.0 with NaOH; autoclave for 20 min

LB plates: Add 15 g of Bacto agar (Difco) to 1 liter of LB; autoclave; pour plates while still molten (~50°C)

LB-Amp plates: When the LB with agar has cooled to about 50°C, add ampicillin to a final concentration of 50 μg/ml, then pour plates

Ampicillin: Make a stock solution of 50 mg/ml in water; filter sterilize. Use at a working concentration of 50 μg/ml

LB top agar, top agarose: For initial rounds of screening, use top agar: add 7 g of Bacto agar per liter of LB, then autoclave in 100 to 200 ml aliquots. For preparation of a high-titer phage stock, use top agarose, with 7 g of agarose per liter of LB, then autoclave

E. Solutions for Immunoscreening

Blocking solution: TBST plus 5% nonfat dry milk

Alkaline phosphatase reaction buffer: 100 mM Tris-HCl (pH 9.5), 100 mM NaCl, 5 mM MgCl$_2$

Nitroblue tetrazolium (NBT): 50 mg/ml in 70% dimethylformamide; store in the dark at 4°C

5-Bromo-4-chloro-3-indoylphosphate (BCIP): 50 mg/ml in dimethylformamide; store in the dark at 4°C

Alkaline phosphatase development solution: Mix 330 μl of NBT with 50 ml of alkaline phosphatase buffer, then add 165 μl of BCIP; mix and use immediately

F. Solutions for Hybridization Screening

Denaturation solution: 0.2 N NaOH, 1.5 M NaCl

Equilibration solution: 1 M Tris-HCl (pH 7.4), 1.5 M NaCl

Prehybridization solution: 50% formamide, 5× SSC, 5× Denhardt's reagent, 0.1% SDS, 20 μl salmon sperm or calf thymus DNA per milliliter

Hybridization solution: Prehybridization solution, with probe added at

10^6 cpm/ml; double-stranded probes should be heated to 100°C for 5 min, and then quickly cooled in ice water immediately prior to addition to the hybridization buffer

SSC: 175.3 g of NaCl, 88.2 g of sodium citrate per liter, pH 7.0

Denhardt's reagent: 5 g of Ficoll, 5 g of polyvinylpyrrolidone, 5 g of bovine serum albumin per 500 ml water (Sambrook *et al.,* 1989)

Wash solutions:

I. 2× SSC, 0.1% SDS, 0.1% sodium pyrophosphate

II. 1× SSC, 0.1% SDS, 0.1% sodium pyrophosphate

III. 0.3× SSC, 0.1% SDS, 0.1% sodium pyrophosphate

G. Solutions for Molecular Biology

TM buffer: 50 mM Tris-HCl (pH 7.5), 10 mM MgSO$_4$

STE buffer: 0.5% SDS, 50 mM Tris-HCl (pH 7.5), 0.4 M EDTA (pH 8.0)

TE buffer: 10 mM Tris-HCl (pH 7.5), 1 mM EDTA (pH 8.0)

PCR buffer (10×): 200 mM Tris-HCl (pH 8.4), 500 mM KCl, 25 mM MgCl$_2$, 1 mg BSA/ml, nuclease free

References

Aebersold, R., Leavitt, J., Saavedra, R. A., Hood, L. E., and Kent, S. B. H. (1987). Internal amino acid sequence analysis of proteins separated by one- or two-dimensional gel electrophoresis after *in situ* protease digestion on nitrocellulose. *Proc. Natl. Acad. Sci. U.S.A.* **84,** 6970–6974.

Ausubel, F. M., Brent, R., Kingston, R. E., Moore, D. D., Seidman, J. G., Smith, J. A., and Struhl, K. (1989 and supplements). "Current Protocols in Molecular Biology." Wiley, New York.

Barbacid, M. (1987). *Ras* genes. *Annu. Rev. Biochem.* **56,** 779–827.

Berger, S. L., and Kimmel, A. R., eds. (1987). "Guide to Molecular Cloning Techniques." Academic Press, London.

Broome, S., and Gilbert, W. (1978). Immunological screening method to detect specific translation products. *Proc. Natl. Acad. Sci. U.S.A.* **75,** 2764–2768.

Chen, E. J., and Seeburg, P. H. (1985). Supercoil sequencing: A fast and simple method for sequencing plasmid DNA. *DNA* **4,** 165–170.

Chen, M. S., Obar, R. A., Schroeder, C. C., Austin, T. W., Poodry, C. A., Wadsworth, S. C., and Vallee, R. B. (1991). Multiple forms of dynamin are encoded by *shibire,* a *Drosophila* gene involved in endocytosis. *Nature (London)* **351,** 583–586.

Cleveland, D. W., Fisher, S. G., Kirschner, M. W., and Laemmli, U. K. (1977). Peptide mapping by limited proteolysis in sodium dodecyl sulfate and analysis by gel electrophoresis. *J. Biol. Chem.* **252,** 1102–1106.

Davis, L. G., Dibner, M. D., and Battey, J. F. (1986). "Basic Methods in Molecular Biology." Elsevier, Amsterdam.

Devereux, J., Haeberli, P., and Smithies, O. (1984). A comprehensive set of sequence analysis programs for the VAX. *Nucleic Acids Res.* **12,** 387–395.

Dingus, J., Obar, R. A., Hyams, J. S., Goedert, M., and Vallee, R. B. (1991). Use of a heat-stable microtubule-associated protein class-specific antibody to investigate the mechanism of microtubule binding. *J. Biol. Chem.* **266,** 18854–18860.

Earnshaw, W. C., and Rothfield, N. (1985). Identification of a family of human centromere proteins using autoimmune sera from patients with scleroderma. *Chromosoma* **91,** 313–321.

Eberwine, J. H., Barchas, J. D., Hewlett, W. A., and Evans, C. J. (1987). Isolation of enzyme cDNA clones by enzyme immunodetection assay: Isolation of a peptide acetyltransferase. *Proc. Natl. Acad. Sci. U.S.A.* **84,** 1449–1453.

Elliott, R. M., and Green, C. D. (1989). A simple purification procedure for lambda gt bacteriophage DNA with hybridization size screening for isolation of longest length cDNA clones. *Anal. Biochem.* **183,** 89–93.

Garner, C. C., and Matus, A. (1988). Different forms of microtubule-associated protein 2 are encoded by separate mRNA transcripts. *J. Cell Biol.* **106,** 779–783.

Gibbons, I. R., Asai, D. J., Ching, N. S., Dolecki, G. J., Mocz, G., Phillipson, C. A., Ren, H., Tang, W.-J. Y., and Gibbons, B. H. (1991a). A PCR procedure to determine the sequence of large polypeptides by rapid walking through a cDNA library. *Proc. Natl. Acad. Sci. U.S.A.* **88,** 8563–8567.

Gibbons, I. R., Gibbons, B. H., Mocz, G., and Asai, D. J. (1991b). Multiple nucleotide-binding sites in the sequence of dynein beta heavy chain. *Nature (London)* **352,** 640–643.

Gibbs, R. A., Nguyen, P.-N., McBride, L. J., Koepf, S. M., and Caskey, C. T. (1989). Identification of mutations leading to the Lesch–Nyhan syndrome by automated direct DNA sequencing of *in vitro* amplified cDNA. *Proc. Natl. Acad. Sci. U.S.A.* **86,** 1919–1923.

Goding, J. W. (1986). "Monoclonal Antibodies: Principles and Practice," 2nd Ed. Academic Press, London.

Gooderham, K. (1984). *Methods Mol. Biol.* **1,** 193–202.

Hammarback, J. A., and Vallee, R. B. (1990). Antibody exchange immunochemistry. *J. Biol. Chem.* **265,** 12763–12766.

Hanks, S. K. (1987). Homology probing: Identification of cDNA clones encoding members of the protein-serine kinase family. *Proc. Natl. Acad. Sci. U.S.A.* **84,** 388–392.

Harlow, E., and Lane, D. (1988). "Antibodies: A Laboratory Manual." Cold Spring Harbor Lab. Press, Cold Spring Harbor, New York.

Holzbaur, E. L. F., Hammarback, J. A., Paschal, B. M., Kravit, N. G., Pfister, K. K., and Vallee, R. B. (1991). Homology of a 150K cytoplasmic dynein-associated polypeptide with the *Drosophila* gene *Glued*. *Nature (London)* **351,** 579–583.

Huynh, T. V., Young, R. A., and Davis, R. W. (1985). Constructing and screening cDNA libraries in λgt10 and λgt11. *In* "DNA Cloning: A Practical Approach" (D. M. Glover, ed.), Vol. 1, pp 49–78. IRL Press, Oxford.

Itoh, N., Slemmon, J. R., Hawke, D. H., Williamson, R., Morita, E., Itakura, K., Roberts, E., Shively, J. E., Crawford, G. D., and Salvaterra, P. M. (1986). Cloning of *Drosophila* choline acetyltransferase cDNA. *Proc. Natl. Acad. Sci. U.S.A.* **83,** 4081–4085.

Kirsch, J., Littauer, U. Z., Schmitt, B., Prior, P., Thomas, L., and Betz, H. (1990). Neuraxin corresponds to a C-terminal fragment of microtubule-associated protein 5 (MAP5). *FEBS Lett.* **262,** 259–262.

Kosik, K. S., Orecchio, L. D., Binder, L., Trojanowski, J. Q., Lee, V. M.-Y., and Lee, G. (1988). Epitopes that span the tau molecule are shared with paired helical filaments. *Neuron* **1,** 817–825.

Kraft, R., Tardiff, J., Krauter, K. S., and Leinwand, L. A. (1988). Using mini-prep plasmid DNA for sequencing double stranded templates with Sequenase. *BioTechniques* **6,** 544–547.

Laemmli, U. K. (1970). Cleavage of structural proteins during the assembly of the head of bacteriophage T4. *Nature (London)* **277,** 680–685.

LeGendre, N. J., and Matsudaira, P. T. (1988). Direct protein microsequencing from Immobilon P transfer membrane. *BioTechniques* **6,** 154–159.

Lewis, S. A., Sherline, P., and Cowan, N. J. (1986a). A cloned cDNA encoding MAP1 detects a single copy gene in mouse and a brain-abundant RNA whose level decreases during development. *J. Cell Biol.* **102,** 2106–2114.

Lewis, S. A., Villasante, A., Sherline, P., and Cowan, N. J. (1986b). Brain-specific expression of MAP2 detected using a cloned cDNA probe. *J. Cell Biol.* **102,** 2098–2105.

Lewis, S. A., Wang, D., and Cowan, N. (1988). Microtubule-associated protein MAP2 shares a microtubule binding motif with tau protein. *Science* **242,** 936–939.

Mayer, M. L. (1991). NMDA receptors cloned at last. *Nature (London)* **354,** 16–17.

McDonald, H. B., Stewart, R. J., and Goldstein, L. S. B. (1990). The kinesin-like ncd protein of *Drosophila* is a minus end-directed microtubule motor. *Cell* **63,** 1159–1165.

Mehra, V., Sweetser, D., and Young, R. A. (1986). Efficient mapping of protein antigenic determinants. *Proc. Natl. Acad. Sci. U.S.A.* **83,** 7013–7017.

Meluh, P. B., and Rose, M. D. (1990). KAR3, a kinesin-related gene required for yeast nuclear fusion. *Cell* **60,** 1029–1041.

Obar, R. A., Dingus, J., Bayley, H., and Vallee, R. B. (1989). The RII subunit of cAMP-dependent protein kinase binds to a common amino-terminal domain in microtubule-associated proteins 2A, 2B, and 2C. *Neuron* **3,** 639–645.

Obar, R. A., Collins, C. A., Hammarback, J. A., Shpetner, H. S., and Vallee, R. B. (1990). Molecular cloning of the microtubule-associated mechanochemical enzyme dynamin reveals homology with a new family of GTP-binding proteins. *Nature (London)* **347,** 256–260.

Ogawa, K. (1991). Four ATP-binding sites in the midregion of the beta heavy chain of dynein. *Nature (London)* **352,** 643–645.

Olmsted, J. B. (1986). Analysis of cytoskeletal structures using blot-purified monospecific antibodies. *In* "Structural and Contractile Proteins, Part C: The Contractile Apparatus and the Cytoskeleton" (R. Vallee, ed.), Methods in Enzymology, Vol. 134, pp. 467–472. Academic Press, Orlando, Florida.

Sambrook, J., Fritsch, E. F., and Maniatis, T. (1989). "Molecular Cloning: A Laboratory Manual," 2nd Ed. Cold Spring Harbor Lab. Press, Cold Spring Harbor, New York.

Sanger, F., Nicklen, S., and Coulson, A. R. (1977). DNA sequencing with chain-terminating inhibitors. *Proc. Natl. Acad. Sci. U.S.A.* **74,** 5463–5467.

Sanger, F., Coulson, A. R., Barrell, B. G., Smith, A. J. M., and Roe, B. A. (1980). Cloning in single-stranded bacteriophage as an aid to rapid DNA sequencing. *J. Mol. Biol.* **143,** 161–178.

Short, M., Fernandez, J. M., Sorge, J. A., and Huse, W. D. (1988). λZAP: A bacteriophage λ expression vector with *in vitro* excision properties. *Nucleic Acids Res.* **16,** 7583–7599.

Singh, H., LeBowitz, J. H., Baldwin, A. S., Jr., and Sharp, P. A. (1988). Molecular cloning of an enhancer binding protein: Isolation by screening of an expression library with a recognition site DNA. *Cell* **52,** 415–423.

Snyder, M., Elledge, S., Sweetser, D., Young, R. A., and Davis, R. W. (1987). "Lambda gt 11: Gene isolation with antibody probes and other applications. *In* "Recombinant DNA," Part E (R. Wu and L. Grossman, eds.), Methods in Enzymology, Vol. 154, pp. 107–128. Academic Press, San Diego.

Southern, E. M. (1975). Detection of specific sequences among DNA fragments separated by gel electrophoresis. *J. Mol. Biol.* **98,** 503–517.

Theurkauf, W. E., and Vallee, R. B. (1982). Molecular characterization of the cAMP-dependent protein kinase bound to microtubule-associated protein 2. *J. Biol. Chem.* **257,** 3284–3290.

Towbin, H., Staehelin, T., and Gordon, J. (1979). Electrophoretic transfer of proteins from polyacrylamide gels to nitrocellulose sheets: Procedure and some applications. *Proc. Natl. Acad. Sci. U.S.A.* **76,** 4350–4354.

Van der Bliek, A. M., and Meyerowitz, E. M. (1991). Dynamin-like protein encoded by the *Drosophila shibire* gene associated with vesicular traffic. *Nature (London)* **351,** 411–414.

Vande Woude, G. F., Oskarsson, M., Enquist, L. W., Nomura, S., Sullivan, M., and Fishinger, P. J. (1979). Cloning of integrated Moloney sarcoma proviral DNA sequences in bacteriophage lambda. *Proc. Natl. Acad. Sci. U.S.A.* **76,** 4464–4468.

Walker, R. A., Salmon, E. D., and Endow, S. A. (1990). The *Drosophila claret* segregation protein is a minus-end directed motor molecule. *Nature (London)* **347,** 780–782.

Weber, K., Pringle, J. R., and Osborn, M. (1972). Measurement of molecular weights by electrophoresis on SDS-acrylamide gel. *In* "Enzyme Structure," Part C (C. Hirs and S. Timasheff, eds.), Methods in Enzymology, Vol. 26, pp. 3–27. Academic Press, New York.

Yamamoto, K. R., Alberts, B. M., Benzinger, R., Lawhorne, L., and Treiber, G. (1970). Rapid bacteriophage sedimentation in the presence of polyethylene glycol and its application to large scale virus purification. *Virology* **40,** 734–744.

Yang, J. T., Saxton, W. M., and Goldstein, L. S. B. (1988). Isolation and characterization of the gene encoding the heavy chain of *Drosophila* kinesin. *Proc. Natl. Acad. Sci. U.S.A.* **85,** 1864–1868.

Yang, J. T., Saxton, W. M., Stewart, R. J., Raff, E. C., and Goldstein, L. S. B. (1990). Evidence that the head of kinesin is sufficient for force generation and motility *in vitro*. *Science* **249,** 42–47.

Young, R., and Davis, R. (1983). Efficient isolation of genes by using antibody probes. *Proc. Natl. Acad. Sci. U.S.A.* **80,** 1194–1198.

CHAPTER 19

Antiidiotypic Antibodies: Methods, Applications, and Critique

Spyros D. Georgatos

Programme of Cell Biology
European Molecular Biology Laboratory
D-6900 Heidelberg, Germany

I. Introduction

The antiidiotypic approach (AIA), although a popular method of analysis in several fields of biological research, has only recently been introduced in cell biology. The usefulness of the AIA can be better appreciated if one looks critically into the spectrum of methods currently available for receptor identification.

• Copurification of a protein ligand with its putative receptor does not provide sufficient evidence for a direct association between these two components.

• *En bloc* solubilization of native ligand–receptor complexes and coimmunoprecipitation by anti-ligand antibodies is not always an option; large macromolecular assemblies, anchored to the cytoskeleton, are often resistant to extraction with nondenaturing detergents.

• Conventional chemical cross-linking has more serious limitations; these concern the relatively narrow windows of working concentrations, the low cross-linking efficiency, and the difficulties encountered in resolving and characterizing large cross-linked products.

• Finally, affinity chromatography and ligand blotting can be used only when the receptor–ligand interaction is strong and of a slow "off rate."

• One way or the other, all of these methods involve use of antibody probes and (in that) are almost as limited as the AIA.

Since its first application in receptor identification (Sege and Peterson, 1978), the AIA has been both praised and criticized. Interestingly, the lack of a consensus has not discouraged further work with antiidiotypic antibodies; dozens of seminal articles have been published during the last decade on the various applications of the AIA. However, because of an ongoing debate, any bias can be misleading. Because the main objective of this series is to provide balanced and reader-friendly clues on novel approaches, the scope of the article will be three-fold: (1) to simplify key concepts on which the AIA is based, (2) to summarize the main experimental steps of the AIA, and (3) to highlight some of the obvious limitations of the AIA. The methods used to generate antiidiotypic antibodies by *in vitro* immunization, and the intricacies of the so-called "third-round" antiidiotypes, will not be covered here; these topics have been discussed by Vaux and Fuller (1991). To obtain a more global picture, the committed reader can consult two comprehensive reviews discussing various aspects of the AIA (Gaulton and Green, 1986; Erlanger *et al.*, 1986), and a short (but pointed) essay discussing the potential pitfalls of using antiidiotypic antibody probes (Meyer, 1990).

II. Structural Basis and Main Principles of Antiidiotypic Approach

What makes the AIA feasible is the intelligent exploitation of a set of structural principles that govern protein–protein interactions, combined with the prudent application of classic immunological techniques. Obviously, none of these principles is written in stone. Therefore, going through the discussion, one should be acutely aware of the many theoretical questions that still remain open with regard to the complexity of protein–protein interactions and the diversity of the immune response.

A. Protein–Chemical Concepts

1. Shape Complementarity

The AIA relies on a fundamental physical principle: that specific protein–protein associations frequently involve a close matching between structures of complementary shape (for a general introduction on this topic, see Colman, 1988). This idea is supported by numerous structural studies on antigen–antibody and protease–inhibitor complexes (see reviews in Janin and Chothia, 1990; Chothia, 1991, Davies *et al.*, 1990). However, shape complementarity is not a static property. In many cases, proteins do have the ability to adapt to each other by subtle conformational changes following an initial low-affinity association (see, e.g., Fermi and Perutz, 1981; Remington *et al.*, 1982; Lesk and Chothia, 1984). From these two points, it follows that interacting surfaces possess some genetically determined complementarity that, nevertheless, can be epigenetically modulated (for an excellent discussion on this topic, see Colman, 1988).

2. Features of Binding Sites

In antibody–antigen complexes, 27–39 residues are involved in the binding (see review in Chothia, 1991); in protease–inhibitor complexes, 10–15 residues of the inhibitors contract 17–29 residues of the proteases (see review in Janin and Chothia, 1990). In both categories of complexes, the area of contact (area buried away from the solvent) is about 1250–1950 Å^2. This parameter is different in other, oligomeric protein complexes (500–10,000 A^2; Miller *et al.*, 1987; Janin *et al.*, 1988).

Significant departures from the principle of spatial complementarity have not been noticed so far among crystallographically analyzed structures. However, whereas some anti-hapten and anti-peptide antibodies possess antigen-binding sites that have the form of deep pockets (Stanfield *et al.*, 1990), in other cases the antigen-binding sites are just shallow troughs (Rose *et al.*, 1990). In the latter, although conformational compatibility (i.e., fitting of "protuberances" into "valleys") is clearly required to establish a close contact between the antigen and the antibody, the degree of three-dimensional complementarity may not be as perfect as in the former case. This distinction is a key point for the rest of the discussion on the AIA, because it underscores and explains some of the limitations of this approach.

Although most interaction sites possess a variable amino acid composition, there are two interesting coincidences concerning the frequency of certain residues in such regions. First, in the antigen-combining sites of antibodies, about half of the contact residues are aromatic (Amit *et al.*, 1986; Padlan *et al.*, 1989; Mian *et al.*, 1991). These bulky amino acids may contribute to the stability of antibody–antigen complexes because of their low conformational entropy and their large surface area (discussed in Davies *et al.*, 1990). Second, the interphases of oligomeric complexes contain a high percentage of arginine residues, which are most probably involved in electrostatic interactions and/or hydrogen bonds (Janin *et al.*, 1988).

3. Specific Protein–Protein Interactions

Shape complementarity allows exact fitting of molecular surfaces to occur, but does not necessarily ensure a stable and specific association. Productive binding correlates best with the potential for noncovalent bonding. Of all possible noncovalent interactions, that is, electrostatic interactions, permanent or induced dipole–dipole interactions, and hydrogen bonds, the latter deserve a special mention. Protein–protein recognition and binding in the structures mentioned above involves an average of 10 hydrogen bonds (Janin and Chothia, 1990). The corresponding number of hydrogen bonds in large oligomeric complexes is lower, but significantly, in the case of smaller oligomers, there is approximately 1 hydrogen bond per 200 $Å^2$ (Janin *et al.*, 1988). Thus, an interphase of 1600–2000 $Å^2$ would contain, again, approximately 8–10 hydrogen bonds.

Although hydrogen bonds do not influence the binding energy as much as hydrophobic interactions (which correlate with the entropy changes on protein–protein association), they contribute significant enthalpic factors and may determine the specificity of protein–protein associations (Colman, 1988). This stems from the fact that hydrogen bonds have two features in common with the covalent bonds that distinguish them from the rest of the noncovalent cohesion forces: stoichiometry and directionality (see Zeegers-Huyskens and Huyskens, 1990).

It is useful to note that hydrogen bonding may provide one of the best physical means for ''replicating'' molecular features of a ligand. A three-dimensional system of hydrogen bond donors and acceptors, arranged as they are at specific angles and distances along interaction sites (Davies *et al.*, 1990; Schultz and Schirmer, 1979), could mediate the precise pairing of a template (in one molecule) with its ''mold'' (in another moelcule). This principle appears to be exploited by ''replicating'' macromolecules, as for example the DNA molecule (discussed in Watson *et al.*, 1987, in the context of DNA replication).

B. Immunochemical and Immunological Concepts

1. Complementarity Determining Regions and Framework Regions

Immunoglobulins consist of distinct structural domains. These domains contain multiple β strands, connected by loop regions, and are folded as β barrels (immunoglobulin fold). Although the sequences of the immunoglobulin domains are relatively constant, the amino-terminal segments of different immunoglobulins, termed the variable domains (V_L for the light chains and V_H for the heavy chains), always differ in sequence. The V domains are not uniformly variable throughout their length; instead, they contain distinct "hypervariable" segments, called the complementarity determining regions (CDRs). These regions (three for each heavy chain and three for each light chain) are localized along the loops that connect the β strands of the V domains. The sequences flanking the CDRs in the V domains are referred to as the framework regions. As a rule, the amino acid residues that contact the antigen belong to the CDRs. However, occasional contacts of the antigen with the framework regions flanking the CDRs have also been found (for an introduction, see Branden and Tooze, 1991).

Considering the structural basis of antibody–antigen recognition, there are so far two alternative views. According to conventional wisdom, the complementarity between the antigen-binding site of the antibody (termed the paratope) and the antibody-binding site of the antigen (termed the epitope) can be attributed to unique conformational features of each CDR system. However, according to more recent studies, there might be only a limited repertoire of conformations for at least five of the six CDRs. These conformations are called canonical structures. To interact with different epitopes, the canonical structures must be modulated. This fine tuning can be accomplished by varying the sequences slightly and, as a result, the surface topography and the relative position of the CDRs (Chothia and Lesk, 1987; Tramontano *et al.*, 1990; Chothia *et al.*, 1989).

2. Idiotopes, Idiotypes, Antiidiotypic and Antimetatypic Antibodies

Antibodies recognizing features of the paratopes of other antibodies can develop after immunization of animals with a purified immunoglobulin (see Fig. 1). This is because the V domains of each immunoglobulin molecule contain specific antigenic determinants (Kunkel *et al.*, 1963; Ouadin and Michel, 1963). These determinants, which are most frequently found along the CDRs, are known as idiotopes and must be differentiated from allotypic, isotypic, and xenotypic antigenic determinants, which occur in other regions of immunoglobulin chains. Idiotopes may be shared by a small group of different immunoglobulin molecules and be encoded by germ-line V genes (public idiotopes); alternatively, idiotopes may be unique to one immunoglobulin molecule, arising from somatic mutation of the germ-line genes (individual or private idiotopes). The sum of idiotopes in a given immunoglobulin molecule defines its idiotype. Antibodies against particular idiotopes are called antiidiotypic antibodies, or antiidiotypes.

Antibodies can also develop to recognize the complex of another immunoglobulin with its antigen, but not the unliganded antibody or the free ligand (see review in Voss, 1990). These antibodies are clearly different from the known antiidiotypes and have been termed antimetatypic antibodies.

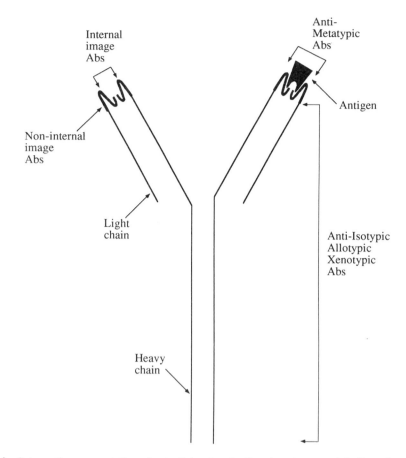

Fig. 1 Schematic representation of potential antigenic sites along immunoglobulin molecules (for a detailed description see text).

3. Network Theory

The discovery of idiotopes provided the basis for the network theory of Jerne (1974, 1984). The fundamental postulate of this hypothesis is that the immune system is balanced by pairs of idiotypic–antiidiotypic antibodies. Jerne further developed the concept of the internal image antiidiotypic antibodies. Such antibodies are thought to occur in the sera of normal individuals and to recognize specific idiotopes of other natural antibodies (principle of connectivity). Because these idiotopes are presumably "replicas" of exogenous antigens, the antiidiotypes are expected to have paratopes that resemble the structure of (nonimmunoglobulin) antigens. Thus, at any instance, the immmune system may possess an "internal image" of the outside antigenic world, inscribed in the paratopes of various antiidiotypic antibodies.

Although the role of antibody networks in the regulation of the immune system is still debated (see, e.g., Coutinho, 1989; Lundkvist *et al.*, 1989), the network theory does provide a framework for classifying antiidiotypic reagents. Thus, antiidiotypic antibodies generated by experimental manipulation (see Section II,B,5) can be divided into two main classes (see Fig. 1): (1) internal image antiidiotypes, that is, antibodies that, by virtue of their complementarity to the CDRs of another antibody, mimic the structure and the biological activity of an antigen recognized by the latter antibody, and (2) noninternal image antiidiotypes, that is, antibodies recognizing partial features of the CDRs of other antibodies, or framework determinants (for reviews see Chothia, 1991; Davies *et al.*, 1990).

4. Antigenicity and Immunogenicity

Early studies have suggested that each protein molecule contains only a limited number of antigenic determinants (Atassi, 1975, 1978, 1984). However, more recent analyses (see reviews in Benjamin *et al.*, 1984; Berzofsky, 1985) indicate that almost all surface-exposed sites on a polypeptide chain are potentially antigenic (i.e., recognizable by antibodies). The intrinsic antigenicity of proteins has been correlated with their hydrophilicity (Hopp and Woods, 1981) and the segmental mobility or flexibility of the polypeptide backbone (Westhof *et al.*, 1984; Tainer *et al.*, 1984). However, both hydrophilicity and mobility are, in final analysis, related to surface accessibility, because the more flexible regions of a protein are usually found on its surface (see Colman, 1988, for discussion). Another feature of antigenic sites (especially in short peptides) is the occurrence of β turns (Wright *et al.*, 1988; see also Schultze-Gahmen *et al.*, 1985).

Not all antigenic sites on a polypeptide chain would be equally effective as immunogens (i.e., able to induce the production of specific antibodies). In practice, the magnitude and the specificity of a humoral response to an antigen depend critically on the genetic background of the immunized animal, on cellular factors (T cell help), on self-tolerance mechanisms, and on the route and scheme of immunogen administration (see reviews in Berzofsy, 1985; Nossal, 1991). For T cell recognition, peptides originating from the processed (proteolyzed) antigen, complexed with major histocompatibility complex (MHC) molecules, must be displayed on the surface of a B lymphocyte and be recognized by a T cell receptor. In general, the particular "epitope" of the processed antigen recognized by the T cell receptor is not identical to the epitope recognized by the B cell. Thus, an effective immunogen should contain, apart from sites for antibody recognition, sites able to elicit T cell help (see review in Berzofsky, 1985; see also Vahlne *et al.*, 1991). Usually, amphipathic regions with alternating hydrophobic and hydrophilic α-helical sequences provide better recognition sites for the T cell receptor (DeLisi and Berzofsky, 1985; see review in Pierce and Margoliash, 1988).

5. "Replication" of Binding Sites onto the Paratopes of Antibodies

On the basis of the above principles, it would seem reasonable to assume that a surface-exposed feature of a molecule can be recognized by and, perhaps, "replicated" onto the paratope of an immunoglobulin molecule. Thus, by immunizing with a purified ligand or a derivative of its presumed receptor-binding site, specific anti-ligand antibodies can be generated. A fraction of these antibodies (Ab1) could possess paratopes complementary to the receptor-binding site. Such a "negative image" of a receptor-binding site can be transformed into a "positive image" by raising secondary, anti-anti-ligand antibodies (antiidiotypic antibodies, or Ab2) against the antigen-combining site of Ab1. If the fidelity of the "replication process' is not compromised between successive immunizations, Ab2 will be analogs of the original ligand that specifically recognize its physiological receptor by virtue of their shape complementarity to it (see Fig. 2a and b). The same method can be used to raise antiidiotypic antibodies against complementary, self-association sites that may occur along protein molecules that have the ability to form homopolymeric assemblies (Fig. 2b and c; for more details see Section V).

The feasibility of successfully "replicating" active sites into the paratopes of idiotypic/antiidiotypic antibodies can be better appreciated by discussing here two outstanding examples:

First, Greene and associates (Bruck *et al.*, 1986) have managed to sequence the CDRs of an antiidiotypic antibody that recognizes a reovirus hemagglutinin cell surface receptor. These studies revealed a distinct homology between the CDR II of the heavy and light chains of the antibody and a segment of the viral hemagglutinin. More recently, the same group (Saragovi *et al.*, 1991) has produced a nonpeptide "mimetic" of the antiidiotype by modeling the structural features of the homologous region. Although primary structure similarity need not be the rule when comparing a particular antiidiotype with the protein ligand which is supposed to mimic, such a finding demonstrates the enormous potential of the AIA.

Second, Billetta *et al.* (1991) have engineered an "internal image antiidiotype" by genetically grafting into one of the CDRs of an immunoglobulin molecule a repetitive tetrapeptide representing part of the *Plasmodium falciparum* circumsporozoite protein. Immunization with the engineered antibody protected animals from parasite invasion, suggesting that ". . .antibody (idiotype) mimicry of an exogenous antigen is possible and may only require a discrete stretch of identity between the two molecules. . ."(Billetta *et al.*, p. 4713, 1991).

Despite the successful applications of the AIA, the general validity of this approach has been challenged: X-ray diffraction analysis of an immunoglobulin–immunoglobulin complex has shown that the molecular features responsible for idiotope–antiidiotope recognition are only *partially* overlapping with the ones involved in idiotypic antibody–antigen recognition (Bentley *et al.*, 1990);

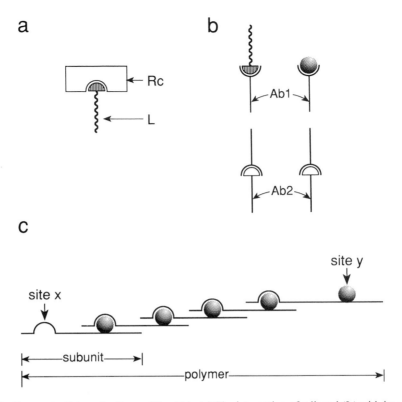

Fig. 2 Two potential applications of the AIA. (a) The interaction of a ligand (L) with its specific receptor (Rc) is diagrammatically shown. Notice that the receptor-binding site of the ligand (hatched region) precisely fits a depression on the receptor molecule. (c) Subunits of a polymer that possess two complementary sites (site x and site y) are shown to "lock" into each other, forming a macromolecular complex. (b) Antibody "replicas" against the receptor-binding site of L, or against site y, are prepared (Ab1). These antibodies are then used to develop antiidiotypic reagents (Ab2) that, by virtue of their complementarity, can recognize the (unknown) receptor Rc and site x, respectively.

furthermore, assessment of the fine specificities of many antibodies raised against a lymphocyte receptor (CD4) and comparison of antibody binding to this molecule with the binding of a natural ligand (the HIV protein gp120) have failed to show *exact* mimicry between the ligand and any of the anti-receptor antibodies (Davies *et al.,* 1992). These results have been interpreted as showing that ". . .production of antibody mimics will occur very rarely or not at all and (that) the anti-idiotype approach is unlikely to be useful. . ." (Davis *et al.,* p. 79, 1992). A balanced interpretation here may be that antibodies do not always "copy" exactly the features of receptor-binding sites and that their specificity has, by definition, its limits. From this angle, the examples discussed above simply confirm that a certain degree of "mismatching" and imprecision is to be expected when a binding site is "replicated" in the paratope of an antibody. Having said that, one can only agree on the idea that genuine internal image antibodies develop rather rarely. This is a notion already made in previous studies on tobacco mosaic virus antigens, showing that internal image antiidiotypes can account for only 15% of the total antiidiotypes (Urbain et al., 1984; also discussed in Gaulton and Green, 1986). According to current views (Chothia, 1991), even this number would appear to be an overestimate.

III. Production and Characterization of Antiidiotypic Antibodies

A. Selection and Characterization of Immunogen

The initial step in all versions of the AIA involves production of an antiserum enriched in antibodies against the receptor-binding site of the ligand of interest. Thus, the first problem that arises concerns the selection of an appropriate immunogen. Choice of suitable immunogens can be greatly facilitated if the sequence and/or the location of the receptor-binding site along the ligand molecule have been previously determined. (This does not require prior purification of the receptor protein, or knowledge of its identity.) Such information enables one to immunize with a fragment of the ligand, or a synthetic peptide, which still contains the intact receptor-binding site. This is a point of importance since the most meaningful results with the AIA have been obtained by immunizing with either small organic molecules, or with small proteins and peptides. Immunization with large proteinaceous ligands carries a risk; irrelevant, yet immunodominant, epitopes can induce the production of "non-Ab1" immunoglobulins. The presence of these contaminants in the resulting antisera considerably complicates the subsequent analysis.

Naturally, a disadvantage of using peptide analogs and protein fragments is that such derivatives may not fold in the same way as the intact ligand (see, e.g., Stanfield et al., 1990). A few rational ways to confront this problem are summarized in Table 1. However, it must be emphasized that the most important property for which a potential immunogen must be screened is the ability to bind specifically to a tissue or cellular preparation that contains the putative receptors. Although binding to a crude cellular fraction containing the receptor does not rigorously prove that the selected immunogen is folding correctly, it nevertheless shows that the ligand and its synthetic or proteolytic derivative share those conformational features that are crucial for receptor recognition.

In the absence of better means, one can always evaluate the conformational "fitness" of synthetic immunogens in a retrospective fashion: if antibodies developed against such derivatives still recognize the native ligand, there is a good chance that (at least a fraction of) the immunogen molecules are folding correctly (methods to assess the ability of antibodies to recognize native

Table I
Criteria for Selecting Peptide Antigens

Property	Method
Direct binding to receptor	Solution binding assays; ligand blotting; affinity chromatography
Correct conformation	Circular dichroism[a]; nuclear magnetic resonance[b]
Biological activity	Appropriate bioassay

[a] For an example see Oas and Kim (1988).
[b] For an example see Wright et al. (1988).

proteins are discussed in Section III,C). It should be noted, however, that according to one school of thought, antibodies developed against peptides may solely recognize the unfolded form of the parent molecule (denatured epitopes, or "unfoldons"; see Laver et al., 1990). This notion, although justified, does not appear to be generally applicable; first, neutralizing antibodies (i.e., antibodies recognizing native viral and other pathogen particles) have been successfully developed employing a variety of peptide antigens (see, e.g., Vahlne et al., 1991; Milich et al., 1988); second, some peptide analogs apparently recognize the genuine receptors for a ligand (for an example see Dyer and Curtiss, 1991); third, other studies reveal that limited proteolysis of large proteins sometimes yields fragments that retain the topology and the physicochemical properties of the native material (see review in Jaenicke, 1991); finally, a variety of observations are consistent with the idea that small synthetic peptides can form stable secondary structures in solution (Lau et al., 1984; Oas and Kim, 1988; Wright et al., 1988).

To bypass modeling of the ligand into smaller derivatives that may fold incorrectly, the AIA can be initiated by first generating monoclonal antibodies against the native ligand, expecting that some of them may represent Ab1 antibodies recognizing an important site of a ligand ("shotgun" approach). An example of this approach is described in Section V.

B. Assessing Antigenicity and Immunogenicity of a Ligand

It is widely realized that, so far, there are no reliable methods to determine accurately the antigenic and immunogenic potential of a protein ligand (see above). However, several computer programs are now available for predicting the secondary structure, the hydrophilicity, and the segmental mobility of proteins or peptides. Similar programs exist for comparing a given protein sequence to sequences frequently found in antigenic sites (see Stern, 1991, for a comprehensive review; see Ada and Skehel, 1985, for other comments).

These probabilistic methods, although nondiagnostic, are helpful for comparing the general properties of different ligand derivatives and can be used to design synthetic immunogens with optimal solution behavior. Solubility to aqueous solvents is an important practical aspect that should be taken into account when designing synthetic immunogens because most of the characterization of the antiidiotypic antibodies involves inhibition and binding experiments that require soluble components. Finally, when the selected immunogen is not predicted to be soluble in aqueous solutions, one should always consider coupling to a hydrophilic carrier, or slight chemical modification that will render it water soluble.

C. "Two-Step" Technique

The general outline of this technique is straightforward. Polyclonal antibodies are first generated against the smallest possible antigen that contains the receptor-binding site of a ligand (whenever this is technically feasible). After characterization and isolation of the relevant antibodies (i.e., Ab1), a second immunization ensues and the antiidiotypic antibodies (Ab2) that develop against the paratope of Ab1 are characterized. When the characterization of these antiidiotypic reagents is completed, one can proceed to probe different cellular fractions to identify the receptor of interest.

Alternatively, the AIA can be initiated by raising monoclonal antibodies against the native form of a ligand. Screening of these antibodies may identify immunoglobulins that behave as Ab1 antibodies. The subsequent steps of such an analysis are not significantly different from the ones already discussed.

Two optimized methods for performing the first (as well as the subsequent immunizations are described in Table II. [These protocols were originally developed by D. Louvard (Institute Pasteur, Paris, France) during a stay at the European Molecular Biology Laboratory (EMBL).] However, it should be noted that the number of immunization schemes currently in use almost matches

Table II
Immunization Schemes

Protocol A
1. At day 0, 100–200 μg of antigen[a] in Freund's complete adjuvant, subcutaneously and in the thigh lymph nodes
2. At day 21, 50–100 μg of antigen in Freund's incomplete adjuvant, intramuscularly
3. At day 37 and every second week thereafter, 50–100 μg of antigen, intramuscularly

Protocol B
1. At day 0, 100–200 μg of antigen in Freund's complete adjuvant, in the thigh lymph nodes and subcutaneously
2. At day 21, 50–100 μg of antigen in Freund's incomplete adjuvant, subscupurarly
3. At day 37 and every second week thereafter, 50–100 μg of antigen intramuscularly and intravenously

[a] Lower doses are used for immunizing mice; the protocols for injecting mice are basically the same as above, except that all injections are done intradermally at the back of the animal.

the number of papers on the AIA published during the last decade. (Techniques for coupling peptide antigens to carrier proteins and methods for preparing the injection cocktails are not included. For details on these topics, the reader should consult a collection of previously published protocols; e.g., ''Methods in Enzymology,'' Vol. 70, 1980).

To measure Ab1 activity in the course of the immune response, one should first screen the antisera against the preparation used as an immunogen and against the native ligand. This can be conveniently done with enzyme-linked immunosorbent assays, (ELISAs), or dot-blot assays (Table III), although some proteins apparently change their conformation on immobilization to a solid matrix. Because short synthetic peptides sometimes are not retained well by nitrocellulose filters, a new procedure that involves immobilization on activated nitrocellulose (by divinyl sulfone-ethylenediamine-glutaraldehyde) can be recommended (for details see Lauritzen *et al.*, 1990).

If the pure ligand cannot be obtained in large quantities, one can also try screening the antisera by modified Western blotting techniques (or ''native'' immunoprecipitation). In this case, crude fractions containing the ligand are

Table III

Technique for Testing the Reactivity of Antibodies against Native Ligands by Dotblot Assays

1. Cut a piece of nitrocellulose filter and a piece of Whatman (Clifton, NJ) 3MM paper, both 12.5 × 9.5 cm
2. Wet the two pieces in double-distilled water
3. Assemble a Bio-Rad (Richmond, CA) dot-blot apparatus and insert the nitrocellulose filter with the paper filter underneath
4. Close the blotter firmly and open an outlet to the atmosphere (for gravity filtration)
5. Apply to each well 50–400 μl of ligand or antigen reconstituted in physiological electrolyte buffer [e.g., phosphate-buffered saline (PBS); absolutely no detergents included]
6. Leave the samples to absorb by gravity onto nitrocellulose filter for 30–60 min at room temperature
7. Connect the apparatus to the vacuum outlet and aspirate the liquid
8. Apply to each well 400 μl of physiological buffer (e.g., PBS)
9. Repeat step 7
10. Apply to each well 400 μl of physiological buffer containing 0.1% (v/v) Tween-20
11. Repeat step 7
12. Ensuring that all of the liquid has been aspirated through the filter, quickly disassemble the apparatus and place the filter on clean 3MM Whatman paper. Move fast and avoid air-drying. Cut strips as required and place into the appropriate ''blocking'' solution (e.g., PBS, 0.1% Tween-20, and 1% gelatin)
13. Proceed as with regular Western blots

[a] Strips of nitrocellulose can be stained after performing the assay in two ways: (1) by incubating in isotonic buffer–0.1% Tween-20 and 2 μl of India ink/ml for 1–15 hr, and destaining with distilled water, or (2) by staining with 0.1% Coomassie Brilliant Blue in 50% methanol for 5–7 min and destaining with 40% methanol and 10% acetic acid.

resolved by sodium dodecyl sulfate-polyacrylamide gel electrophoresis (SDS-PAGE) and the proteins are transferred to nitrocellulose filters. Although most proteins are completely unfolded on electrophoresis, it has been found that numerous ligands (including enzymes and DNA-binding proteins) can be efficiently renatured after removal of the SDS from the blots and washing under physiological conditions (see, e.g., Ferrell and Martin, 1989; Northwood et al., 1991; Georgatos et al., 1987; Djabali et al., 1991; Daniel et al., 1983). Four of these protocols for renaturing proteins in situ are described in Table IV. Three technical points are pertinent to the above methods. First, proteins immobilized on filters should not be allowed to air dry before assaying. The surface tension regime in the air–liquid interphase can severely affect protein conformation and result in complete denaturation. Second, treatment of the protein sample before electrophoresis (or of the blot) by heat, acid, or alkali should be absolutely avoided to maintain the integrity of sensitive amino acid residues. Third, reduced as well as nonreduced samples should always be tested in parallel,

Table IV
Methods for Renaturing Proteins after SDS-PAGE

Method A[a] for renaturing proteins on blots
1. Wash nitrocellulose blots with 10–20% (v/v) 2-propanol, at room temperature, for 2–10 min
2. Rinse the blots five times with distilled water (make sure that the nitrocellulose sheets are rewetted properly)
3. Incubate the blots in a solution containing 155 mM NaCl, 20 mM Tris-HCl (pH 7.3), 0.1% (v/v) Tween-20, at room temperature, for 1 hr
4. Bathe the blots in a solution containing 155 mM NaCl, 20 mM Tris-HCl (pH 7.3), 1 mM MgCl$_2$, 1 mM dithiothreitol (DTT), 0.1 phenylmethylsulfonyl fluoride (PMSF), 0.1% Tween-20, and 0.1–1.0% (w/v) gelatin (preheated at 100°C for 15 min), for 15–18 hr at room temperature with gentle shaking

Method B[b] for renaturing proteins on blots
1. Bathe the blots in 7 M guanidine-HCl, 50 mM Tris, 50 mM DTT, 2 mM ethylenediaminetetraacetic acid (EDTA) (pH 8.3), for 1 hr at room temperature
2. Incubate blots in 100 mM NaCl, 50 mM Tris, 2 mM DTT, 2 mM EDTA, 1% (w/v) bovine serum albumin (BSA), 0.1% (w/v) Nonidet P-40 (pH 7.5) for 12–18 hr at 4°C

Method C[c] for renaturing proteins on blots
Incubate blot in 50 mM Tris-HCl, 2 mM CaCl$_2$, 50 mg of BSA/ml, and 90 mM NaCl (pH 8.0) for 30 min at 37°C

Method D[d] for renaturing proteins on polyacrylamide gels
1. Wash the gel with 20% 2-propanol in 50 mM Tris (pH 8.0)
2. Treat the gel with 6 M guanidine-HCl, 50 mM Tris (pH 8.0), 5 mM 2-mercaptoethanol
3. Incubate the gel with buffer containing 0.04% Tween-40, for 14–16 hr at 4°C

[a] After Georgatos et al. (1987).

[b] After Ferrell and Martin (1989). The guanidine-HCl treatment is used to denature the protein completely before renaturation, preventing misfolding from wrongly folded and incompletely denatured intermediates.

[c] After Daniel et al. (1983).

[d] After Northwood et al. (1991).

because reduction sometimes results in subunit dissociation or irreversible denaturation.

A method to examine whether an antiserum contains anti-ligand antibodies that recognize the native form of cytoplasmically exposed cellular proteins (suggested to the author by R. Duden and T. E. Kries, EMBL) is to microinject the antibodies into living cells. Under conditions of nonsaturation (not infrequent *in vivo*), the antibodies may bind to the native ligand. The partitioning of the microinjected antibodies can be assessed by staining with appropriate fluorescently labeled antibodies and visualization of the specimen by indirect immunofluorescence.

To complete the characterization of the primary antisera, one should examine whether an excess of soluble immunogen can inhibit the binding of Ab1 to the native ligand, and whether the antisera (that presumably contain Ab1) can inhibit the binding of the native ligand to a preparation containing the putative receptors. If the above analyses suggest the existence of genuine Ab1 activity in the antiserum, this activity must be purified with an immunogen affinity matrix. Although the methods for eluting antibodies from columns containing the immobilized immunogen are more or less established, one point of caution must be discussed here. Too harsh an elution condition (e.g., acid elution, Potassium thiocyanate (KSCN) elution, etc.) may result in loss of antibody activity. On the other hand, too mild an elution scheme may leave the high-affinity antibodies uneluted and dissociate only the low-affinity antibodies. Tsang and Wilkins (1991) have compared 13 different elution methods and reached the conclusion that the best results are obtained by a cocktail containing high concentrations of $MgCl_2$ and ethylene glycol at neutral pH.

The purified Ab1 can be directly injected into a second group of animals for generating antiidiotypic antibodies. The methods used for performing the second immunization and the follow-up are exactly as in the previous case. However, when a monoclonal Ab1 is available, one has the option of immunizing inbred animals, to eliminate allotypic and xenotypic reactivities. It is also useful to monitor the responses of the immunized animals on a weekly or fortnightly basis, because in a number of cases the titer of Ab2 has been seen to vary markedly during the immune response (Schechter *et al.*, 1984; Marriott *et al.*, 1987).

To identify antiidiotypic antibody activities, the secondary antisera should be screened against Ab1. To facilitate analysis, especially when different species of outbred animals are used, two alternative methods may be considered. First, the immune sera can be depleted of xenotypic, isotypic, and allotypic reactivities by absorbing them through a matrix containing total immunoglobulins of the animal species used for immunization. Second, to examine whether the antisera contain antibodies that map near or at the antigen-combining site of Ab1, the purified primary antibodies can be digested with proteases to produce Fab fragments. The purified Fabs can then be immobilized on nitrocellulose filters and tested for binding an activity present in the secondary antisera. This analysis, although not yet proving the anti-idiotypic character of the antibodies, provides a practical way to identify potentially interesting antisera.

To evaluate the antiidiotypy of the secondary antibodies, one should examine whether they inhibit the reaction of Ab1 (or Ab1 Fabs) with the ligand, or whether their reaction with Ab1 is inhibited by an excess of soluble ligand. These tests are usually feasible and sufficiently diagnostic. However, to show that the antiidiotypes constitute "internal image antibodies" and not "anti-framework antibodies" is considerably more demanding. A first indication can be obtained by examining the ability of the anti-idiotypes to block *in vitro* binding of the ligands to its putative receptor (see, e.g., Pain *et al.*, 1988, 1990). However, unequivocal evidence for this requires either bioassays, or criteria beyond the cross-reactivity of the antibodies (see Section IV and V).

In some cases the binding of the antiidiotypes to the putative receptors and the corresponding idiotypes was found to be uninhibitable by the ligand. An explanation that has been proposed to explain this paradox is that antiidiotypes occasionally develop against the "reverse" aspects of the CDRs of the idiotypic antibodies (Erlanger, 1985), and therefore possess the "anti-conformation" of a receptor-binding site (see Fig. 3).

D. "One-Step" (Autoimmune) Technique

Animals immunized with neurotransmitter analogs, peptide hormones, or synthetic peptides develop anti-ligand (idiotypic) antibodies that can occasionally act themselves as autoantigens, provoking a secondary antiidiotypic response (see, e.g., Cleveland *et al.*, 1983; Schechter *et al.*, 1982, 1984; Rivas *et al.*, 1988; Djabali *et al.*, 1991; Jennings and Cotton, 1990; Lombes *et al.*, 1989). Exploiting such responses (which may also develop during ascites production in nonirradiated animals; S. Fuller, personal communication) makes feasible the generation of anti-receptor antibodies in one step.

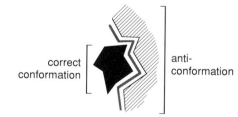

Fig. 3 "Replication" of a paratope by antiidiotypic antibodies. A hypothetical interface of the CDR system of Ab1 is depicted as a gray zig-zag line. Typical Ab2 will have the conformation shown in the black area on the left and they will not bind to Ab1 complexed with antigen; "replication" of the hatched "anti-conformation" generates atypical antiidiotypes that may still react with Ab1 when antigen is bound to it (see text for details).

Most of the methods employed in the one-step approach are similar to those outlined above for the two-step technique. However, because in many instances the sera obtained from immunized animals contain both Ab1 and Ab2 antibodies, an additional step is necessary to separate and characterize these activities. This extra step usually involves affinity purification of Ab1, using a ligand affinity matrix.

IV. Receptor Identification by Antiidiotypic Antibodies

To identify receptors or partners of a ligand by antiidiotypic reagents, conventional immunochemical and immunohistochemical techniques are usually employed. These include Western blotting of cellular fractions that show ligand-binding activity, immunoprecipitation, and in some instances, immunodecoration of the receptor *in situ*. In respect to the latter, it must be emphasized here that antiidiotypic antibodies are not expected to react with the putative receptor when it is occupied by the ligand under *in vivo* conditions. Thus, in fixed specimens, a genuine "internal image antiidiotype" may not decorate the organelles that contain the receptor protein (see Section V,B, below). This, of course, must be distinguished from a trivial lack of reactivity due to the fixation method or other variables. If the antiidiotypes do decorate the receptor *in situ*, there are two possibilities: either a significant subpopulation of the receptor sites are unliganded *in vivo*, or the antiidiotype reacts with a site distinct from the ligand-binding site of the receptor (and therefore does not qualify as true internal image antibody). A way to differentiate between these two alternatives can be to repeat the immunostaining with the antiidiotypic antibodies, using cells preincubated with an excess of ligand (in the case of extracellular ligands), or after inducing receptor saturation by overexpressing the ligand in living cells. However, this may not be always possible.

Internal image antiidiotypes are, by construction, expected to be monospecific. Such monospecificity is sometimes apparent even when the crude antiidiotypic serum is used to probe different cellular fractions. However, in theory, it is perfectly possible for an antiidiotypic antibody to react with more than one protein species, if a common site along the ligand is recognized by multiple receptors *in vivo*. These different receptors may either be structurally related (receptor family), or unrelated molecules with similar ligand-binding pockets. Hence, any cross-reacting proteins that do not have the expected characteristics of the putatuve receptor should not be discarded as background but rather be considered as equally plausible candidates for a function shared with the receptor of interest.

To characterize a putative receptor identified by screening with antiidiotypic antibodies, some investigators have raised an independent polyclonal antibody against the antiidiotype-cross-reacting protein (Pain *et al.*, 1990). This is a useful approach, because it overcomes the limitations in immunolocalizing the receptor by the antiidiotypic antibodies (see above) and because such an antibody can be further exploited in immunoprecipitation experiments to isolate native receptor–ligand complexes. It is also obvious that, as soon as a candidate has been identified by virtue of its cross-reaction with the antiidiotypes, this molecule must be recharacterized as a ligand-binding protein. It should be emphasized here that receptor characterization is not completed merely by demonstrating a reaction of a cellular protein with antiidiotypic antibodies. As with any method used for receptor identification, the complete characterization of a receptor requires evidence that the ligand and the candidate protein associate *in vivo* and that a function in which the ligand is involved can be affected by (genetic) elimination or inactivation of the candidate factor identified by the AIA (see, Murakami *et al.*, 1990). Although seldom achieved, demonstration that the antiidiotypes possess biological activity functionally mimicking the ligand (Schechter *et al.*, 1982, 1984; Sege and Peterson, 1978), constitutes strong evidence that the AIA has been sucessful.

V. Applications: Use of Antiidiotypic Antibodies to Study Molecular Interactions of Intermediate Filament Proteins

Intermediate filaments (IFs) constitute major cytoskeletal components of the eukaryotic cytoplasm. These elements are nonrandomly distributed inside the cell and, in most cases, are organized as radial networks extending from the plasma membrane to the nuclear surface (Goldman *et al.*, 1985). The fibrous, 10-nm thick, ropelike backbone of IFs is built from laterally and longitudinally associated subunits, the so-called intermediate filament proteins (IFPs). Depending on cell type and developmental state, IFs can be assembled from different IFPs (or combinations of IFPs). Data indicate the existence of more than 40 IFPs and the list continues to grow. Despite their sequence differences, these proteins share a common domain substructure (N-terminal "head," middle "rod," and C-terminal "tail" domains) and can be grouped into six categories (types I–VI; for a review see Stewart, 1990).

To understand the roles of the different IFs, two questions are of mechanistic importance: first, how subunits self-associate in an orderly fashion to form a polymeric structure and, second, how the individual IFs are interacting with other cellular structures to build highly organized filamentous networks. In the final analysis, both questions can be reduced to a problem of protein–protein association that implies site-specific recognition and binding. Because structures that are indistinguishable from the native IFs can be assembled *in vitro* from isolated IFPs, it seems reasonable to assume that the blueprint for the self-association reactions that lead to filament formation is contained with the sequences of the subunit molecules. This does not appear to be the case when one considers the assembly of IF networks; such elaborate systems cannot be reproduced *in vitro* by the mere mixing of isolated IFPs. Thus, to understand the regulation of the IFs in a cellular context, one should examine the molecular interactions between IFPs and other cellular factors. Described below are two instances in which we have tried to address some aspects of these problems by employing the AIA.

A. Identification of Self-Association Sites in Intermediate Filament Proteins

To initiate this analysis, we have selected a monoclonal antibody (anti-Ct) that had been shown to affect native IFs when microinjected into living fibroblastic cells. To develop anti-Ct, a fusion peptide that represents the 100 C-terminal amino acid residues of the IFP vimentin was used as an antigen (for details see Kouklis, 1990). Because of its *in vivo* effects, we reasoned that anti-Ct may block self-association sites of vimentin subunits (see Fig. 2c for a schematic explanation) and somehow interfere with the normal remodeling of IFs.

The epitope of anti-Ct was mapped by immunoblotting of vimentin fragments between residues 364 and 416 of this protein (Kouklis *et al.*, 1991). Further work also showed that anti-Ct recognizes a similar epitope present in the IFPs desmin and peripherin (all type III subunits), but not in type IV and type V IFPs (neurofilament and lamin subunits). We termed the common epitope of type III proteins the ε *epitope* (Fig. 4).

To identify sites complementary to the ε epitope within the type III IFP molecules, we then immunized rabbits with purified anti-Ct IgG. In this manner we obtained two potentially antiidiotypic antisera. As shown in Fig. 5C, one of these antisera (antiserum 28) contained antibodies recognizing desmin and vimentin. Other experiments (Kouklis *et al.*, 1991) demonstrated that peripherin was also recognizable by this antiserum. Thus, immunization with anti-Ct ap-

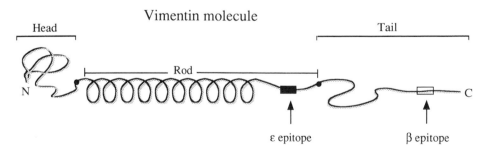

Fig. 4 Diagrammatic representation of the domain substructure of vimentin. The scheme depicts its N-terminal domain (head), its helical middle domain (rod), and its C-terminal domain (tail). The sites recognized by the anti-Ct antibody (ε epitope) and the corresponding antiidiotype 28 (β epitope) are indicated by arrows.

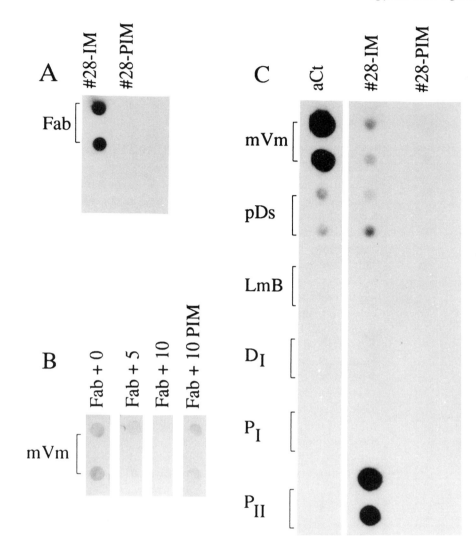

Fig. 5 Characterization of antiidiotypic antibodies to anti-Ct. (A) Reaction of antiserum 28 with purified Fab fragments of anti-Ct. Fab fragments were produced by papain digestion of purified anti-Ct IgG and isolated by absorbing out the Fc fraction of the digest by Sepharose 4B--protein A. Aliquots (2 μg) of these preparations were spotted onto nitrocellulose filters (see Table III) and probed either with antiserum 28 (rabbit serum obtained after immunization of rabbits with purified anti-Ct IgG), or the corresponding preimmune serum at a dilution of 1:200. The reaction was developed by ^{125}I-labeled protein A. (B) Inhibition of the binding of Fab fragments of anti-Ct to vimentin by antiserum 28. Aliquots (3 μg) of purified vimentin were spotted on replica nitrocellulose filters and incubated with 450 μg of anti-Ct Fabs/ml plus 0,5, and 10 μl of the immune serum or 10 μl of the corresponding preimmune system. The reaction was developed by alkaline phosphatase–goat anti-mouse antibodies (light and heavy chain specific). (C) Reaction of antiserum 28 with IFPs and synthetic peptides. Approximately 3 μg of mouse vimentin (mVm), porcine desmin (pDs), rat lamin

we obtained two potentially antiidiotypic antisera. As shown in Fig. 5C, one of these antisera (antiserum 28) contained antibodies recognizing desmin and vimentin. Other experiments (Kouklis *et al.*, 1991) demonstrated that peripherin was also recognizable by this antiserum. Thus, immunization with anti-Ct appeared to elicit anti-anti-Ct antibodies that recognized all three IFPs that are reactive with anti-Ct. Interestingly, when we tested various peptide preparations, antiserum 28 reacted strongly with a peptide (PII) representing a 30 C-terminal residues of peripherin, that is, a part of its tail domain. PII contains a segment highly conserved among vimentin, desmin, and peripherin (T-I/V-E-T-R-D-G-X-V), whereas the rest of its sequence is different among the three IFPs. We assumed that this sequence motif constitutes an epitope common among type III IFPs and termed it the β *epitope* (Fig. 4).

After appropriate analysis was carried out, we found that the antiserum 28 antibodies had the following properties: (1) they reacted with anti-Ct Fabs (Fig. 5A); (2) they inhibited the reaction of anti-Ct Fabs with vimentin (Fig. 5B); and (3) they reacted, after affinity purification over a PII column, with type III IFPs and anti-Ct IgG, but not with other IFPs (as, e.g., nuclear lamin B and neurofilament L protein) or an irrelevant mouse IgG (for original data see Kouklis *et al.*, 1991). Based on these features, we tentatively concluded that the antiserum 28 antibodies represent true antiidiotypic antibodies that recognize both the paratope of anti-Ct, and a conserved sequence motif in the tail domains of different type III IFPs.

Implicit in the above interpretation is that the ε and β epitopes bind to each other. A first indication in favor of this had been that injection of anti-Ct into 3T3 cells changes the organization of vimentin filaments, resulting in the formation of thick fibers, a retraction of IFs away from the cell membrane, and, finally, a collapse of IFs into a honeycomb structure (Fig. 6). To confirm it, we examined the possibility for a direct interaction between the ε and β sites. First, by means of ligand blotting, we showed that PII binds to intact vimentin, the rod domain of vimentin, and a peptide that lacks the N-terminal head domain of vimentin. We also demonstrated that neurofilament L subunits (that do not react with anti-Ct) do not bind PII. Second, we showed that the reaction of anti-Ct to the rod fragment of vimentin is competed by soluble PII (Kouklis *et al.*, 1991). These two pieces of evidence clearly suggested that a segment represented in the PII sequence binds at or near the ε epitope of vimentin.

B (LmB), and ~28 μg of the synthetic peptides DI, PI, and PII (corresponding to different parts of tail domains of chicken desmin and rat peripherin, respectively) were immobilized on nitrocellulose filters. These blots were probed with anti-Ct (aCt; 20 μg/ml), antiserum 28 (#28-IM), or preimmune serum 28 (#28-PIM), both of the latter at a dilution of 1:150. The reactions were developed by either ^{125}I-labeled goat–anti-mouse antibodies (aCt), or by ^{125}I-labeled protein A (#28), respectively. (Reproduced from the *Journal of Cell Biology*, Rockefeller Press.)

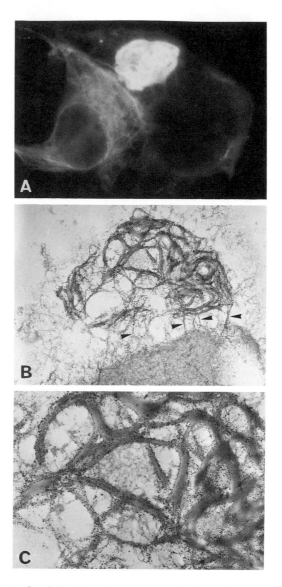

Fig. 6 *In vivo* effects of anti-Ct. 3T3 cells were injected with purified anti-Ct and analyzed either by indirect immunofluorescence (using a polyclonal anti-vimentin antibody) as in (A), or by immunoelectron microscopy, as in (B) and (C). Note the mass of immunoreactive material that develops 1 day postinjection and the reticular pattern of anastomosed thick fibrils that are detected by electron microscopy. (Reproduced from the *Journal of Cell Biology*, Rockefeller Press.)

To examine the effects of blocking the ε–β interaction on filament assembly, we proceeded with *in vitro* reconstitution experiments. As depicted in Fig. 7, when vimentin subunits were mixed with increasing quantities of PII and then induced to polymerize by the addition of salt, we observed the gradual conversion of regular 10-nm filaments into thick fibrils (~40 nm and increasing), as the amounts of PII increased. Thus, competition between the β site contained in the PII peptide and the endogenous β site of vimentin, and, conceivably, "titration" of the natural ε site on vimentin by the synthetic peptide, reproduced the thick filament morphologies we had detected by injecting the anti-Ct antibody (which is also expected to block the ε site *in vivo*). These alterations in the diameter of vimentin IFs fitted well other data obtained with the type III IFP desmin. Specifically, Kaufmann *et al.* (1985) had previously observed that removal of 27 C-terminal residues from desmin subunits results in filaments with a "bundled" appearance.

From all of the above, it follows that the application of the AIA in the self-association of IFPs has already yielded useful information, unveiling a potential interaction between two specific sites of type III subunits that may play an important role in the regulation of the lateral packing of IFs. Further experiments are now in progress to examine whether this specific mechanism is operative under *in vivo* conditions.

B. Exploring Interactions of Cytoplasmic Intermediate Filament Proteins with Lamin B

Previous *in vitro* work has established that vimentin, desmin, and peripherin bind specifically (through their C-terminal domains) to type B lamin proteins under *in vitro* conditions (Georgatos and Blobel, 1987; Georgatos *et al.*, 1987; Djabali *et al.*, 1991). The type B lamins, themselves members of the IF superfamily, are located along the inner side of the inner nuclear membrane and are organized (together with the type A lamins) as an elaborate polymeric meshwork termed the nuclear lamina (Aebi *et al.*, 1986). The *in vitro* binding of vimentin, desmin, and peripherin to the type B lamins suggests that cytoplasmic IFs may be attached to the nuclear lamina, providing a physical link that may be important for conveying signals from the cell surface to the nucleus.

A topological paradox associated with this interaction has been that the nuclear lamina and the cytoplasmic IFs are normally separated by a double-membrane barrier (nuclear membranes). Assuming that the lamin B–cytoplasmic IF interaction does occur under physiological conditions, three *in vivo* scenarios can explain the *in vitro* binding data. First, cytoplasmic IFs may cross the nuclear pores (as thinner protofilaments) and directly connect to the nuclear lamina. Second, these interactions may take place only during mitosis (when the lamin Bs become cytoplasmic after nuclear envelope breakdown). Third, the detected interaction between lamin B and type III IFPs could represent a distant "evolutionary memory" of another association between nonnu-

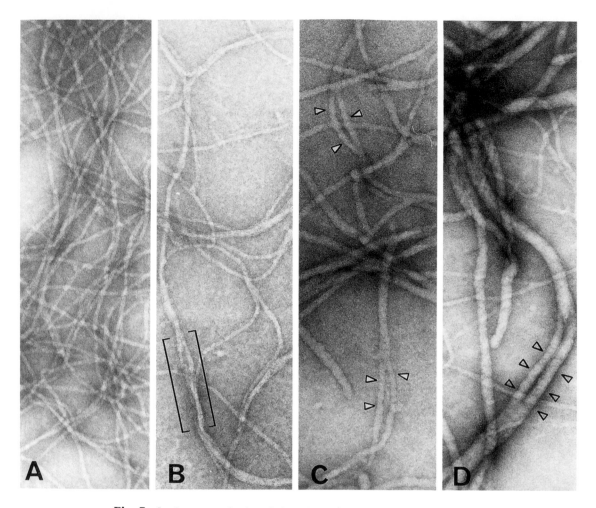

Fig. 7 *In vitro* reconstitution of vimentin IFs in the presence of the peptide PII. Purified vimentin protofilaments (400 μg/ml) were mixed with 0 (A), 100 (B), 500 (C), and 1500 (D) μg of PII/ml in 10 mM Tris-HCl, pH 7.4. Samples were preincubated for 60 min at room temperature before addition of concentrated KCl to 150 mM. After another 60 min at 37°C, the samples were processed for negative staining and electron microscopy. Note the gradual increase in the filament diameter. Arrowheads and bracket indicate points of filament unraveling. (Reproduced from the *Journal of Cell Biology,* Rockefeller Press.)

clear lamin B-like proteins and cytoplasmic IFs. The latter possibility (which does not exclude the other two interpretations) is favored by information showing that such lamin B-like proteins do exist in specialized areas of the plasma membrane (Cartaud *et al.,* 1989, 1990).

To investigate the specificity of lamin B–type III IF interactions, and to

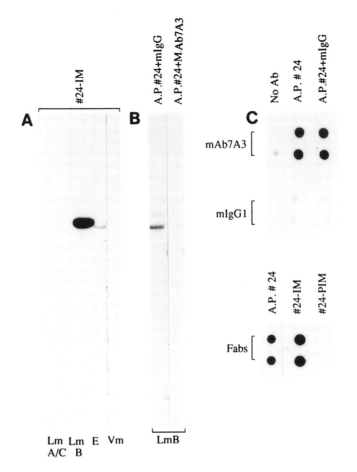

Fig. 8 Characterization of antiidiotypic antibodies against MAb 7A3. (A) Equivalent amounts of rat lamins A/C (Lm A/C), lamin B (LmB), a karyoskeletal extract of bird erythrocyte nuclear envelopes (E), and purified mouse vimentin (Vm) were probed by antiserum 24 (#24) developed against purified 7A3 IgG (1:200 dilution). Note the strong reaction with rat lamin B, the weak cross-reaction with bird lamin B (of lower molecular weight), and the lack of a reaction with the lamins A/C (which are structurally similar to lamin B). (B) Inhibition of the antiidiotype–lamin B reaction by soluble idiotype (7A3 antibody). Antibody 24, affinity purified in a lamin B–agarose column at a concentration of 5 µg/ml, was used to probe two replica blots of purified rat lamin B, in the presence of nonimmune mouse IgG (A.P. #24 + mIgG), or of purified 7A3 IgG (A.P. #24 + MAb7A3), both at a concentration of 200 µg/ml. Note the lack of reaction in the second case. (C) Binding of 7A3 to its anti-idiotype. In the upper panel, 3 µg of purified MAb 7A3 and 3 µg of an irrelevant mouse IgG₁ were applied to nitrocellulose and incubated with buffer alone (No Ab), affinity-purified antibody 24 (A.P. #24, at 1 µg/ml), or a mixture of affinity-purified antibody 24 and nonimmune mouse IgG (1 and 50 µg/ml, respectively). In the lower panel, 2 µg of 7A3 Fabs were applied to nitrocellulose filters and probed with affinity-purified antibody 24 (A.P. #24, 2.5 µg/ml), antiserum 24 (#24-IM, diluted 1:200), or preimmune serum 24 (#24-PIM, diluted 1:200). All of the reactions were developed by ^{125}I-labeled protein A. (Reproduced from the *Journal of Biological Chemistry,* American Society for Biochemistry and Molecular Biology, Inc.)

develop probes that could identify such lamin B-like factors, we have employed a monoclonal antibody (7A3), previously prepared by Matteoni and Kreis (1987), and shown to collapse the vimentin IFs of living cells. Epitope mapping of this antibody showed that its binding site is localized within the C-terminal tail domain of vimentin. Further experiments also showed that 7A3 recognizes the type III IFP peripherin and a synthetic peptide (PI) modeled after the proximal half of its C-terminal domain (Papamarcaki et al., 1991). PI is known to contain the peripherin lamin B-binding site (Djabali et al., 1991). Monoclonal antibody 7A3 inhibited the binding of PI to lamin B, suggesting that the corresponding epitope is located at or near the lamin B-binding site of peripherin (Papamarcaki et al., 1991).

When rabbits were immunized with purified 7A3 IgG, we obtained antiserum 24, which strongly reacts with lamin B but not with lamin A or other IF proteins (Fig. 8A). Addition of soluble 7A3 to antiserum 24 antibodies (affinity purified over lamin B) abolished their reaction with lamin B (Fig. 8B), whereas affinity-purified antiserum 24 IgG bound to 7A3 Fabs (Fig. 8C). These data are consistent with the idea that the anti-lamin B activity antiserum 24 comprises a genuine

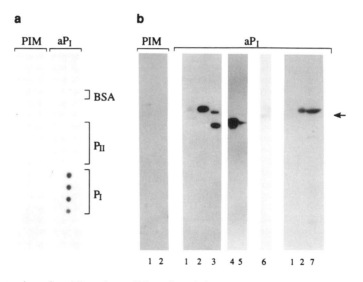

Fig. 9 Detection of an idiotypic–antiidiotypic pair in the same serum. (a) Peptides PI and PII (0.45, 0.9, 1.35, and 1.8 μg), as well as 24 μg of BSA, were immobilized on nitrocellulose. These preparations were probed with anti-PI (aPI), an antiserum obtained by immunizing rabbits with PI, or with the corresponding preimmune serum (PIM). (b) Western blots of purified rat liver lamins A/C (lanes 1), purified rat liver lamin B (lanes 2), a karyoskeletal extract of bird erythrocyte nuclear envelopes (lane 3), purified mouse neuroblastoma peripherin (lane 4), purified mouse vimentin (lane 5), purified porcine neurofilament L protein (lane 6), and whole-rat liver nuclear envelopes (lane 7) were probed with either the anti-PI antiserum (aPI; diluted from 1:150 to 1:400), or the preimmune serum (PIM; diluted 1:150). All reactions were developed by ^{125}I-labeled protein A. Arrow points to the position of rat lamin B. (Reproduced from Cell, MIT Press).

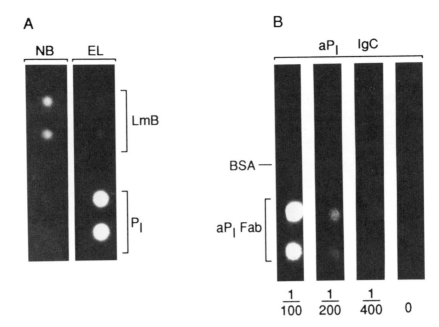

Fig. 10 Separation of the anti-lamin B and the anti-peripherin/PI antibodies. A sample of aPI serum was applied to a PI–agarose column. The resultant column flow-through (NB) and the fraction eluted by low pH (EL) were used to probe dot blots of rat lamin B (LmB; 10 μg) or PI (PI; 2.7 μg). Note the separation of the two activities. (B) Demonstration of antibody–antibody complexation in the aPI antiserum. Fabs of total aPI IgG (~1–1.5 μg) and BSA (50 μg) were immobilized on nitrocellulose and probed with aPI (at the indicated dilutions), or buffer alone (0). The reactions were developed by ^{125}I-labeled protein A. Note the presence of an antibody–antibody reaction. This is due to binding of anti-lamin B (idiotypic) to the anti-peripherin (antiidiotypic) antibodies, because absorption of the aPI Fabs to PI–agarose and probing with the NB fraction shown in (A) significantly quenches the reactions (not shown here). [Reproduced (in reverse contrast) from *Cell*, MIT Press.]

antiidiotypic antibody reacting with the paratope of 7A3 and recognizing a "receptor" for type III IFs, that is, the protein lamin B. This interpretation is supported by the fact that antiserum 24 antibodies inhibited the binding of lamin B to PI (Papamarcaki *et al.,* 1991).

In another twist of the same project, we have immunized rabbits with the PI peptide (Djabali *et al.,* 1991). When the immune sera were tested, we detected two distinct reactivites. One antibody population reacted with peripherin, vimentin, and desmin, whereas another antibody population was specifically reacting with only one nuclear envelope protein, which could be identified as lamin B (Fig. 9). Because the two classes of antibodies were separable by affinity chromatography over a PI–agarose column (Fig. 10A), it seemed unlikely that the reactivities detected were due to a common epitope in lamin B and type III IFPs. Binding of the two antibodies to the corresponding antigens was shown to be inhibited as lamin B and PI bound to each other *in vitro*. Furthermore,

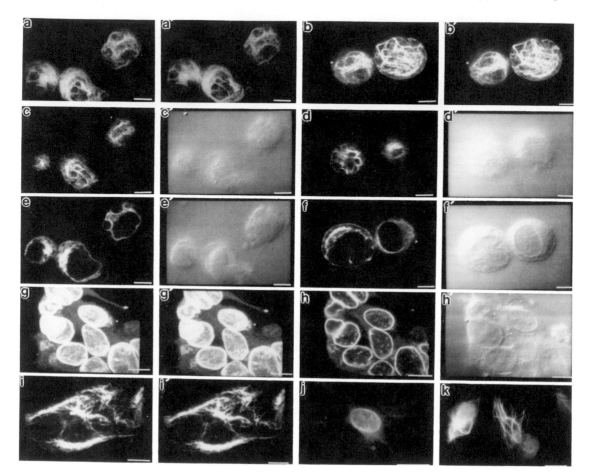

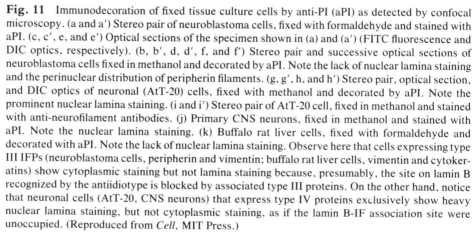

Fig. 11 Immunodecoration of fixed tissue culture cells by anti-PI (aPI) as detected by confocal microscopy. (a and a') Stereo pair of neuroblastoma cells, fixed with formaldehyde and stained with aPI. (c, c', e, and e') Optical sections of the specimen shown in (a) and (a') (FITC fluorescence and DIC optics, respectively). (b, b', d, d', f, and f') Stereo pair and successive optical sections of neuroblastoma cells fixed in methanol and decorated by aPI. Note the lack of nuclear lamina staining and the perinuclear distribution of peripherin filaments. (g, g', h, and h') Stereo pair, optical section, and DIC optics of neuronal (AtT-20) cells, fixed with methanol and decorated by aPI. Note the prominent nuclear lamina staining. (i and i') Stereo pair of AtT-20 cell, fixed in methanol and stained with anti-neurofilament antibodies. (j) Primary CNS neurons, fixed in methanol and stained with aPI. Note the nuclear lamina staining. (k) Buffalo rat liver cells, fixed with formaldehyde and decorated with aPI. Note the lack of nuclear lamina staining. Observe here that cells expressing type III IFPs (neuroblastoma cells, peripherin and vimentin; buffalo rat liver cells, vimentin and cytokeratins) show cytoplasmic staining but not lamina staining because, presumably, the site on lamin B recognized by the antiidiotype is blocked by associated type III proteins. On the other hand, notice that neuronal cells (AtT-20, CNS neurons) that express type IV proteins exclusively show heavy nuclear lamina staining, but not cytoplasmic staining, as if the lamin B-IF association site were unoccupied. (Reproduced from *Cell,* MIT Press.)

binding of the anti–lamin B antibodies to the anti–PI/peripherin antibodies could be detected *in vitro* (Fig. 10B). Finally, the anti–lamin B antibodies were found not to decorate the nuclear lamina of tissue culture cells expressing type III IFPs, but to stain heavily the nuclear lamina of cells expressing type IV IFPs (Fig. 11).

The interpretation we have proposed to explain these data is that, on immunization with PI, the animals developed anti-PI/peripherin antibodies, some of which replicated the lamin B-binding site of type III IFPs. Such antiPI/peripherin antibodies probably acted as autoantigens to elicit antiidiotypic antibodies that recognize the peripherin "receptor" lamin B. Thus, for the type III IF–lamin B system, the two-step and one-step methods produced identical results. In more recent experiments we have also been able to produce monoclonal antibodies recognizing lamin B by immunizing mice with PI (K. Djabali and S.D.G., unpublished observations). It is remarkable that all the antiidiotypes that we have developed were able to differentiate between type A and type B lamins, which belong to the same subfamily and share a number of structural and biochemical features. In future, these reagents should be suitable for *in vivo* studies and for identifying other IF-binding proteins.

VI. Conclusions

The AIA offers a powerful screening method for identifying potential receptors or partners of biologically significant ligands. However, there are pitfalls and, to minimize the possibility of misleading results, the following points need to be considered.

1. Before initiating the AIA, the interaction between the ligand of interest and its putative partners should be characterized as completely as possible, using conventional binding assays.

2. The antiidiotypy of the reagents employed for receptor identification should be experimentally investigated and not "inferred" from indirect observations.

3. One should realize that not all antiidiotypic antibodies can serve as probes for identifying unknown receptors and binding patterns; true internal image antiidiotypes are rarely obtained.

4. Finally, the AIA should not be viewed as a "wonder method" that bypasses conventional biochemical, ultrastructural, and genetic analysis, but rather as a screening method complementary to all of these approaches.

Acknowledgments

I would like to thank Gunter Blobel (Rockefeller University, New York, NY) for support and for suggesting the writing of this article. I would also like to thank John Tooze, Kai Simons, Thomas

Kreis, George Simos, and especially Stephen Fuller (all at EMBL, Heidelberg, Germany) for their numerous critical comments and for stimulating discussions. The experimental work presented here was carried out by Karima Djabali (College de France, Paris, France), Panos D. Kouklis (EMBL), Andreas Merdes (EMBL), and Thomais Papamarcaki (University of Ioannina, Ioannina, Greece). This article is dedicated to Elias Brountzos (Antigoni Metaxa General Hospital, Pireas, Greece), who has been a constant source of courage and inspiration during the last 10 years.

References

Ada, G., and Skehel, J. J. (1985). Are peptides good antigens? *Nature (London)* **316,** 764–765.

Aebi, U., Cohn, J., Buhle, L., and Gerace, L. (1986). The nuclear lamina is a meshwork of intermediate-type filaments. *Nature (London)* **323,** 560–564.

Amit, A. G., Mariuzza, R. A., Phillips, S. E. V., and Roljak, R. J. (1986). Three dimensional structure of an antigen–antibody complex at 2.8 Å resolution. *Science* **233,** 747–753.

Atassi, M. Z. (1975). Antigenic structure of myoglobin: The complete immunochemical anatomy of a protein and conclusions relating to antigenic structures of proteins. *Immunochemistry* **12,** 423–438.

Atassi, M. Z. (1978). Precise determination of the entire antigenic structure of lysozyme. *Immunochemistry* **15,** 909–936.

Atassi, M. Z. (1984). Antigenic structures of proteins. *Eur. J. Biochem.* **145,** 1–24.

Benjamin, D. C., Berzofsky, J. A., East, I. J., Gurd, F. R. N., Hannum, C., Leach, S. J., Margoliash, E., Michael J. G., Miller, A., Prager, E. M., Reichlin, M., Sercarz, E. E., Smith-Gill, S. J., Todd, P. E., and Wilson, A. C. (1984). The antigenic structure of proteins: A reappraisal. *Annu. Rev. Immunol.* **2,** 67–101.

Bentley, G. A., Boulot, G., Riottot, M. M., and Poljak, R. J. (1990). Three-dimensional structure of an idiotope–anti-idiotype complex. *Nature (London)* **348,** 254–257.

Berzofsky, J. A. (1985). Intrinsic and extrinsic factors in protein antigenic structure. *Science* **229,** 932–940.

Billetta, R., Hollingdale, M. R., and Zanetti, M. (1991). Immunogenicity of an engineered internal image antibody. *Proc. Natl. Acad. Sci. U.S.A.* **88,** 4713–1417.

Branden, C., and Tooze, J. (1991). Recognition of foreign molecules by the immune system. *In* "Introduction to Protein Structure," pp. 179–199. Garland, New York.

Bruck, C., Co, M. S., Slaoui, M., Gaulton, G. N., Smith, T., Fields, B. N., Mullins, J. I., and Greene, M. I. (1986). Nucleic acid sequence of an internal image-bearing monoclonal anti-idiotype and its comparison to the sequence of the external antigen. *Proc. Natl. Acad. Sci. U.S.A.* **83,** 6578–6582.

Cartaud, A., Courvalin, J. C., Ludosky, M. A., and Cartaud, J. (1989). Presence of a protein antigenically related to nuclear lamin B in the post-synaptic membrane of *Torpedo marmorata* electrolyte. *J. Cell Biol.* **109,** 1745–1752.

Cartaud, A. Ludosky, M. A., Courvalin, J. C., and Cartaud, J. (1990). A protein antigenically related to nuclear lamin B mediates the association of intermediate filaments with desmosomes. *J. Cell Biol.* **111,** 581–588.

Chothia, C. (1991). Antigen recognition. *Curr. Opinion Struct. Biol.* **1,** 53–59.

Chothia, C., and Lesk, A. M. (1987). Canonical structures for the hypervariable of immunoglobulins. *J. Mol. Biol.* **196,** 901–917.

Chothia, C., Lesk, A. M., Tramontano, A., Levitt, M., Smith-Gill, S. J., Air, G., Sheriff, S., Padlan, E. A., Davies, D., Tulip, W. R., Colman, P. M., Spinelli, S., Alzari, P. M., and Poljak, R. J. (1989). Conformations of immunoglobulin hypervariable regions. *Nature (London)* **342,** 877–883.

Cleveland, W. L., Wassermann, N. H., Sarangarajan, R., Penn, A. S., and Erlanger, B. F. (1983). Monoclonal antibodies to the acetylcholine receptor by a normally functioning auto-anti-idiotypic mechanism. *Nature (London)* **305,** 56–57.

Colman, P. M. (1988). Structure of antibody–antigen complexes: Implications for immune recognition. *Adv. Immunol.* **43**, 99–132.

Coutinho, A. (1989). Beyond clonal selection and network. *Immunol. Rev.* **110**, 63–87.

Daniel, T. O., Schneider, W. J., Goldstein, J. L., and Brown, M. S. (1983). Visualization of lipoprotein receptors by ligand blotting. *J. Biol. Chem.* **258**, 4606–4611.

Davies, D. R., Padlan, E. A., and Sheriff, S. (1990). Antibody–antigen complexes. *Annu. Rev. Biochem.* **59**, 439–473.

Davis, S. J., Schockmel, G. A., Somoza, C., Buck, D. W., Healey, D. G., Rieber, E. P., Reiter, C., and Williams, A. (1992). Antibody and HIV-1 gp120 recognition of CD4 undermines the concept of mimicry between antibodies and receptors. *Nature (London)* **358**, 76–79.

DeLisi, C., and Berzofsky, J. A. (1985). T-cell antigenic sites tend to be amphipathic structures. *Proc. Natl. Acad. Sci. U.S.A.* **82**, 7084–7052.

Djabali, K., Portier, M.-M., Gros, F., Blobel, G., and Georgatos, S. D. (1991). Network antibodies identify nuclear lamin B as a physiological attachment site for peripherin intermediate filaments. *Cell* **64**, 109–121.

Dyer, C. A., and Curtiss, L. K. (1991). A synthetic peptide mimic of plasma apolipoprotein E that binds the LDL receptor. *J. Biol. Chem.* **266**, 22803–22806.

Erlanger, B. F. (1985). Anti-idiotypic antibodies: What do they recognize? *Immunol. Today* **6**, 10–11.

Erlanger, B. F., Cleveland, W. L., Wassermann, N. H., Ku, H. H., Hill, B. L., Sarangarajan, R., Rajagopalan, R., Cayanis, E., Edelman, I. S., and Penn, A. S. (1986). Auto-anti-idiotypy: A basis for autoimmunity and a strategy for anti-receptor antibodies. *Immunol. Rev.* **94**, 23–37.

Fermi, G., and Perutz, M. F. (1981). "Haemoglobin and Myoglobin: Atlas of Molecular Structures in Biology." Clarendon, Oxford.

Ferrell, J. E., and Martin, G. S. (1989). Thrombin stimulates the activities of multiple previously unidentified protein-kinases in platelets. *J. Biol. Chem.* **264**, 20723–20729.

Gaulton, G. N., and Green, M. I. (1986). Idiotypic mimicry of biological receptors. *Annu. Rev. Immunol.* **4**, 253–280.

Georgatos, S. D., and Blobel, G. (1987). Lamin B constitutes an intermediate filament attachment site at the nuclear envelope. *J. Cell Biol.* **105**, 117–125.

Georgatos, S. D., Weber, K., Geisler, N., and Blobel, G. (1987). Binding of two desmin derivatives to the plasma membrane and the nuclear envelope of avian erythrocytes: Evidence for a conserved site-specificity in intermediate filament–membrane interactions. *Proc. Natl. Acad. Sci. U.S.A.* **84**, 6780–6784.

Goldman, R. D., Goldman, A., Green, K., Jones, J., Lieska, N., and Yang, H.-Y. (1985). Intermediate filaments: Possible functions as cytoskeletal connecting links between the nucleus and the cell surface. *Ann. N.Y. Acad. Sci.* **455**, 1–17.

Hopp, T. P., and Woods, K. R. (1981). Prediction of protein antigenic determinants from amino-acid sequences. *Proc. Natl. Acad. Sci. U.S.A.* **78**, 3824–3828.

Jaenicke, R. (1991). Protein stability and protein folding. *Ciba Found. Symp.* **161**, 206–221.

Janin, J., and Chothia, C. (1990). The structure of protein–protein recognition sites. *J. Biol. Chem.* **265**, 16027–16030.

Janin, J., Miller, S., and Chothia, C. (1988). Surface, subunit interfaces and interior of oligomeric proteins. *J. Mol. Biol.* **204**, 155–164.

Jennings, I. G., and Cotton, R. G. H. (1990). Structural similarities among enzyme pterin binding sites as demonstrated by a monoclonal anti-idiotypic antibody. *J. Biol. Chem.* **265**, 1885–1889.

Jerne, N. K. (1974). Towards a network theory of the immune system. *Ann. Immunol. (Paris)* **125C**, 373–389.

Jerne, N. K. (1984). The generative grammar of the immune system. *EMBO J.* **4**, 847–882.

Kaufmann, E., Weber, K., and Geisler, N. (1985). Intermediate filament forming ability of desmin lacking either the amino-terminal 67 or the carboxy-terminal 27 residues. *J. Mol. Biol.* **185**, 733–742.

Kouklis, P. D. (1990). Charakterisierung und Anwendung monoklonaler Antikörper gegen Maus-Vimentin. Ph. D Thesis, Univ. of Heidelberg, Heidelberg.

Kouklis, P. D., Papamarcaki, T., Merdes, A., and Georgatos, S. D. (1991). A potential role for the COOH-terminal domain in the lateral packing of type III intermediate filaments. *J. Cell Biol.* **114,** 773–786.

Kunkel, H. G., Mannick, M., and Williams, R. C. (1963). Individual antigenic specificities of isolated antibodies. *Science* **140,** 1218–1223.

Lau, S. Y. M., Taneja, A. K., and Hodges, R. S. (1984). Synthesis of a model protein of defined secondary and quaternary structure. *J. Biol. Chem.* **259,** 13253–13261.

Lauritzen, E., Masson, M., Rubin, I., and Holm, A. (1990). Dot immunobinding and immunoblotting of picogram and nanogram quantities of small peptides on activated nitrocellulose. *J. Immunol. Methods* **131,** 257–267.

Laver, W. G., Air, G. M., Webster, R. G., and Smith-Gill, S. J. (1990). Epitopes on protein antigens: Misconceptions and realities. *Cell* **61,** 553–556.

Lesk, A. M., and Chothia, C. (1984). Mechanisms of domain closure in proteins. *J. Mol. Biol.* **174,** 175–191.

Lombes, M., Edelman, I. S., and Erlanger, B. F. (1989). Internal image properties of a monoclonal auto-anti-idiotypic antibody and its binding to aldosterone receptors. *J. Biol. Chem.* **264,** 2528–2536.

Lundkvist, I. A., Coutinho, A., Varela, F., and Holmberg, D. (1989). Evidence for a functional idiotypic network among natural antibodies in normal mice. *Proc. Natl. Acad. Sci. U.S.A.* **86,** 5074–5078.

Marriott, S. J., Roeder, D. J., and Consigli, R. A. (1987). Anti-idiotypic antibodies to a polyoma virus monoclonal antibody recognise cell surface components of mouse kidney cells and prevent polyoma virus infection. *J. Virol.* **61,** 2747–2753.

Matteoni, R., and Kreis, T. E. (1987). Translocation and clustering of endosomes and lysosomes depends on microtubules. *J. Cell Biol.* **105,** 1253–1265.

Meyer, D. I. (1990). Mimics or gimmicks? *Nature (London)* **347,** 424–425.

Mian, I. S., Bradwell, A. R., and Olson, A. J. (1991). Structure, function and properties of antibody binding sites. *J. Mol. Biol.* **217,** 133–151.

Milich, D. R., McLachlan, A., Hughes, J. L., Moriarty, A., and Thornton, G. B. (1988). Multivalent Hepatitis B virus synthetic vaccine. *In* "Vaccines '88" (H. Ginsberg, F. Brown, R. A. Lerner, and R. M. Chanock eds.), pp. 13–18. Cold Spring Harbor Lab. Press, Cold Spring Harbor, New York.

Miller, S., Lesk, A. R., Janin, J., and Chothia, C. (1987). The accessible surface-area and stability of oligomeric proteins. *Nature (London)* **328,** 834–836.

Murakami, H., Blobel, G., and Pain, D. (1990). Isolation and characterization of the gene for a yeast mitochondrial import receptor. *Nature (London)* **347,** 488–491.

Northwood, I. C., Gonzalez, F. A., Wartmann, M., Raden, D. L., and Davis, R. J. (1991). Isolation and characterization of two growth factor-stimulated protein kinases that phosphorylate the epidermal growth factor receptor at threonine 669. *J. Biol. Chem.* **266,** 15266–15276.

Nossal, G. J. V. (1991). Molecular and cellular aspects of immunological tolerance. *Eur. J. Biochem.* **202,** 729–737.

Oas, T. G., and Kim, P. S. (1988). A peptide model of a protein folding intermediate. *Nature (London)* **336,** 42–48.

Ouadin, J., and Michel, M. (1963). Une nouvelle form d'allotypie des globulins du serum de lapin apparement liee "a la function et al specificite" anticorps. *C. R. Hebd. Seances Acad. Sci.* **257,** 805–808.

Padlan, E. A., Silverton, E. W., Sheriff, S., Cohen, G. H., Smith-Gill, S. J., and Davies, D. R. (1989). Structure of an antibody–antigen complex: Crystal structure of the HYHEL-10 Fab-lysozyme complex. *Proc. Natl. Acad. Sci. U.S.A.* **86,** 5938–5942.

Pain, D., Kanwar, Y. S., and Blobel, G. (1988). Identification of a receptor for protein import into chloroplasts and its localization to envelope contact zones. *Nature (London)* **331,** 232–237.

Pain, D., Murakami H., and Blobel, G. (1990). Identification of a receptor for protein import into mitochondria. *Nature (London)* **347**, 444–449.

Papamarcaki, T., Kouklis, P. D., Kreis, T. E., and Georgatos, S. D. (1991). The "lamin B-fold." *J. Biol. Chem.* **266**, 21247–21251.

Pierce, S. K., and Margoliash, E. (1988). Antigen processing: An interim report. *Trends Biochem. Sci.* **13**, 27–29.

Remington, S. J., Wiegand, G., and Huber, R. (1982). Crystallographic refinement and atomic models of 2 different forms of citrate synthase at 2.7 Å and 1.7 Å resolution. *J. Mol. Biol.* **158**, 111–152.

Rivas, C. I., Vera, J. C., and Maccioni, R. B. (1988). Anti-idiotypic antibodies that react with microtubule-associated proteins are present in sera of rabbits immunized with synthetic peptides from tubulin's regulatory domain. *Proc. Natl. Acad. Sci. U.S.A.* **85**, 6092–6096.

Rose, D. R., Strong, R. K., Margolies, M. N., Gefter, M. L., and Petsko, G. A. (1990). Crystal structure of the antigen-binding fragment of the murine anti-arsonate monoclonal antibody 36-71 at 2.9 Å resolution. *Proc. Natl. Acad. Sci. U.S.A.* **87**, 338–342.

Saragovi, H. U., Fitzpatrick, D., Raktabutr, A., Nakanishi, H., Kahn, M., and Greene, M. I. (1991). Design and synthesis of a mimetic from an antibody complementarity-determining region. *Science* **253**, 792–753.

Schechter, Y., Maron, R., Elias, D., and Cohen, I. R. (1982). Autoantibodies to insulin receptor spontaneously develop as anti-idiotypes in mice immunized with insulin. *Science* **216**, 542–545.

Schechter, Y., Elias, D., Maron, R., and Cohen, I. R. (1984). Mouse antibodies to the insulin receptor developing spontaneously as anti-idiotypes. *J. Biol. Chem.* **259**, 6411–6415.

Schultz, G. E., and Schirmer, R. H. (1979). Noncovalent forces determining protein structure. *In* "Principles of Protein Structure." (C. R. Cantor, ed.), pp. 27–44. Springer-Verlag, New York.

Schultze-Gahmen, U., Prinz, H., Glatter, U., and Beyreuther, K. (1985). Towards assignment of secondary structures by anti-peptide antibodies. Specificity of the immune response to a beta-turn. *EMBO J.* **4**, 1731–1737.

Sege, K., and Peterson, P. A. (1978). Use of anti-idiotypic antibodies as cell-surface receptor probes. *Proc. Natl. Acad. Sci. U.S.A.* **75**, 2443–2447.

Stanfield, R. L., Fieser, T. M., Lerner, R. A., and Wilson, I. A. (1990). Crystal structures of an antibody to a peptide and its complex with peptide antigen at 2.8 Å resolution. *Science* **248**, 712–719.

Stern, P. S. (1991). Predicting antigenic sites on proteins. *Trends Biotechnol.* **9**, 163–169.

Stewart, M. (1990). Intermediate filaments: Structure, assembly and molecular interactions. *Curr. Opinion Cell Biol.* **2**, 91–100.

Tainer, J. A., Getzoff, E. D., Alexander, H., Houghten, R. A., Olson, A. J., Lerner, R. A., and Hendrickson, W. A. (1984). The reactivity of anti-peptide antibodies is a function of the atomic mobility of sites in a protein. *Nature (London)* **312**, 127–134.

Tramontano, A., Chothia, C., and Lesk, A. M. (1990). Framework residue-71 is a major determinant of the position and conformation of the 2nd hypervariable region in the VH domains of immunoglobulins. *J. Mol. Biol.* **215**, 175–182.

Tsang, V. C. W., and Wilkins, P. P. (1991). Optimum dissociating condition for immunoaffinity and preferential isolation of antibodies with high specific activity. *J. Immunol. Methods* **138**, 291–299.

Urbain, J., Slaoui, M., Mariame, B., and Leo, O. (1984). Idiotypy and internal images. *In* "Idiotypy in Biology and Medicine" (H. Kohler, J. Urbain, and P. A. Cazenave, eds.), Academic Press, San Diego.

Vahlne, A., Horal, P., Eriksson, K., Jeansson, S., Rymo, L., Hedstrom, K. G., Czerkinsky, C., Holmgren, J., and Svennerholm, B. (1991). Immunizations of monkeys with synthetic peptides disclose conserved areas on gp120 of human immunodeficiency virus type 1 associated with cross-neutralizing antibodies and T-cell recognition. *Proc. Natl. Acad. Sci. U.S.A.* **88**, 10744–10748.

Vaux, D., and Fuller, S. D. (1991). The use of antiidiotype antibodies for the characterization of protein–protein interactions. *Methods Cell Biol.* **34**, 1–38.

Voss, E. W. (1990). Antibody and immunoglobulin: Structure and function. *In* "Immunology: Clinical, Fundamental, and Therapeutic Aspects" (B. P. Ram, M. C. Harris, and P. Tyle, eds.), pp. 49–88. VCH, New York.

Watson, J. D., Hopkins, N. H., Roberts, J. W., Steitz, J. A., and Weiner, A. M. (1987). The replication of DNA. *In* "Molecular Biology of the Gene," Vol. I, pp. 283. Benjamin/Cumming, Menlo Park, California.

Westhof, E., Altschuh, D., Moras, D., Bloomer, A. C., Mondragon, A., Klug, A., and Regenmortel, M. H. V. V. (1984). Correlation between segmental mobility and the location of antigenic determinants in proteins. *Nature* (*London*) **311**, 123–126.

Wright, P. E., Dyson, H. J., and Lerner, R. A. (1988). Conformation of peptide fragments of proteins in aqueous solution: Implications for initiation of protein folding. *Biochemistry* **27**, 7167–7175.

Zeegers-Huyskens, T., and Huyskens, P. (1990). Intermolecular forces. *In* "Intramolecular Forces" (P. L. Huyskens, W. A. Luck, and T. Zeegers-Huyskens, eds.), pp. 1–30. Springer-Verlag, Berlin.

Appendix

David J. Asai

Department of Biological Sciences
Purdue University
West Lafayette, Indiana 47907

Table I
Some Properties of Human Immunoglobulins[a]

Property	IgG	IgA	IgM	IgD	IgE
Sedimentation coefficient	7S	7–11S	19S	6–7S	8S
Molecular weight	150,000	150,000–170,000 and dimer	800,000–900,000	185,000	200,000
Number of four-chain units	1	1 or 2	5	1	1
Concentration in normal serum	8–16 mg/ml	1–4 mg/ml	0.5–2 mg/ml	0–0.4 mg/ml	17–450 ng/ml
Percentage of total immunoglobulin	75–85	5–10	5–10	0–1	<1
Percentage carbohydrate content	3	8	12	13	12

[a] The data in this table are from Barrett (1974) and Roitt (1988). These references are gratefully acknowledged.

Table II
Immunoglobulin Binding to Protein A and Protein G[a]

Immunoglobulin	Protein A	Protein G	Immunoglobulin	Protein A	Protein G
Mouse			IgG$_{2b}$	±	+
IgG$_1$	±	+	IgG$_{2c}$	+ +	+ +
IgG$_{2a}$	+ +	+ +	IgM	−	−(?)
IgG$_{2b}$	+ +	+ +	Rabbit		
IgG$_3$	+ +	+ +	IgG	+ +	+ +
IgM	−	−	IgM	−	?
Rat			Human		
IgG$_1$	+	+	IgG$_1$	+ +	+ +
IgG$_{2a}$	−	+ +	IgG$_2$	+ +	+ +

(*continues*)

Table II (*Continued*)

Immunoglobulin	Protein A	Protein G	Immunoglobulin	Protein A	Protein G
IgG$_3$	−	++	IgM	−	?
IgG$_4$	++	++	Sheep		
IgM	−	−	IgG$_1$	−	++
Guinea pig			IgG$_2$	++	++
IgG$_1$	+	?	Horse		
IgG$_2$	++	?	IgG$_a$	+	++
Pig			IgG$_b$	+	++
IgG$_1$	++	?	IgG$_c$	+	++
IgG$_2$	++	?	IgG(T)	−	+
IgM	±	?	IgM	−	?
Goat			Bovine		
IgG$_1$	+	++	IgG$_1$	−	++
IgG$_2$	++	++	IgG$_2$	++	++

[a] This table is from Beltz and Burd (1989), who consolidated the data of Björck and Kronvall (1984) and Langone (1982). These sources are gratefully acknowledged.

References

Barrett, J. T. (1974). "Textbook of Immunology," 2nd Ed. Mosby, St. Louis.

Beltz, B. S., and Burd, G. D. (1989). "Immunocytochemical Techniques." Blackwell, Oxford.

Björck, L., and Kronvall, G. (1984). Purification and some properties of streptococcal protein G, a novel IgG-binding protein. *J. Immunol.* **133,** 969–974.

Langone, J. J. (1982). Protein A of *Staphylococcus aureus* and related immunoglobulin receptors produced by streptococci and penumonococci. *Adv. Immunol.* **32,** 157–252.

Roitt, I. (1988). "Essential Immunology," 6th Ed. Blackwell, Oxford.

INDEX

VOLUMES IN SERIES

Founding Series Editor
DAVID M. PRESCOTT

Volume 1 (1964)
Methods in Cell Physiology
Edited by David M. Prescott

Volume 2 (1966)
Methods in Cell Physiology
Edited by David M. Prescott

Volume 3 (1968)
Methods in Cell Physiology
Edited by David M. Prescott

Volume 4 (1970)
Methods in Cell Physiology
Edited by David M. Prescott

Volume 5 (1972)
Methods in Cell Physiology
Edited by David M. Prescott

Volume 6 (1973)
Methods in Cell Physiology
Edited by David M. Prescott

Volume 7 (1973)
Methods in Cell Biology
Edited by David M. Prescott

Volume 8 (1974)
Methods in Cell Biology
Edited by David M. Prescott

Volume 9 (1975)
Methods in Cell Biology
Edited by David M. Prescott

Volume 10 (1975)
Methods in Cell Biology
Edited by David M. Prescott

Volume 11 (1975)
Yeast Cells
Edited by David M. Prescott

Volume 12 (1975)
Yeast Cells
Edited by David M. Prescott

Volume 13 (1976)
Methods in Cell Biology
Edited by David M. Prescott

Volume 14 (1976)
Methods in Cell Biology
Edited by David M. Prescott

Volume 15 (1977)
Methods in Cell Biology
Edited by David M. Prescott

Volume 16 (1977)
Chromatin and Chromosomal Protein Research I
Edited by Gary Stein, Janet Stein, and Lewis J. Kleinsmith

Volume 17 (1978)
Chromatin and Chromosomal Protein Research II
Edited by Gary Stein, Janet Stein, and Lewis J. Kleinsmith

Volume 18 (1978)
Chromatin and Chromosomal Protein Research III
Edited by Gary Stein, Janet Stein, and Lewis J. Kleinsmith

Volume 19 (1978)
Chromatin and Chromosomal Protein Research IV
Edited by Gary Stein, Janet Stein, and Lewis J. Kleinsmith

Volume 20 (1978)
Methods in Cell Biology
Edited by David M. Prescott

Advisory Board Chairman
KEITH R. PORTER

Volume 21A (1980)
Normal Human Tissue and Cell Culture, Part A: Respiratory, Cardiovascular, and Integumentary Systems
Edited by Curtis C. Harris, Benjamin F. Trump, and Gary D. Stoner

Volume 21B (1980)
Normal Human Tissue and Cell Culture, Part B: Endocrine, Urogenital, and Gastrointestinal Systems
Edited by Curtis C. Harris, Benjamin F. Trump, and Gary D. Stoner

Volume 22 (1981)
Three-Dimensional Ultrastructure in Biology
Edited by James N. Turner

Volume 23 (1981)
Basic Mechanisms of Cellular Secretion
Edited by Arthur R. Hand and Constance Oliver

Volume 24 (1982)
The Cytoskeleton, Part A: Cytoskeletal Proteins, Isolation and Characterization
Edited by Leslie Wilson

Volume 25 (1982)
The Cytoskeleton, Part B: Biological Systems and *in Vitro* Models
Edited by Leslie Wilson

Volume 26 (1982)
Prenatal Diagnosis: Cell Biological Approaches
Edited by Samuel A. Latt and Gretchen J. Darlington

Series Editor
LESLIE WILSON

Volume 27 (1986)
Echinoderm Gametes and Embryos
Edited by Thomas E. Schroeder

Volume 28 (1987)
***Dictyostelium discoideum:* Molecular Approaches to Cell Biology**
Edited by James A. Spudich

Volume 29 (1989)
Fluorescence Microscopy of Living Cells in Culture, Part A: Fluorescent Analogs, Labeling Cells, and Basic Microscopy
Edited by Yu-Li Wang and D. Lansing Taylor